Symbols of the Elements and Their Atomic Weights

Name	Symbol	Atomic Number	Atomic Weight*	Name	Symbol	Atomic Number	Atomic Weight*
Actinium	Ac	89	(227)	Mercury	Hg	80	200.6
Aluminum	Al	13	27.0	Molybdenum	Mo	42	95.9
Americium	Am	95	(243)	Neodymium	Nd	60	144.2
Antimony	Sb	51	121.8	Neon	Ne	10	20.2
Argon	Ar	18	39.9	Neptunium	Np	93	(237)
Arsenic	As	33	74.9	Nickel	Ni	28	58.7
Astatine	At	85	(210)	Niobium	Nb	41	92.9
Barium	Ba	56	137.3	Nitrogen	N	7	14.01
Berkelium	Bk	97	(247)	Nobelium	No	102	(254)
Beryllium	Be	4	9.01	Osmium	Os	76	190.2
Bismuth	Bi	83	209.0	Oxygen	O	8	16.00
Boron	B	5	10.8	Palladium	Pd	46	106.4
Bromine	Br	35	79.9	Phosphorus	P	15	31.0
Cadmium	Cd	48	112.4	Platinum	Pt	78	195.1
Calcium	Ca	20	40.1	Plutonium	Pu	94	(242)
Californium	Cf	98	(251)	Polonium	Po	84	(210)
Carbon	C	6	12.01	Potassium	K	19	39.1
Cerium	Ce	58	140.1	Praseodymium	Pr	59	140.9
Cesium	Cs	55	132.9	Promethium	Pm	61	(147)
Chlorine	Cl	17	35.5	Protactinium	Pa	91	(231)
Chromium	Cr	24	52.0	Radium	Ra	88	(226)
Cobalt	Co	27	58.9	Radon	Rn	86	(222)
Copper	Cu	29	63.5	Rhenium	Re	75	186.2
Curium	Cm	96	(247)	Rhodium	Rh	45	102.9
Dysprosium	Dy	66	162.5	Rubidium	Rb	37	85.5
Einsteinium	Es	99	(254)	Ruthenium	Ru	44	101.1
Erbium	Er	68	167.3	Samarium	Sm	62	150.4
Europium	Eu	63	152.0	Scandium	Sc	21	45.0
Fermium	Fm	100	(253)	Selenium	Se	34	79.0
Fluorine	F	9	19.0	Silicon	Si	14	28.1
Francium	Fr	87	(223)	Silver	Ag	47	107.9
Gadolinium	Gd	64	157.3	Sodium	Na	11	23.0
Gallium	Ga	31	69.7	Strontium	Sr	38	87.6
Germanium	Ge	32	72.6	Sulfur	S	16	32.1
Gold	Au	79	197.0	Tantalum	Ta	73	180.9
Hafnium	Hf	72	178.5	Technetium	Tc	43	(99)
Helium	He	2	4.00	Tellurium	Te	52	127.6
Holmium	Ho	67	164.9	Terbium	Tb	65	158.9
Hydrogen	H	1	1.008	Thallium	Tl	81	204.4
Indium	In	49	114.8	Thorium	Th	90	232.0
Iodine	I	53	126.9	Thulium	Tm	69	168.9
Iridium	Ir	77	192.2	Tin	Sn	50	118.7
Iron	Fe	26	55.8	Titanium	Ti	22	47.9
Krypton	Kr	36	83.8	Tungsten	W	74	183.9
Lanthanum	La	57	138.9	Uranium	U	92	238.0
Lawrencium	Lr	103	(257)	Vanadium	V	23	50.9
Lead	Pb	82	207.2	Xenon	Xe	54	131.3
Lithium	Li	3	6.94	Ytterbium	Yb	70	173.0
Lutetium	Lu	71	175.0	Yttrium	Y	39	88.9
Magnesium	Mg	12	24.3	Zinc	Zn	30	65.4
Manganese	Mn	25	54.9	Zirconium	Zr	40	91.2
Mendelevium	Md	101	(256)				

*The values given in parentheses are mass numbers of the principal isotopes of unstable elements.

FOUNDATIONS

FOUNDATIONS

FOUNDATIONS

FOUNDATIONS

FOUNDATIONS

Department of Chemistry, University of South Florida, Tampa, Florida

OF CHEMISTRY

Jesse S. Binford, Jr.

MACMILLAN PUBLISHING CO., INC.
New York
COLLIER MACMILLAN PUBLISHERS
London

Macmillan Publishing Co., Inc.
866 Third Avenue, New York, New York 10022

Collier Macmillan Canada, Ltd.

Library of Congress Cataloging in Publication Data

Binford, Jesse S
  Foundations of chemistry.

  Includes index.
  1. Chemistry.   I. Title.
QD31.2.B57   540   75-45278
ISBN 0-02-309880-5

Printing: 1 2 3 4 5 6 7 8   Year: 7 8 9 0 1 2

TO LOLITA

PREFACE

This book was written because no satisfactory text was available to prepare college students for general chemistry. To avoid lowering the standards of the general chemistry course and help the poorly prepared student, we developed a new one-term course that lies somewhere between high school and college levels. In it we emphasize the topics students find most difficult in general chemistry, but we treat these subjects in a more fundamental manner than a college course usually does.

For which students is this course necessary, and why do they take it? We have found that most are majoring in the life sciences (nursing, premedicine, and biology), a significant number are in engineering, and a large number are still undecided. Their reasons for taking the course are that they did not have chemistry in high school (40%), or they think their high school course was weak (30%). Most of the others think they have been out of high school too long or simply need a science course. A large proportion (80%) have definite plans to take general chemistry.

Students usually do unsatisfactory work in general chemistry because of (1) an inability to use systematic methods, (2) an inability to apply simple arithmetic and algebra to chemical problems, (3) an inability to understand concepts and models, and (4) a lack of interest in chemistry. The intent of *Foundations of Chemistry* is to overcome these deficiencies. The text has been tested since 1971. The first year it was little more than a weekly syllabus and was used in conjunction with commercial textbooks. Revised and expanded as the result of student reaction in 1972, and again in 1974, it has developed into a textbook. The final revision resulted in the elimination of considerable extraneous material and brought the text to its present form. The emphasis has always been (1) basic explanation of the concepts that are often misunderstood in college chemistry, (2) problem solving, (3) practical everyday examples and applications, and (4) systemization of chemical facts. The last-mentioned is particularly important, because many essential topics for beginning students are almost pure memory work. The more systematically these subjects are presented, the easier they are to assimilate. As a result of student requests, many examples with detailed solutions and exercises with answers have been added. Furthermore, the problems at the end of each chapter are paired, and the answers to one complete set are provided.

In my own course, which does not have laboratory sessions associated with it, I supplement this text with required film and videotape demonstrations. There are weekly problem assignments and weekly quizzes supplemented by audiotape problem solutions and videotape quiz reviews. *Foundations of Chemistry* is not meant to be a book for independent study but rather one that requires encouragement and leadership from the teacher.

We have found the course to be a success in providing the necessary foundations for general chemistry. Our experience with several hundred students who have now taken both the "foundations" course and the first term of general chemistry indicates that the students make one letter grade higher than predicted on their placement examinations. Also, their average grade in general chemistry is the same as the class average, although comparison of performance on placement examinations would predict a lower average.

This text uses the "one-factor" method, or dimensional-analysis, to solve problems almost to the exclusion of other methods. The word "one-factor" is used to emphasize the superiority of the method. More understanding is required in using this method because the student must go back to the physical reality or to the definition in each problem and not just blindly use memorized formulas or take a 50-50 chance on a supposed proportion. However, more and more algebra is introduced in later chapters and by the last chapter, algebraic formulas are used almost exclusively.

Special thanks are given to Conard Fernelius and Mike Holloway, who made a number of helpful suggestions early in the project. Special credit is due my wife, Lolita Fritz Binford, who created the original artwork and who gave her help and encouragement throughout the project.

J. S. B.

CONTENTS

175   **8**

SOLUTIONS

195   **9**

EQUILIBRIUM

ACIDS AND BASES **10** 219

ACID CONSTANTS, TITRATION, AND INDICATORS **11** 241

1

METHODS, PROPERTIES, AND THE PERIODIC TABLE

Science is a human enterprise. Its most significant achievements are results of the systematic application of ordinary logic to knowledge gained through observation.

A conversation with a three-year-old will give examples of the scientific method. Turn the pages of a magazine and ask the child to point out the pictures of animals. The child has observed cats, birds, and horses and might readily identify them as animals. But a picture of a dog might get a negative response. "That's not an animal. That's a dog!" If you then ask what the child knows about animals, you might be answered, "Well, they are made of meat." The child has not only accumulated information through observation but has organized it and found regularities. If you then point out that dogs are made of meat, the child may learn from this generalization that dogs are animals. Most examples of the learning process are applications of the scientific method.

Often a scientific discovery is said to be an accident. But, as in the case of the child "discovering" that dogs are animals, the scientist merely remembers something almost forgotten. One could say the scientist suddenly recognizes a relationship that could have been seen earlier, because all the information was there; however, the connection between the facts is noticed for the first time when something falls into place. Unfortunately such events do not usually occur until a considerable amount of work and mental effort have been expended, often in wrong directions.

1.1 THE SCIENTIFIC METHOD

A complete sequence of actions which comprise the scientific method are as follows: (1) obtaining facts by **observation,** that is, collecting data, (2) **organization** of these facts in a systematic way, (3) deriving from this organization of facts a "law," which is so-called not merely because it has stood the test of attempts to disprove it but mainly because it can be used in the **prediction** of facts not yet observed.

There is a hierarchy of terms expressing the degree of confidence in the correctness of the organization of data that may eventually be accepted as a law. The preliminary idea is usually called a **hypothesis,** and several hypotheses may be proposed for the same set of data, just for the sake of argument. A single new fact will frequently destroy a given hypothesis. After the hypothesis has been around for a while and has been found resistant to disproof, it is promoted to the level of **theory.** A theory is accepted by most scientists as an adequate explanation of the facts but usually with some reserve. The ultimate expression of acceptance is to describe a theory as a **law,** which signifies that it is generally accepted by scientists. The acceptance of the law does not usually occur until it has demonstrated repeatedly its powers of *prediction* of yet undiscovered knowledge. Some scientific ideas which, because of their abstractness are difficult to use for prediction, such as Einstein's theory of relativity, will be described as theories for many years. There was a famous hypothesis made by Avogadro in 1811 that explained certain aspects of chemical reactions between gases (Section 4.1). It was not accepted for such a long time (about 50 years) that it is still ironically referred to as "Avogadro's hypothesis."

1.2 AN EXAMPLE OF THE SCIENTIFIC METHOD—ARCHIMEDES' LAW OF FLOATING BODIES

An interesting set of experiments* has been devised by psychologists for children in which the children try to reconstruct Archimedes' law using the equipment shown in Figure 1.1. A child is given a bucket of water and several objects and asked to classify them on the basis of floating or sinking. Then the child is asked to summarize the observations and to devise a law if possible.

To describe an object it is necessary to list its properties. A **property** is a characteristic of an object that distinguishes it from other objects, such as color, size, and weight. As a result of this experiment, children develop the idea that the properties of weight and size considered separately are not very useful in characterizing the

---

*This is a part of the life-long study of the Swiss psychologist Piaget, who began as a zoologist. His interest in the development of intelligence in children and his observations on his own three children have resulted in important new evidence that IQ is not "fixed at birth" but depends on stimulation of the very young. See Bärbel Inhelder and Jean Piaget: *The Growth of Logical Thinking from Childhood to Adolescence.* Basic Books, New York, 1958.

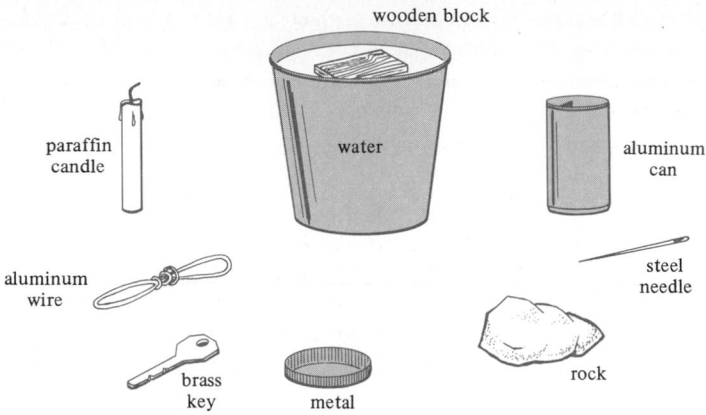

Figure 1.1. Archimedes' law of floating bodies. Why do some objects float and others sink?

phenomenon of floating; however, they discover the property of weight for equal volumes is very helpful.

The kinds of results obtained by the children are reported in the following paragraphs according to age groups. The psychologists can obtain evidence for new learning theories in this way, and you will be able to see the necessity for each of the steps in the scientific method. You will also see how an inferior hypothesis must be rejected when confronted by a single fact with which it is not compatible.

Children who are *four years of age* fail to see any relation between the properties of the objects and whether they float. In other words their powers of observation are too limited at this age for this kind of experiment.

Children at *five or six years of age* give as reasons for an object not floating that it is "big," "heavy," "long," or "small." Every object that sinks is described in terms of a similar property of another object that sinks. However, they are not bothered by the contradictions that, although the "heavy" block of wood floats as well as the "small" cork, the "heavy" rock sinks as does the "small" key.

Children of about *seven or eight years of age* typically classify the objects in three categories: (1) those that float (wood, candle), (2) those that sink (key, rock), and (3) those that sink or float depending on whether they are pushed through the surface and are filled with water (cover, aluminum wire, needle, can). Furthermore, they may hypothesize that *small–light* objects float while *small–heavy* objects sink and *large–light* objects float while *large–heavy* objects sink.

At later stages the psychologists recognize that some children will use knowledge acquired in school on this subject. They have identified these children through their questioning and have not included them among their reported interviews.

For the *twelve-year-old* Fran, the following encounter was recorded.

Fran does not manage to discover the law, but neither does he accept any of the earlier hypotheses. He classifies the objects presented correctly but hesitates before the aluminum wire.—Why are you hesitating?—Because of the lightness, but no, that has no effect.— Why?—The lightness has no effect. It depends upon the sort of matter; for example the wood can be heavy and it floats. And for the cover: I thought of the surface.—The surface plays a role?—Maybe, the surface that covers the water, but that doesn't mean anything. . . . Thus he discards all of his hypotheses without finding a solution.

Ala, who is the same age as Fran, makes the following hypotheses:

Why do you think this key will sink?—Because it is heavier than water.—This little key is heavier than water? (The bucket is pointed out.)—I mean the same quantity of water would be less heavy than the key.—What do you mean?—You would put them (metal or water) in containers which contain the same amount and weigh them.

Finally, Wur, who is *fourteen years old,* suggests the following very enlightening hypothesis:

I take a wooden cube and a plastic cube which I fill with water. (The cubes are the same size [and the plastic cube is hollow].) I weigh them and the difference can be seen on the scale according to whether an object is heavier or lighter than water.

In this way he has eliminated volume as a variable by basing the hypothesis on weight per unit volume, taking the equal size cubes as the unit of volume.

The usual way of stating **Archimedes' law** is as follows: *The force causing an object to float on water (the buoyant force) is equal to the weight of the water it displaces.* The law of Archimedes will be put to the test every time a new ship is launched on the waters. A quantitative example of the law that shows further predictive powers will be given in the next chapter (Section 2.7).

The degree to which young children use the scientific method in the development of their own intelligence is remarkable. Although it would not be practical in a text of this kind to insist that you relive the processes of thought that led to all the scientific concepts described, we hope that the concepts will be presented in such a way that you will want to make use of the scientific method yourself to solve chemical problems in your everyday life. It is after all, a natural method for the human species.

Exercise 1.   If you introduce the needle into the water (Figure 1.1) end first, it will sink to the bottom. However, if you carefully lay it horizontally on the water surface, it will float. The same is true of a straight piece of aluminum wire. Does this disprove Archimedes' law? If not, can you propose a hypothesis to account for the new observation? Can you propose new experiments to test your hypothesis?

1.3   THE SCIENCE OF CHEMISTRY

Chemistry is the science of matter—its properties; its composition; its structure; its synthesis; its behavior under changing conditions of pressure, temperature, light, and electrical forces; its interactions with other kinds of matter, and its changes in energy during these interactions.

Matter is that which has mass and volume. To have volume means that it occupies space to the exclusion of other forms of matter. The property that best describes mass is its **inertia,** which is its tendency to remain at rest when you attempt to move it or its tendency to remain in motion when you attempt to stop it. The greater the force needed to overcome the inertia of an object, the greater is its mass. Another important property of mass is the **weight** that is associated with it. The weight of matter is the force gravity exerts upon it.

The distinction between weight and mass is subtle, but in this time of weightless astronauts drifting around on your television screen it is obvious that such a difference exists. For example, the astronaut would have to use the property of inertia to obtain the mass of an object in his space ship. On earth, however, you can say two objects have the same mass if their gravitational attractions to earth (weights) are the same. An ideal piece of equipment for this comparison is a one-pan balance like the one shown in Figure 1.2. The beam (the horizontal bar which runs from front to rear at the top) is balanced when the pan is unloaded, or empty, because of the large counterbalance at the rear. The pan will descend when an object to be "massed" is placed upon it. By removing enough rings, which are on the same end of the beam as the pan and object, to regain balance, you can say that the mass of the rings removed and that of the object are equal. All that is needed to define mass quantitatively is a standard mass to which the rings can be compared. One such standard is maintained in the National Bureau of Standards in Washington, D.C. Because all the masses discussed in the remainder of this text will be subject to the gravitational force of the earth, "weight" will be considered to mean "mass" and "weighing" to mean "massing."

Volume can always be expressed in cubic length. The volume of a cube, shown in Figure 1.3, is the cube of the length of an edge, that is, the length of an edge multiplied by itself three times. The volume of any irregularly shaped object could be expressed in terms of small cubes that occupy the same volume. In order to express

Figure 1.2   The one-pan balance.

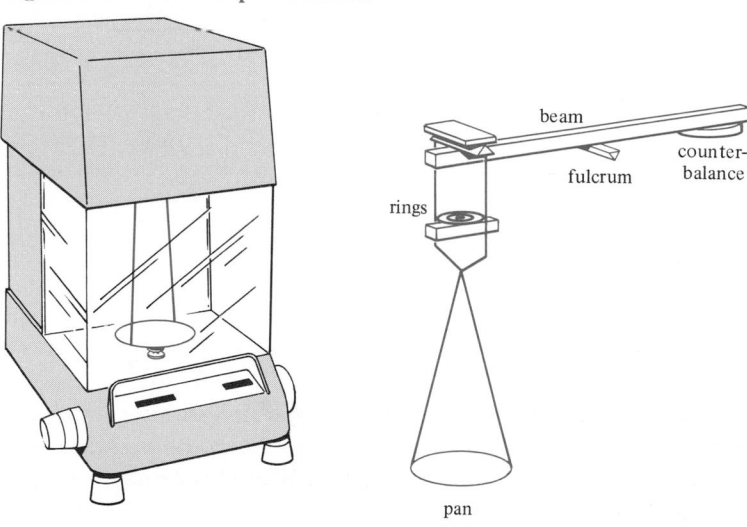

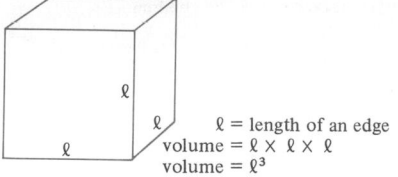

ℓ = length of an edge
volume = $\ell \times \ell \times \ell$
volume = ℓ^3

Figure 1.3. The volume of a cube.

volume quantitatively, all that is needed is a standard of length. This standard could be the length of some object maintained at the Bureau of Standards. The best standard, however, is provided by a certain number of wavelengths of light (Section 13.2), which not only is accurately measurable but also does not change with temperature and other conditions.

The Forms of Matter   Chemists have developed a special vocabulary to describe the different forms of matter that are observed. The outline in Figure 1.4 will help you see the interrelations of the various forms.

Figure 1.4.   The forms of matter.

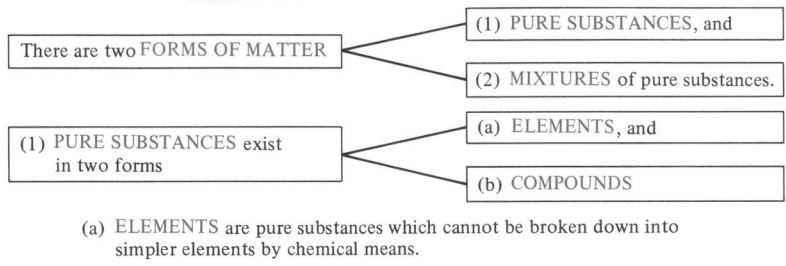

(a) ELEMENTS are pure substances which cannot be broken down into simpler elements by chemical means.

(b) COMPOUNDS are chemical combinations of two or more different elements.

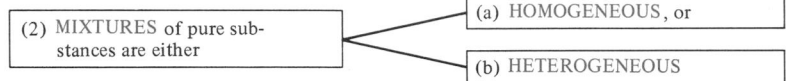

(a) HOMOGENEOUS MIXTURES, also called SOLUTIONS, have no visible discontinuities between substances.

(b) HETEROGENEOUS MIXTURES have visible boundaries of separation between substances.

Separation of Mixtures by Physical Means   Physical changes are very common in nature. A **physical change** does *not* involve the formation of new compounds or the breakdown of existing compounds, and it is easily reversed. Most simple substances are capable of existing in three **phases: solid, liquid,** and **gas,** which are called the **states of matter.** Changes from one state to another are examples of physical change. By simply increasing the temperature, matter may be changed from solid $\longrightarrow$ liquid $\longrightarrow$ gas, provided that chemical decomposition does not occur. By decreasing the temperature, the changes may be reversed: gas $\longrightarrow$ liquid $\longrightarrow$ solid.

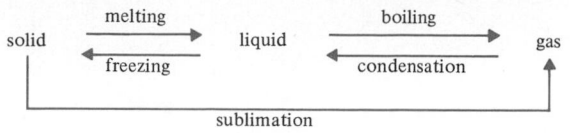

Figure 1.5. Changes in physical state.

Most changes in state are very easily reversed. A summary of the most common changes in state is shown in Figure 1.5, which gives the name used to describe the change. The temperature of every particular phase change is exactly reproducible. The freezing point of water, for example, is so reproducible that it serves as a basis for the temperature scales. The quantitative aspects of temperature are discussed in Section 2.3.

Another example of physical change is forming a solution. This physical change can be used to separate heterogeneous mixtures. For example, salt and pepper, a **heterogeneous** mixture, can be separated by adding water, which forms a solution (a **homogeneous** mixture) with the salt but not with the pepper. After separating the solution from the pepper by **filtration,** a physical process illustrated in Figure 1.6a, the salt and water can be separated by changing the temperature as shown in Figure 1.6b. If the temperature is raised enough, the water boils away leaving the salt behind. As the water is lost, the boiling temperature rises. Such a change in temperature while a liquid is boiling is good evidence that the liquid is a solution and not a pure substance.

The process of separation illustrated in Figure 1.6b is called **distillation** and the equipment in which it is carried out is called a **still.** A similar separation could be accomplished by evaporation of water at temperatures below its boiling point (Section 7.5).

Decreasing the temperature can also produce physical changes useful in separating mixtures. If the temperature of a salt solution is lowered, two things may happen. **Recrystallization** may occur if the water cannot hold all the salt at the lower temperature. (This is a physical change which is the opposite of dissolving.) The salt becomes solid again, and can be separated from the remaining solution by filtration. Frequently two different solids in a liquid solution can be separated from one another by recrystallization because the two solids have different tendencies to remain dissolved. The other possibility is that the temperature will reach the freezing point of water and pure ice will form. The ice can then be separated from the remaining solution by filtration.

> **Environmental Note:** Pure distilled water is quite expensive because of the energy required for purification. The U.S. Department of the Interior sponsored a research project in Florida using solar energy to purify water by distillation. The heat requirements are so great that if all the solar energy falling on a given area of salt water for an entire day were used in distillation it would provide only 0.4 in. of water on agricultural land of equal area. The best efficiency attained so far would provide only 0.14 in.
>
> Another project sponsored by the government in Texas depends on freezing the salt water and collecting the ice. The energy requirements for freezing water are not as great as for boiling it.

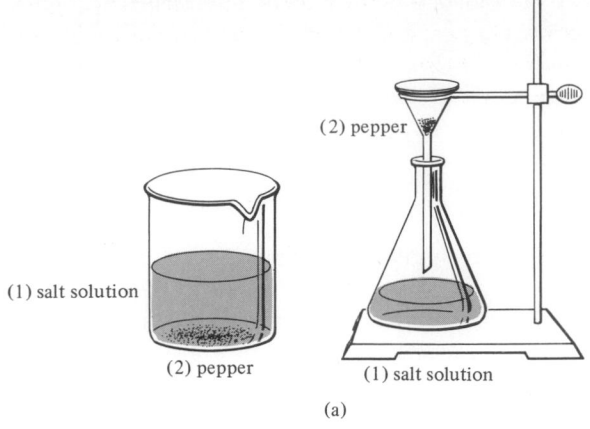

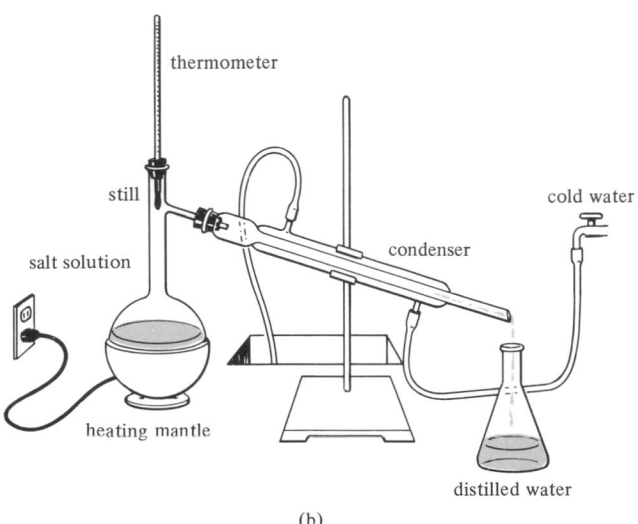

Figure 1.6. Methods of physical separation. (a) Filtration. (b) Distillation.

Preparation of
Elements by
Chemical Means

When substances are subjected to energy in the form of heat, electricity, or light or when they are mixed with other substances, they frequently undergo changes in which new substances are formed. These changes are called **chemical changes.** The energy involved in chemical changes is greater than the energy associated with physical changes. One of the new substances formed may be an **element,** which means that this substance cannot be broken down into simpler substances by similar treatment.

The following processes are examples of chemical change: The white crystalline solid potassium chlorate is heated in a test tube. The solid melts at sufficiently high temperatures, and a gas begins to bubble out of it. This gas will cause a glowing match to burst into flame and can be identified as oxygen. When the substance in the test tube is cooled in the presence of the oxygen, a white solid is formed; however this solid does not have the properties of potassium chlorate. If this white solid is remelted and an electric current such as that from an automobile battery is passed through it in the absence of air, two new substances are formed. A greenish-yellow gas called chlorine is produced at the positive electrode, and a vapor is produced at the negative electrode which, on cooling, becomes a soft metallic solid called potassium.

The processes that produce oxygen, chlorine, and potassium from potassium chlorate are chemical changes. All further attempts to simplify the composition of oxygen, chlorine, potassium, and other substances like them (elements), by heating them, subjecting them to an electric current, interacting them with other substances, or by any other chemical change have been unsuccessful. Therefore, chemists conclude that these substances are elements and not compounds.

1.4 PERIODIC ARRANGEMENT OF THE ELEMENTS

One of the most interesting and rewarding organizations of information in the scientific world is the periodic table based on the properties of the elements. The form of this table was discovered by the Russian chemist Mendeleev in 1869. The periodic table now contains all the elements known to man arranged in their proper sequence. All the compounds in the world are made of these elements. There is good reason to believe that the modern periodic table is complete in the sense that no new elements will be discovered occupying positions between those already occupied.

An important part of the periodic table is shown in Table 1.1. Many of these elements are probably familiar to you, the carbon that you use in a pencil, the sulfur that you use to dust roses, and the oxygen and nitrogen that are mixed together in the air you breathe. Other elements like sodium and chlorine are so reactive that they are rarely used in pure form by the nonchemist. Still others like francium and astatine are so rare that they are not used in their elemental forms even by chemists.

By arranging all the elements in the order of increasing weight, that is by "the relative weights of the ultimate particles" of each element, as John Dalton stated it in 1808 (Section 4.3), there is a definite periodic repetition of certain chemical properties. With rare exceptions, reading Table 1.1 in the usual way from left to right and from top to bottom, the elements are arranged in the order of increasing relative weights. The vertical columns are called **groups** and each new horizontal row is called a **period.**

It is important to learn the symbol for each element not only because it is simpler than using the whole name but also because it is accepted internationally. The scientists from all those countries which are members of IUPAC (International Union of Pure and Applied Chemistry) have agreed on what symbol will represent which

Symbols in the
Periodic Table

TABLE 1.1* The first 18 elements of the periodic table with the complete first two and last two principal groups

		Principal Groups							
		I	II	III	IV	V	VI	VII	0
Periods	1	H hydrogen							He helium
	2	Li lithium	Be beryllium	B boron	C carbon	N nitrogen	O oxygen	F fluorine	Ne neon
	3	Na sodium	Mg magnesium	Al aluminum	Si silicon	P phosphorus	S sulfur	Cl chlorine	Ar argon
	4	K potassium	Ca calcium					Br bromine	Kr krypton
	5	Rb rubidium	Sr strontium					I iodine	Xe xenon
	6	Cs cesium	Ba barium					At astatine	Rn radon
	7	Fr francium	Ra radium						

* The complete periodic table is shown inside the front cover of the textbook.

element. Even in those countries where English is not used, you would be able to recognize the symbols for the elements. The periodic table is the alphabet of chemistry and it is an alphabet in which position is all important.

Group Properties

All of the elements within the same group have similar properties. Each of the principal groups is discussed on the following pages, starting on the left side of the periodic table and moving across to the right. Within each group there will be elements from different periods, period 1, period 2, period 3,. . . .

One important property of an element is the kinds of compounds it forms. Elements in the same group form similar compounds. For example, oxygen and sulfur form compounds with hydrogen having two parts hydrogen and one part of either oxygen or sulfur: H_2O or H_2S. Note that the subscript refers to the element it follows. The meaning of these subscripts in terms of John Dalton's "ultimate particles" will be discussed in Chapter 4.

1.5 NAMING OF BINARY COMPOUNDS

A binary compound is a chemical combination of two different elements. All of the compounds listed in Section 1.4 are binary, such as LiH, Na_2O, and Al_2O_3. Some compounds have common names that give no indication of their chemical composition, such as water and ammonia. It is impractical to have a common name for every

1. Like most metals the alkali metals (a) have a shiny lustrous appearance (when freshly cut), (b) are good conductors of electricity, (c) are malleable (can be pounded into sheets), and (d) are ductile (can be drawn into wires).

2. Alkali metals are all solids but they have low melting points; for example, Na melts at a temperature slightly below the boiling point of water and Cs is a liquid on a warm day.

3. Alkali metals react with water to produce hydrogen gas.

			Compounds with hydrogen, HYDRIDES (all solids)
P1	H	A gas and nonmetal, not group I	
P2	Li		*LiH, lithium hydride
P3	Na	Lighter than water	NaH, sodium hydride
P4	K		KH, potassium hydride
P5	Rb		RbH, rubidium hydride
P6	Cs	Heavier than water	CsH, cesium hydride
P7	Fr	Radioactive, not found in nature	These hydrides react with water to produce hydrogen gas.

Compounds with oxygen, OXIDES (all solids)

$\dagger Li_2O$, lithium oxide

Na_2O, sodium oxide

K_2O, potassium oxide

Rb_2O, rubidium oxide

Cs_2O, cesium oxide

Compounds with halogens, HALIDES

LiCl,	lithium chloride	A salt††
NaCl,	sodium chloride	Table salt
KCl,	potassium chloride	Sodium-free dietary salt
RbCl,	rubidium chloride	
CsCl,	cesium chloride	
KI,	potassium iodide	In iodized salt
NaF, etc.	sodium fluoride	Treatment of teeth

Group II Alkaline Earth Metals	Group III	Group IV

P2	Be
P3	Mg
P4	Ca
P5	Sr
P6	Ba
P7	Ra

Group III

Every element below and to the left of the metalloids is a metal.

A metalloid B

A metal Al

HYDRIDE

AlH_3 A solid

OXIDE

Al_2O_3 sapphire, ruby, etc.

HALIDES

$AlCl_3$

$AlBr_3$

etc.

Group IV

A nonmetal C

A metalloid Si

HYDRIDE

CH_4 A gas

This compound, called *methane*, is the principal constituent of "natural gas."

HYDRIDES (All solids)	OXIDES (All solids)	HALIDES (All solids)
BeH_2	BeO	BeF_2
MgH_2	MgO	$MgCl_2$
CaH_2	CaO	$CaCl_2$
SrH_2	SrO	$SrBr_2$
BaH_2	BaO	BaI_2
RaH_2	RaO	$RaCl_2$

*The naming of compounds is discussed in Section 1.5.

† To be read "Two parts lithium and one part oxygen." See Section 4.1.

†† All metal halides are classified as salts.

Group V	Group VI	Group VII Halogens or "salt formers"

The following elements are all nonmetals. They are all located above and to the right of the metalloids.

N	A gas, 78% of air
P	A brittle white solid

O	A gas, 21% of air
S	A brittle yellow solid

P2	F	A pale yellow *gas*
P3	Cl	A greenish-yellow *gas*
P4	Br	A red-brown *liquid*
P5	I	A dark violet *solid*
P6	At	Radioactive

These elements are very reactive, toxic, pungent substances.

HYDRIDES

NH_3 ammonia

PH_3 phosphine

HYDRIDES

H_2O water

H_2S hydrogen sulfide

HYDRIDES

HF   hydrogen fluoride

HCl   hydrogen chloride

HBr   hydrogen bromide

HI   hydrogen iodide

Group 0
Inert Gases

Lighter than air.

P1	He
P2	Ne

Heavier than air.

P3	Ar
P4	Kr
P5	Xe
P6	Rn

Found in underground reservoirs.

1% of air is argon.

HYDRIDES:	None
OXIDES:	None
FLUORIDES:	With Kr and Xe only, unstable.

The extreme stability of the inert gases is a clue to the stability of the compounds which the other elements form.

compound, however, and systems of nomenclature have been devised that simplify the problem of naming compounds. In the case of binary compounds, the general rule is to begin by naming the first element. This is followed by a prefix based on the name of the second element to which is added the suffix -ide.

EXAMPLE 1.   Give names for the following binary compounds: (a) HI, (b) $MgBr_2$, (c) Al_2O_3.

Solution: (a) hydrogen iodide, (b) magnesium bromide, (c) aluminum oxide.

The combination of elements in some binary compounds is not unique. For example, the two oxides of carbon, CO and CO_2 are both very common substances. In such cases Greek prefixes for appropriate numbers must be used. Thus the oxide with one part oxygen is called carbon monoxide, and the oxide with two parts oxygen is called carbon dioxide. A more complete list of prefixes is

one	mono-	four	tetra-
two	di-	five	penta-
three	tri-	six	hexa-

The series of nitrogen oxides, all of which exist under certain conditions, is named as follows.

N_2O_5	dinitrogen pentoxide	N_2O_3	dinitrogen trioxide
N_2O_4	dinitrogen tetroxide	NO	nitrogen oxide
NO_2	nitrogen dioxide	N_2O	dinitrogen oxide

EXAMPLE 2. Name the following compounds using the appropriate numerical Greek prefix: (a) PCl_3 and PCl_5, (b) CS_2, (c) SiO_2.

Solution: (a) phosphorus trichloride and phosphorus pentachloride (b) carbon disulfide (c) silicon dioxide.

1.6 CONCLUSION AND SUMMARY OF THE SCIENTIFIC METHOD

Because the scientific method appears to be a step by step process, many people get the impression that it is a slow, methodical procedure and not really very exciting or logical. To be sure the scientific method can be slow—the accumulation of a large number of bits of information, classifying them, and trying to explain them. However, the application of the scientific method can be rapid and very exciting. In fact, the search for knowledge and the development of explanations can become a game. What is the minimum number of observations necessary to give a reasonably complete overview of a subject? Is there some critical experiment that will determine whether or not a suggested explanation is really valid? How adventurous is one prepared to be in devising explanations? Many an investigator has proposed explanations which were not acceptable to his contemporaries but which in the end came to complete acceptance.

Observation and organization are important aspects of the scientific method, but the most fruitful aspect is *prediction*. Sometimes this takes the form of a mathematical equation with which an unknown numerical value can be calculated. Other times novel experiments are suggested that will open new areas of knowledge.

The ability to predict knowledge is the feature of new scientific work most convincing to other scientists. Consider the case of the periodic table. It was not until several of Mendeleev's predictions came true that much interest was stimulated by his table. Mendeleev had daringly predicted the existence of several new elements because there were places in his table that none of the known elements seemed to fit. What was more striking, he predicted their properties and suggested where they might be found. According to his predictions several missing elements were discovered—gallium in 1875, scandium in 1879, and germanium in 1887. By 1895, textbooks in chemistry were beginning to publish the periodic table, a 26-year delay which should serve as fair warning to all students reading this text.

To summarize, the scientific method consists of *observation, organization,* and *prediction*. But underlying all scientific work is the fundamental belief that what is unknown about nature will be no more difficult to discover than that which is known, and that nature's basic simplicity and orderliness will yield itself to human enterprise.

PROBLEMS*

1. The amount of sand in a truckload is often expressed in "yards." How would you use a yard length to express volume?

2. Which of the following is *not* convertible to cubic feet? (a) quart, (b) cubic mile, (c) acre, (d) fifth of a gallon

3. Criticize the following statement: All heavy objects sink in water.

4. Criticize the following statement: All light objects float on water.

5. Explain the following in terms of weight and mass. (a) A bicyclist coasts downhill at increasing speed and (b) coasts up the other side. *Hint:* In which part does *weight* cause motion and in which part does *inertia* cause motion?

6. Explain the following in terms of weight and mass: A bicyclist finds it difficult (a) to start from a standstill in high gear and (b) to pedal slowly up a steep hill; however (c) he/she finds it extremely easy to pedal slowly on level ground.

7. Without referring to the periodic table, give the name and location by principal group number and period number of each of the following elements:
(a) Na   (d) C
(b) Cl   (e) S
(c) He

8. Same as problem 7.

(a) K   (d) Si
(b) I   (e) O
(c) Ne

9. Without referring to the periodic table, give the symbol and location by principal group number and period number for each of the following elements:
(a) radium   (d) phosphorus
(b) argon   (e) hydrogen
(c) lithium

10. Same as problem 9.

(a) calcium   (d) nitrogen
(b) krypton   (e) bromine
(c) aluminum

11. Describe each of the following changes as being either physical or chemical:
(a) Salt and water are separated by freezing part of the water.

(b) Gasoline combines with oxygen to form water and carbon dioxide.

12. Describe the following changes as being either physical or chemical:
(a) A pure, red, solid substance is heated and produces a colorless gas and a silvery metallic liquid.
(b) Salt and water are separated by distillation of the water.

13. Identify the following mixtures as being either homogeneous or heterogeneous:
(a) sugar and water
(b) a vein of gold bearing quartz

(c) air (unpolluted)

14. Same as problem 13.

(a) gasoline and water
(b) a "sterling silver" bracelet (which contains silver and copper)
(c) a concrete sidewalk

*Odd and even number problems are paired. Answers to even numbered problems will be found on pages 311–320.

15. When ice is formed from liquid water the physical change is described as "freezing." Give correct names for each one of the following physical changes.
    (a) It rains.

    (b) Snow disappears on a very cold day (below 32°F).
    (c) Liquid air (which is made by cooling air to −320°F) disappears after being poured into an empty beaker at room temperature. *Hint:* See Figure 1.5.

16. Same as problem 15.

    (a) Water disappears from an overheated automobile radiator.
    (b) Dry ice (carbon dioxide) disappears when dropped into water.
    (c) Moisture forms on the surface of a cold drink can.

17. In naming the following compounds it is not necessary to use the Greek prefixes indicating number. Give the correct name for each.
    (a) CaO   (c) $RaCl_2$
    (b) Na_2S (d) AlH_3

18. Same as problem 17.

    (a) NaBr   (c) Al_2S_3
    (b) $MgCl_2$ (d) CaH_2

19. Name the following compounds using the proper Greek prefixes to indicate number.
    (a) CCl_4
    (b) $SiBr_4$
    (c) SO_3

20. Same as problem 19.

    (a) CF_4
    (b) NI_3
    (c) SO_2

2

MEASUREMENT AND QUANTITATIVE PROPERTIES

A large part of chemistry is written in the language of numbers. Much concerning nature is revealed to the person who is willing to make careful measurements, and those measurements are usually expressed in terms of numbers. Whether the measurements are of mass, volume, distance, or any of the more esoteric quantities in chemistry, there is almost invariably a scale of graduated amounts from which the numbers can be read or estimated.

Before turning our attention to the precision of these scales let us look at the units of measurement in which scales are marked. The system used throughout the scientific world is the metric system.

2.1 THE METRIC SYSTEM OF MEASUREMENT AND POWERS OF TEN

The language that the chemist uses to express himself is often different from our everyday language, but if you wish to understand him you must learn his language. In choosing a language, the chemist, and all scientists, keep two things in mind, *precision* and *simplicity*.

Although the English system of measurement *is precise, it is not simple*. The length of the foot is precisely known, but to change from feet to inches or feet to miles peculiar conversion factors must be used. The relationship between length and volume is even more complex. Few can remember how to convert cubic feet to quarts. Furthermore, no simple relationship exists between mass and volume units. In the metric system, however, one unit can be converted to another by simply multiplying by a power of ten.

Powers of ten are shown by exponents. The numbers . . . , $\frac{1}{100}$, $\frac{1}{10}$, 1, 10, 100, . . . can be written . . . , 10^{-2}, 10^{-1}, 10^0, 10^1 (or 10), 10^2, ... You will observe that the exponent in the second set of numbers represents the number of zeros in the first set of numbers. If the number is less than 1, the exponent is negative; if it is greater than 1, the exponent is positive.

The basic metric units are listed in Table 2.1. The letter shown in the parenthesis is the standard abbreviation for the unit indicated.

TABLE 2.1 The basic units of measurement in the metric system*

length	meter (m)
volume	liter (l)
mass (weight)	gram (g)

In SI units, or International System of Units, the base unit of mass is the kilogram (See Table 2.3). There is no base unit for volume. A derived unit, cubic meter (m^3), is used instead.

Several prefixes may be added to these units to derive new units. The most common of these prefixes are listed in Table 2.2.

TABLE 2.2 The common prefixes used in the metric system

kilo-(k)	meaning 1000 or 10^3
centi-(c)	meaning $\frac{1}{100}$ or 10^{-2}
milli-(m)	meaning $\frac{1}{1000}$ or 10^{-3}
micro-(μ)	meaning $\frac{1}{1,000,000}$ or 10^{-6}
nano-(n)	meaning $\frac{1}{1,000,000,000}$ or 10^{-9}
pico-(p)	meaning $\frac{1}{1,000,000,000,000}$ or 10^{-12}

Each of the prefixes in Table 2.2 represents a power of ten. You should become familiar with the exponential notation shown in this table. Examples 1 through 6 will show you how to use it in multiplication and division and how to interpret negative exponents.

EXAMPLE 1.   What power of 10 is 1000? Use the exponential notation to write this power.

Solution: 1000 is the third power (or "cube") of 10, because, 10 times itself *three* times is equal to 1000. In exponential notation this gives

$$10 \times 10 \times 10 = 10^3$$

You will notice that the power of ten is the same as the number of zeros in 1000.

EXAMPLE 2.   What power of 10 is $\frac{1}{100}$? Use the negative exponential notation to write this power.

Solution: Use the same method as that in Example 1 to rewrite the denominator as a power of 10.

$$\frac{1}{100} = \frac{1}{10} \times \frac{1}{10} = \frac{1}{10^2}$$

For convenience and by definition a power may be moved from denominator to numerator (and *vice-versa*) by reversing the sign of the exponent

$$\frac{1}{10^2} = 10^{-2}$$

In words, you may say that "$\frac{1}{100}$ is the negative two power of 10."

The rules for handling exponents in multiplication and division of powers of ten are as follows:

Multiplication and Division of Powers of Ten

1. To *multiply* powers of ten, *add* their exponents.
2. To *divide* powers of ten, *change the sign* of the exponent in the denominator and *add* it to the exponent in the numerator.
3. By definition, $10^0 = 1$.

EXAMPLE 3.   What is the product of $10^3 \times 10^4$?

Solution: $$10^{3+4} = 10^7$$

EXAMPLE 4.   What is the product of $10^3 \times 10^{-3}$?

Solution: $$10^{3-3} = 10^0 = 1$$

EXAMPLE 5.   What is the quotient of $10^6/10^2$?

Solution: $$10^6 \times 10^{-2} = 10^{6-2} = 10^4$$

EXAMPLE 6.   What is the quotient of $10^{-7}/10^{-2}$?

Solution: $$10^{-7} \times 10^2 = 10^{-7+2} = 10^{-5}$$

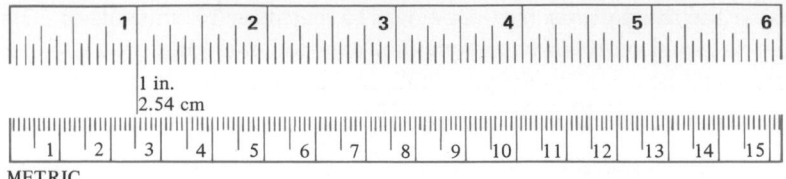

Figure 2.1.   Comparison of English and metric units of length.

In Table 2.3 you will find a list of the most important units derived from the metric system; some useful interrelationships among length, volume, and mass units; and some English equivalents.

TABLE 2.3   Derived metric units, useful interrelationships, and English equivalents

1 kg = 1000 g	2.54000 . . . cm $\equiv$ 1 in. (by definition)
1 cm = $\frac{1}{100}$ m	454 g = 1 lb
1 ml = $\frac{1}{1000}$ ℓ	1 liter = 1.06 qt

1 ml H_2O (at 4°C and standard atmospheric pressure) has mass = 1 g

1 Å = 10^{-8} cm (Angstrom, a unit of length)

One of the psychological problems of adjustment to the metric system will be the initial unfamiliarity with estimating quantities by hand and eye. In Figure 2.1 a 6-in. ruler and a centimeter scale are shown together in their true proportions.

In Figure 2.2 a 10-cm cube is compared with a quart bottle of milk. Since the volume is 1000 cm³, which is the same as 1000 ml, it is exactly equal to 1 ℓ. For

Figure 2.2.   Comparison of English and metric units of volume and weight.

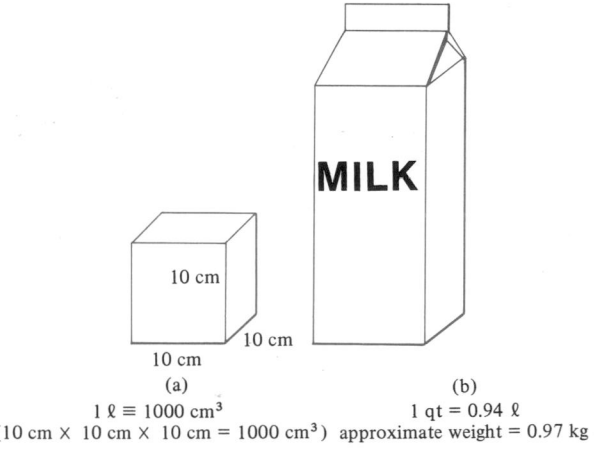

(a)
1 ℓ $\equiv$ 1000 cm³
(10 cm × 10 cm × 10 cm = 1000 cm³)

(b)
1 qt = 0.94 ℓ
approximate weight = 0.97 kg

purposes of comparison, a quart container is shown next to it on the same scale. The quart and the liter have very nearly the same volume (see Table 2.3) and the weight of a quart of milk is very nearly 1 kg. Whole milk is slightly heavier than an equal volume of water.

2.2 CONVERTING UNITS BY THE ONE-FACTOR METHOD

Conversion of one metric unit to another is a very simple and very important process. The method used here is one you may use for many types of calculations throughout this book. It is an example of dimensional analysis, which we shall call the **one-factor method.** It can be used whenever the unknown quantity is *proportional* to the known quantity, either directly or indirectly.* In this method you begin with a known numerical value, complete with units, and multiply it by a "one-factor" which has units so chosen as to convert the original units to whatever new unit is desired. (This is equivalent to multiplying by 1.) You may proceed by making a fraction containing equal quantities of the original unit and the new unit. For example, if the unit you wish to cancel is kilometer, you would use a one-factor that has kilometer in the denominator. The simplest equality to remember for this unit is

$$1 \text{ km} = 1000 \text{ m}$$

The corresponding one-factor would be written as follows.

$$\frac{1000}{1} \frac{m}{km}$$

Multiplying a number containing kilometer by this factor will be like multiplying by 1. It will not change its value, but it will eliminate the kilometer as a unit by cancellation. The process can be repeated chain fashion as often as desired as shown in Examples 1 through 4.

EXAMPLE 1. How many centimeters (cm) are 100 km?

Solution: Both centimeter and kilometer are units of length so they must be proportional to each other. For example if the number of kilometers were doubled the number of centimeters would be doubled. Therefore you may begin a solution of the problem by writing,

$$x \text{ cm} = (100 \text{ km}) \cdots$$

where x is the unknown number. Then you may generate the appropriate one-factors by using equalities that are familiar to you. In this example 100 km is the numerical value of a certain distance. An

*A discussion of proportionality is found in the Appendix.

appropriate "one-factor" is $\dfrac{1000}{1}\dfrac{m}{km}$ since 1000 m = 1 km and the kilometer in 100 km will cancel. This multiplication alone would convert kilometer to meter, but multiplication by another "one-factor," $\dfrac{1}{10^{-2}}\dfrac{cm}{m}$, will cancel the meter and convert the distance to centimeter. This can all be done in a single chain as shown.

$$x\ cm = (100\ \cancel{km})\left(\frac{1000}{1}\ \frac{\cancel{m}}{\cancel{km}}\right)\left(\frac{1}{10^{-2}}\ \frac{cm}{\cancel{m}}\right)$$
$$= (10^2)(10^3)(10^2)\ cm$$
$$= 10^7\ cm$$

EXAMPLE 2. Express 10 ml in terms of a liter.

Solution: Since milliliter and liter are both units of volume, they are directly proportional to one another. You may begin by writing the following:

$$x\ l = (10\ ml)\ \cdots$$

A one-factor is needed with milliliter in the denominator to cancel milliliter already in the numerator. You may use the equality 1 ml = 10^{-3} l, which comes from the meaning of the prefix, 1 ml = $\dfrac{1}{1000}$ l.

$$x\ l = (10\ \cancel{ml})\left(\frac{10^{-3}}{1}\ \frac{l}{\cancel{ml}}\right)$$
$$= (10)(10^{-3})\ l$$
$$= 10^{-2}\ l$$

EXAMPLE 3. How many centimeters are in 2 ft?

Solution: The number of centimeters is proportional to the number of feet; therefore, you may write

$$x\ cm = (2\ ft)\ \cdots$$

A one-factor is needed with feet in its denominator and, since 1 ft = 12 in., you may continue as follows.

$$x\ cm = (2\ \cancel{ft})\left(\frac{12}{1}\ \frac{in.}{\cancel{ft}}\right)\cdots$$

From Table 2.3 you may use the equality, 2.54 cm = 1 in., to provide a one-factor with inch in the denominator to cancel with inch already in the numerator.

$$x\ cm = (2\ \cancel{ft})\left(\frac{12}{1}\ \frac{\cancel{in.}}{\cancel{ft}}\right)\left(\frac{2.54}{1}\ \frac{cm}{\cancel{in.}}\right)$$
$$= (2)(12)(2.54)\ cm$$
$$= 60.96\ cm$$

EXAMPLE 4. How many liters are in 1 m³?

Solution: In this example 1 m³ is the numerical value of a certain volume. Since the number of liters is directly proportional to the number of cubic meters (m³) you should begin with 1 m³.

A useful one-factor is $\dfrac{1 \; cm^3}{1 \; ml}$ since 1 cm³ = 1 ml and this factor will provide a bridge between cubic meter and liter. Also, since 1 cm = 10^{-2} m, $\dfrac{1 \; cm}{10^{-2} \; m}$, and its cube $\dfrac{1^3 \; cm^3}{(10^{-2})^3 \; m^3}$ are also one-factors. The final one-factor needed, $\dfrac{10^{-3} \; \ell}{1 \; ml}$, comes from the equality 1 ml = 10^{-3} ℓ.

$$x \; \ell = (1 \; \cancel{m^3}) \left(\frac{1^3 \; \cancel{cm^3}}{(10^{-2})^3 \; \cancel{m^3}} \right) \left(\frac{1 \; \cancel{ml}}{1 \; \cancel{cm^3}} \right) \left(\frac{10^{-3} \; \ell}{1 \; \cancel{ml}} \right)$$

After like units in numerator and denominator are cancelled, the only unit that remains is liter. The numbers should be grouped together for computation.

$$x \; \ell = \frac{10^{-3}}{(10^{-2})^3} \; \ell = \frac{10^{-3}}{10^{-6}} \; \ell = 10^3 \; \ell$$

or

$$1 \; m^3 = 1000 \; \ell$$

Note that in this power of ten calculation, $(10^{-2})^3 = (10^{-2})(10^{-2})(10^{-2}) = 10^{-2-2-2} = 10^{-6}$.

Cubic meters might be a convenient unit for measuring volumes of gases, which have very little density compared with solids and liquids. One m³ of air at ordinary pressures and temperatures weighs about 1.2 kg (one liter of liquid water weighs 1 kg).

The desirability of making calculations accurately and quickly in chemistry cannot be overemphasized. Most troubles in chemistry courses are a result of this inability. As a student you need some method that will make clear at every step of the way where you have been and where you need to go. The *one-factor method* is just such a method, and you should keep practicing it until you can make calculations with ease and confidence. Never make a calculation until the problem is set up and the chain has been extended to its practical limits.

The metric system has so many advantages that in retrospect it is difficult to understand why this country kept the English system after the revolution. After all we did abandon shilling, pence, and pound. Thomas Jefferson tried to convince the Congress that a change would be desirable during his presidency, but he failed. The National Bureau of Standards has a quiet program attempting to bring about such a change. In time we may be the only English-speaking country using the English system. England has switched to the metric system. Australia is well on its way toward a complete change (Figure 2.3). Only Brunei, Burma, Liberia, Yemen, and the United States have not adopted the metric system.

Figure 2.3. A part of the Australian government's campaign to switch to the metric system.

"A pint's a pound the world around", is an old saying of dubious validity.* Soon, however, a liter of water will be known to weigh a kilogram in a much larger portion of the world than ever before.

2.3 TEMPERATURE SCALES

In the metric system of measurement new units are generated by multiplying base units by different powers of ten. This is not the case with the Celsius or centigrade scale of temperature. However, the Celsius scale is widely used throughout the world (Figure 2.3). It is used instead of the Fahrenheit scale by chemists in the United States. These two scales are best defined by describing the construction and operation of thermometers.

If the temperature of a material is increased, the volume generally increases. This effect is commonly used in designing thermometers. The mercury (Hg) thermometer, for example, is constructed by filling a reservoir bulb at one end of a length of glass capillary tubing with mercury (see Figure 2.4). Its use as a thermometer depends on the fact that mercury expands more than the glass when both are heated, and the small increase in volume of the mercury in the bulb causes a significant change in the height of the mercury column in the fine bore of the capillary. The space above the column of mercury is evacuated.

A mark may be made on the glass at the mercury level while the thermometer is immersed in a water and ice mixture and given the value 0°C (Celsius). A similar mark may be made for a water and steam mixture and called 100°C. Then we have the basis for a temperature scale. The distance between these marks may be divided into 100 equal divisions to represent intermediate temperatures, and divisions of the same size may be added above 100°C and below 0°C to measure higher and lower temperatures.

---

* Sixteen fluid ounces are in 1 pt and 16 oz are in 1 lb, but a fluid ounce is a measure of volume, not weight. A fluid ounce of water does not weigh 1 oz but 1.04 oz!

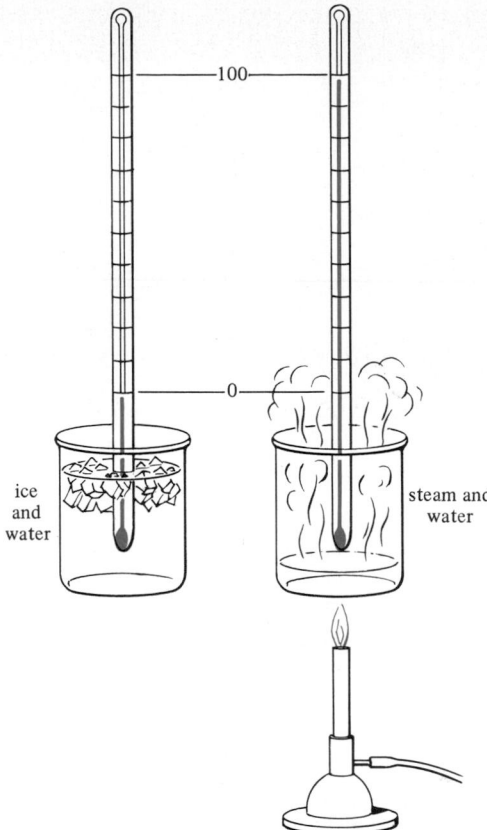

Figure 2.4. Thermometer, Celsius scale.

The choice of zero on this scale is quite arbitrary, and other choices are equally valid. For example on the Fahrenheit scale, $0°F = -17.8°C$.

The relationship between the Fahrenheit and Celsius scales is shown graphically in Figure 2.5. A straight line is established by two points: $(100°C, 212°F)$ and $(0°C, 32°F)$. Then any point on the line represents corresponding temperatures on the Celsius and Fahrenheit scales. Such straight line relationships are called **linear** relationships.

Exercise 1. The highest temperature container that can be held comfortably against your cheek is said to be about 55°C. Using the graph in Figure 2.5, estimate what this temperature is on the Fahrenheit scale.

Answer: (131°F)

A formula that gives an exact relationship between Celsius and Fahrenheit is

$$°F = (°C \times \tfrac{180}{100}) + 32$$
$$°F = (°C \times \tfrac{9}{5}) + 32$$

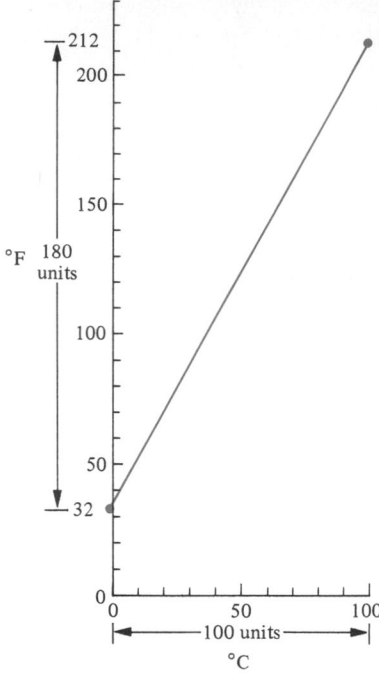

Figure 2.5. Graphical relationship between the Celsius and Fahrenheit scales of temperature.

This formula uses the fact that 100 units on the Celsius scale is equal to 180 units on the Fahrenheit scale (see Figure 2.5). Thus degrees Celsius must be multiplied by $\frac{180}{100}$ or $\frac{9}{5}$. Also since °F = 32 when °C = 0 (substitute these numbers into the formula), the number 32 must be added to give the correct relationship. This formula is applicable to any temperature, including negative temperatures.

EXAMPLE 1. Calculate the temperature in degrees Celsius corresponding to 131°F.

Solution:

$$131 = (°C \times \tfrac{9}{5}) + 32$$
$$131 - 32 = °C \times \tfrac{9}{5}$$
$$(131 - 32) \times \tfrac{5}{9} = °C$$
$$99 \times \tfrac{5}{9} = °C$$
$$55 = °C$$

Exercise 2. Show that $-40°C = -40°F$. This is the *only* temperature that is identical on the two scales.

Exercise 3. Show that $37.0°C = 98.6°F$. This is normal body temperature.

The Swedish astronomer Celsius chose the points 0°C and 100°C on his scale simply for the ease of reproducing these temperatures in laboratories around the world. The German physicist, Fahrenheit, who invented the mercury thermometer, had a physiological basis for 100°F; he probably believed 100°F was normal body temperature. He attempted, by mixing various salts with snow and water, to reach the lowest possible temperature. The lowest temperature that he reached was called 0°F. Lord Kelvin, an English physicist, carried this idea still further. He observed the behavior of gases as their temperatures are lowered (Section 3.5) and established an absolute zero of temperature, which he argued was the lowest possible temperature. On his scale 0°C is 273°K, called "degrees Kelvin" or "degrees absolute." In general you can calculate *absolute temperature* as follows:

$$°K = °C + 273$$

EXAMPLE 2.   Normal room temperature is considered to be 25°C. What is this in degrees Kelvin?

Solution:

$$°K = °C + 273$$
$$= 25 + 273$$
$$= 298$$

EXERCISE 4.   Calculate −40°C in °K.

Answer: (233°K)

2.4   SIGNIFICANT FIGURES AND ERRORS

You have now been introduced to most of the units of measurement that chemists use in their everyday work. When the chemist counts these units on the scale of whatever piece of equipment he happens to be using, he gets a number represented by a certain figure, which he writes down. Of course he tries to estimate the smallest possible fraction of the scale unit. The idea of **significant figures** is based on the fact that it is possible to read a scale just so accurately and no more.

Suppose that you wish to measure the three dimensions, length (l), width (w), and height (h), of the metal bar in Figure 2.6 using the centimeter scale provided. What is the error in measuring each one of these dimensions, and how many significant figures are in each? If you now calculate the volume of the bar from these dimensions, how great is the error in the calculated value and how many significant figures should be used to express it?

In reading scales you can generally make a visual estimate of 0.1 the smallest unit, if the unit is not too small. Since each centimeter is marked into ten units in Figure 2.6, each one of these is 0.1 cm. This means that the best visual estimate that can be made is 0.1 × 0.1 cm or 0.01 cm. This is considered the probable error in every measurement, regardless of its total length. For example in measuring the height, $h = 0.37$ cm, but it is possibly as large as 0.38 cm or as small as 0.36 cm. Such a value is often written as 0.37 ± 0.01. Since both the 3 and the 7 are known with some

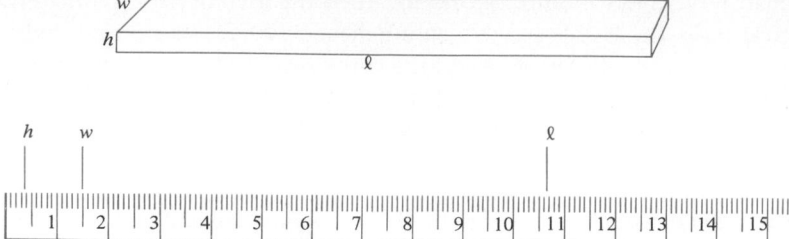

Figure 2.6.   Errors in measurement of a metal bar of length (*l*), width (*w*), and height (*h*).

degree of certainty, this number is said to have two significant figures. There would still be two significant figures if the uncertainty were as large as 0.37 ± 0.04. Note that zeros *before* the decimal point are never considered significant.

The **percentage error** depends on the size of the measurement in the following way:

$$\% \text{ error} = \frac{\text{error in the quantity}}{\text{the quantity}} \times 100\%$$

The per cent error in the measured height we have chosen as an example may then be calculated.

$$\% \text{ error} = \frac{0.01 \text{ cm}}{0.37 \text{ cm}} \times 100\%$$

$$= \frac{1}{0.37}\%$$

$$= 2.7\%$$

$$= 3\% \text{ (rounded off)}$$

A table of all the measured values follows.

Dimension	Estimated Value (cm)	Number of Significant Figures	Probable Error (cm)	Error (%)
h	0.37	2	0.01	3
w	1.49	3	0.01	0.7
l	10.65	4	0.01	0.09

You will see from these examples that *zeros are* significant when located between two nonzero digits. They are also significant when they express a degree of certainty at the end of a number.

Now, to calculate the volume you may use the following formula:

$$\text{volume} = l \times w \times h$$
$$\text{volume} = 10.65 \times 1.49 \times 0.37 \text{ cm}^3$$

The product of these numbers appears in the display of an eight-digit hand calculator shown in Figure 2.7, and of course it is the same product that would be obtained by hand.

How many of these digits are you justified in writing down, or, in other words, how many significant figures are in the product? The upper and lower limits are obtained below by combining the largest dimensions with one another, the intermediate ones with one another and the smallest ones with one another.

$$\text{upper limit} = 10.66 \times 1.50 \times 0.38 = 6.076200 \text{ cm}^3$$
$$\text{intermediate value} = 10.65 \times 1.49 \times 0.37 = 5.871345 \text{ cm}^3$$
$$\text{lower limit} = 10.64 \times 1.48 \times 0.36 = 5.668992 \text{ cm}^3$$

Rounded off, these values are 6.1, 5.9, and 5.7, respectively, or $5.9 \pm 0.2 \text{ cm}^3$. Therefore, the calculated volume has two significant figures, the error is 0.2 cm³ and the per cent error is calculated as follows:

$$\% \text{ error} = \frac{0.2}{5.9} \frac{\text{cm}}{\text{cm}} \times 100 = 3.4$$
$$= 3\% \text{ (rounded off)}$$

The per cent error is approximately the same as that in the height (h) which was the dimension having the largest per cent error. Similarly, the number of significant figures is the same as that of the dimension with the fewest number of significant figures.

Before looking at errors in problems of division, consider that property of matter which combines both volume and weight.

Figure 2.7. Display on an eight-digit calculator for the product 10.65 × 1.49 × 0.37.

When Archimedes' law of floating bodies was explained, you saw that it was the weight per unit volume of the object compared with that of water which determined whether or not the object floated (Section 1.2). Although weight and volume are important properties in comparing objects with one another, neither property taken separately gives you any idea of what material the object is made. Golden Gate Bridge is certainly larger and heavier than London Bridge, but this tells you nothing of their composition. On the other hand, the weight per unit volume is characteristic of the substance of which objects are composed. This ratio, which is obtained by dividing the weight (mass) of a sample by its volume, is called its density.

DEFINITION:
$$density = \frac{weight}{volume}$$

A convenient choice of density units for liquid and solid substances is grams per cubic centimeter ($g\ cm^{-3}$). This unit is sometimes written "g per cm^3," or "g/cm^3." The same rule applies to units that applies to exponents of powers of 10 that are moved from denominator to numerator or vice-versa: *change the sign of the exponent*. The densities of a number of different materials are shown in Table 2.4.

TABLE 2.4 Densities of various materials

Material	Density ($g\ cm^{-3}$)
balsa wood	0.11–0.14
cork	0.22–0.26
gasoline	0.66–0.69
butter	0.86–0.87
human fat	0.92
water	1.00
sea water	1.03
aluminum (Al)	2.70
titanium (Ti)	4.5
iron (Fe)	7.9
mercury (Hg)	13.6
osmium (Os)	22.5

Golden Gate Bridge, which is made of steel, has a density of $7.8\ g\ cm^{-3}$ whereas London Bridge, which is made of granite, has a density of $2.7\ g\ cm^{-3}$. Since the density of water is $1.000\ g\ cm^{-3}$ at 4°C (and very nearly the same for all other temperatures at which it is a liquid) both Golden Gate and London Bridges would sink if dropped into water. On the other hand, a paraffin candle, which has a density of $0.89\ g\ cm^{-3}$, and a block of pine wood, which varies in density from 0.4 to $0.6\ g\ cm^{-3}$ depending on the species, both float on water because their densities are less than $1.0\ g\ cm^{-3}$.

The property of density may be used to distinguish one element from another. The metals in principal group I of the periodic table have very low densities for solids. For the first three, lithium (Li), sodium (Na), and potassium (K), the densities are 0.53, 0.97, and $0.86\ g\ cm^{-3}$, respectively. Since these densities are less than that of water, they all float on water (Section 1.4). As you move to the right in the table, the density of the elements in their solid state increases, reaching a maximum toward the middle of a period. These regularities in observed properties are examples of the periodic nature of the arrangement.

EXAMPLE 1. Suppose that you wish to know the volume of the Hope diamond and that you know its weight is 44 carats, to the nearest carat. The diamond itself, Figure 2.8 is mounted in a setting surrounded by smaller diamonds, which you need not consider. The Hope diamond is blue in color* but it is quite pure and has the same density as any other diamond, 3.51 g cm⁻³.

Solution: From the density and its weight in carats you can calculate the volume of the diamond. The carat is equal to 0.200 g. Since the volume of the diamond in cubic centimeters is directly proportional to its weight in carats, you may begin in the following way.

$$x \text{ cm}^3 = (44 \text{ carats}) \cdots$$

A one-factor is needed with carat in the denominator to cancel carat already in the numerator. You may use the equality, 1 carat = 0.200 g.

$$x \text{ cm}^3 = (44 \text{ carat}) \left(\frac{0.200}{1} \frac{g}{\text{carat}} \right) \cdots$$

To obtain a one-factor with gram in the denominator, which will cancel with gram already in the numerator, use the fact that 1 cm³ of diamond is *equivalent to*† 3.51 g. You will obtain cubic centimeter in the numerator, which is also needed.

$$x \text{ cm}^3 = (44 \text{ carats}) \left(\frac{0.200}{1} \frac{g}{\text{carat}} \right) \left(\frac{1}{3.51} \frac{\text{cm}^3}{g} \right)$$

$$= \frac{(44)(0.200)}{3.51} \text{ cm}^3$$

Figure 2.8. The Hope diamond weighs 44 carats. What is its volume?

*It is very likely that the Hope diamond was cut from the French Blue diamond (68 carats) which was stolen from the Royal Treasury of France in 1792, and never recovered. The most prominent owner of the "Blue" was Marie Antoinette. The Hope diamond was purchased for $180,000 in 1947 and given to the Smithsonian Institution in Washington, D.C.

† See Appendix for the meaning of "equivalent to."

If you are working this problem without the aid of hand calculator or slide rule, you will probably proceed in the following way.

$$(44)(0.200) = 8.8$$

$$
\begin{array}{r}
2.5071\cdots \\
351\overline{)880.0000} \\
702 \\
\hline
1780 \\
1755 \\
\hline
2500 \\
2457 \\
\hline
430 \\
351 \\
\hline
79
\end{array}
$$

Summarizing

$$\text{volume} = \frac{(44)(0.200)}{3.51} = 2.5071\ldots\,\text{cm}^3$$

The same question arises as before: How many significant figures are justified? To obtain a partial answer to this question the calculation may be repeated using the presumed uncertainty in your knowledge of the diamond's weight, ± 1 carat. After all, it may weigh as little as 43 carats and as much as 45 carats. Using the smallest weight you have,

$$\text{volume} = \frac{(43)(0.200)}{3.51} = 2.4501\ldots\,\text{cm}^3$$

and using the largest,

$$\text{volume} = \frac{(45)(0.200)}{3.51} = 2.5641\ldots\,\text{cm}^3$$

All you know about the volume is that it is between 2.45 and 2.56 cm^3, or 2.51 ± 0.06 cm^3. With this much uncertainty in the third significant figure, it is probably better to express the value as 2.5 cm^3 with two significant figures. (If the uncertainty were less than 0.05 cm^3 it would be better to express the volume as 2.51 cm^3.)

As a general **rule** for significant figures **in multiplication and division,** *the answer obtained when numbers are multiplied or divided has the same number of significant figures as the number used in the calculation which has the fewest number of significant figures.*

In the preceding example you are perfectly justified in using a slide rule, which gives three significant figures, and you would be wasting time if you made the calculation on paper. Still, if you have not learned to use a slide rule yet and do not have access to a calculator, you can save time by rounding off to a smaller number of significant figures, where permissible, before multiplying or dividing. In the example in this section the same answer would be obtained if all numbers were written as two significant figures, and the calculation would be simplified, as shown.

$$\text{volume} = \frac{(44)(0.20)}{3.5} = \frac{8.8}{3.5}$$

$$\text{volume} = 2.5 \text{ cm}^3$$

$$
\begin{array}{r}
2.5 \\
35\overline{)88.0} \\
70 \\
\hline
180 \\
175 \\
\hline
5
\end{array}
$$

In the Section 2.5 you will see that it is often desirable to express numbers as "powers of ten" in calculations, and you will learn how significant figures are involved in using this technique.

2.5 SCIENTIFIC NOTATION: MORE ON POWERS OF TEN

The **powers of ten method** or **scientific notation** is used by chemists for two reasons. First, it indicates the degree of precision of the numbers used in a calculation. Second, it avoids errors arising from the use of "bulky" numbers which have a large quantity of zeros before or after the decimal point. The rules are few and easy to remember once put into use.

RULE I. *Move the decimal point in each number to the position behind the first nonzero digit, giving a number between 1 and 10.*

Case A: If the number is greater than 1, count the number of moves and multiply by this power of 10, using a *positive* exponent.
For example

$$3067.5 = 3.0675 \times 10^3$$

(Note also that, $3000 = 3 \times 1000 = 3 \times 10^3$.)
Case B: If the number is less than one, do just as before, but use a *negative* exponent.
For example

$$0.0491 = 4.91 \times 10^{-2}$$

(Note also that, $0.04 = 4 \times \dfrac{1}{100} = 4 \times 10^{-2}$.)

RULE II. *Multiply and divide the numbers and their powers of 10 **separately**.*

Case A: In *multiplication, add* the exponents of the powers of 10.
For example

$$(3.0675 \times 10^3)(4.91 \times 10^{-2}) = 15.10 \times 10^{3+(-2)}$$
$$= 15.10 \times 10^1 \quad \text{or} \quad 1.510 \times 10^2$$

Case B: In division, subtract exponents.
For example

$$\frac{3.0675 \times 10^3}{4.91 \times 10^{-2}} = 0.624 \times 10^{3-(-2)} = 0.624 \times 10^5 \quad \text{or} \quad 6.24 \times 10^4$$

Scientific notation permits you to indicate whether a zero is a significant figure or not. All nonzero digits are significant, but zeros are not significant when they merely indicate where the decimal point should be. It is customary to write an extra zero to the left of the decimal point of a number less than one. This zero emphasizes the decimal point and helps avoid errors, but it is not significant.

EXERCISES 1. How many significant figures are indicated in each number below?
(a) 3067.5 *and* 3.0675 × 10³
(b) 0.0491 *and* 4.91 × 10⁻²
(c) 23,000,000
(d) 2.3 × 10⁷
(e) 2.30 × 10⁷

Answers:

(e) 3 significant figures (definitely)
(d) 2 significant figures (definitely)
(c) 2 significant figures (probably, but possibly 8)
(b) 3 significant figures
(a) 5 significant figures

In the last number, the zero is considered significant because it was not needed to locate the decimal point and, thus, must indicate a greater degree of precision.

When you use the hand calculator it is often necessary to employ scientific notation. Any number as large as 100,000,000 or smaller than 0.00000001 cannot be displayed on an eight-digit calculator, and such numbers are quite common in chemistry. In Figure 2.9, the number, 6.023 × 10²³, is displayed on a calculator capable of scientific notation. This facility is indicated by the presence of EE on the keyboard. Without the facility of powers of ten on your keyboard you would find it necessary to make the power of ten calculation separately as shown in Rule II. Significant figures must be determined separately with either type of calculator, using the rule in Section 2.4.

Scientific notation is ideal for slide rule calculations. Since the slide rule does not distinguish between different powers of ten of the same set of significant figures and does not indicate where the decimal goes in the final answer, one needs an approxi-

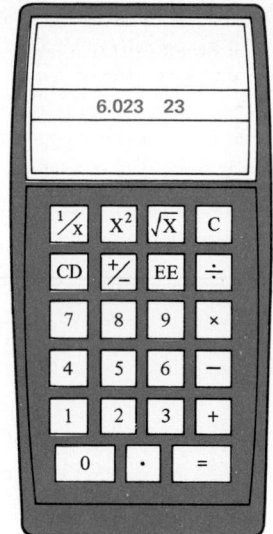

Figure 2.9. Display in scientific notation of the number 6.023×10^{23}.

mation to the answer, preferably before turning to the slide rule. This is easily done using scientific notation.

For example, take the product

$$3067.5 \times .0491 = (3.0675 \times 10^3)(4.91 \times 10^{-2})$$

An approximation is easily obtained by rounding off the numbers to one significant figure.

$$(3 \times 10^3)(5 \times 10^{-2}) = (3 \times 5)(10^3 \times 10^{-2})$$
$$= 15 \times 10^1$$

Thus it is known before picking up the slide rule that the product is approximately 150. The slide rule calculation increases the significant figures and the approximation removes any doubt about where the decimal goes.

Every chemist has made calculations without benefit of calculator, slide rule, paper, pencil, or even lights. Frequently this is as good a calculation as he needs, or the best that is justified considering the uncertainty of the numbers he is working with. Every chemistry student should learn to appreciate the importance of calculations that give only one significant figure and, of course, the correct power of ten.

Try the following exercises using scientific notation and using the rule for significant figures. Note that sometimes a calculated value beginning with the integer 1 is given an "extra" significant figure.

Exercises 2. (a) $(2.30 \times 10^7) \times (1.65 \times 10^{-3})$

(b) $\dfrac{7.81 \times 10^{-3}}{6.4298 \times 10^6}$

(c) $(0.000212) \times (0.000063)$

(d) $\dfrac{0.000063}{0.000212}$

(e) $(6.9273 \times 10^2) \times (8.96 \times 10^{-3}) \times (5 \times 10^6)$

Answers:

(a) 3.80×10^4

(b) 1.215×10^{-9} or 1.22×10^{-9}

(c) 1.34×10^{-8} or 1.3×10^{-8}

(d) 3.0×10^{-1}

(e) 3×10^7

2.6 RULES FOR DECIMAL PLACES IN ADDITION AND SUBTRACTION

When the estimated values of two quantities must be added or subtracted to give the approximate value of a third quantity, it is the absolute errors that are carried through the operation. For example, suppose a boy wants to weigh his cat as accurately as possible with the bathroom scales, which he can read to the nearest 0.2 lb. The method he uses is to weigh himself and the cat and then himself alone.

weight of boy and cat	84.4 ± 0.2 lb
weight of boy alone	-76.2 ± 0.2 lb
weight of cat	8.2 ± 0.4 lb

The largest error expected would be 0.4 lb obtained by adding the two errors $0.2 + 0.2$ lb, and the weight of the cat is correctly written with one decimal place.

Scientists try to design experiments so that they will not have to subtract two values that are almost the same size from one another, because the derived value has fewer significant figures than the original values. Although the per cent errors in the weighing operations are both small (approx. $\frac{0.2}{80} \times 100 = 0.3\%$) the possible per cent error in the calculated weight of the cat is more than ten times as large (approx. $\frac{0.4}{8} \times 100 = 5\%$).

A similar decimal point rule applies in addition. Suppose the cat is found to weigh 8.26 ± 0.02 pounds using some other more accurate scale and method. What would be the best way of reporting the combined weight of cat and boy?

weight of boy alone	76.2 ± 0.2 lb
weight of cat	8.26 ± 0.02 lb
weight of cat and boy	84.46 ± 0.22 lb
which should be written	84.5 ± 0.2 lb

The **rule** that applies for use of significant figures **in addition and subtraction** is that *the sum or difference can have no more decimal places than either of the values being added or subtracted.*

EXAMPLE 1. A road map gives the distance from Tampa to Chicago as 1159 mi. If the distance from the University of South Florida (USF) on the northern outskirts of Tampa to downtown is 10.3 mi, how far is it from USF to Chicago?

Solution: Most road maps report mileage figures to the nearest mile; therefore, it would not be correct to write 1159 mi as 1159.0 mi. The calculation is made as follows:

$$
\begin{array}{ll}
1159 \text{ mi} & \text{distance from Chicago to downtown Tampa} \\
-10.3 \text{ mi} & \text{distance from USF to downtown Tampa} \\
\hline
1149 \text{ mi} & \text{distance from USF to Chicago}
\end{array}
$$

Since the distance from Chicago to downtown Tampa has no decimal places, the distance from USF to Chicago can have no decimal places.

In order to add or subtract numbers having different powers of ten, it is necessary to change all numbers to the same power.

Using the Same Power of Ten in Addition and Subtraction

EXAMPLE 2. A horse fly weighing 1.62×10^{-1} g lands on a brick having the accurately known weight of 1.7723×10^3 g. What is their combined weight?

Solution: To change 1.62×10^{-1} g to something times 10^3 g, it is necessary to multiply by 10^4. This is permissible if you also multiply by 10^{-4} because the overall effect will be that of multiplying by 1. (*Note:* $10^4 \times 10^{-4} = 10^0 = 1$.) Proceed as follows.

$$
\begin{aligned}
1.62 \times 10^{-1} \text{ g} &= (1.62 \times 10^{-4}) \times (10^{-1} \times 10^4) \text{ g} \\
&= 0.000162 \times 10^3 \text{ g}
\end{aligned}
$$

The result can now be added to the weight of the brick.

$$
\begin{array}{ll}
1.7723 \times 10^3 \text{ g} & \\
+0.000162 \times 10^3 \text{ g} & \\
\hline
1.772462 \times 10^3 \text{ g} & \text{(incorrect)} \\
1.7725 \times 10^3 \text{ g} & \text{(correct, rounded off to four places)}
\end{array}
$$

The same result would be obtained if both numbers are written as something times 10^{-1}.

2.7 OTHER EXAMPLES

EXAMPLE 1. Determination of significant figures from per cent accuracy. If the per cent accuracy of a method of measurement is known, it can be used to determine the number of significant figures. Consider for example the

Figure 2.10. Per cent accuracy and significant figures. [© *1977 Chicago Tribune-N.Y. News
Syndicate, and created by Chester Gould.*]

case of E. Kent Hardly in Figure 2.10. If the method of measuring his
fortune is only 99% accurate, aren't more figures being reported to him
than are actually significant?

Solution: The quantity $1,736,427.80 is reported as if it had either 8 or 9 significant
figures. If it is correct to the nearest $0.01, it has 9 significant figures; if
it has been rounded off to the nearest $0.10, it has 8 significant figures.
Even if it has been rounded off, the percentage error is calculated as
follows.

$$\% \text{ error in a quantity} = \frac{\text{error in the quantity}}{\text{the quantity}} \times 100$$

$$\% \text{ error} = \frac{\$0.10}{\$1,736,427.80} \times 100 = \frac{\$1.0 \times 10^{-1}}{\$1.7364278 \times 10^{6}} \times 10^{2}$$

$$= \frac{1.0}{1.7364278} \times \frac{10^{-1} \times 10^{2}}{10^{6}} = 0.6 \times 10^{-5} \%$$

But this error is much too small for the accuracy of the method, which is 99% and therefore has an error of 1%. What is the error in the quantity if its per cent error is 1%?

$$1\% \text{ of } \$1,736,427.80 = \frac{1}{100} \times \$1,736,427.80$$

$$= \frac{1}{10^2} \times \$1.7364278 \times 10^6 = \$1.7364278 \times \frac{10^6}{10^2}$$

$$= \$1.7364278 \times 10^4 \approx \$2 \times 10^4 \quad \text{(rounded off)}$$

The size of Mr. Hardly's fortune then, including the possible error, is

$$\$1,736,427.80 \pm \$2 \times 10^4$$

or, expressing both numbers times the same power of 10,

$$\$1.7364278 \times 10^6 \pm \$0.02 \times 10^6$$

Since there is an error of two units in the third significant figure, the number is best written with three significant figures as

$$\$1.74 \times 10^6, \text{ or if you prefer } \$1.74 \text{ million}$$

In other words it would have been more honest for Mr. Hardly's money counters to report his fortune as shown, and to admit an uncertainty of $20,000 in their estimate.

EXAMPLE 2. *The identification of an element from its density.* Sometimes a substance can be identified by its qualitative properties: color, odor, stability when heated, behavior when mixed with other substances, and so on. Other times a substance may have qualitative properties that are so similar to those of other substances that a quantitative approach must be taken.

An interesting problem concerning properties of elements was given to Archimedes (287–212 B.C.), according to legend, by the King of Syracuse. The method used to solve the problem is still in use today. The problem was to determine whether the King's new crown was made of pure Au (gold) or whether the gold had been alloyed with a baser metal, such as Ag (silver) or Cu (copper)—without damaging the crown. If the crown were as dense as pure Au ($d = 19.3$ g cm^{-3}), it could be assumed that it was not an alloy, because Au was the densest metal known. If it were alloyed with Ag ($d = 10.5$ g cm^{-3}) or Cu ($d = 8.92$ g cm^{-3}), or any other metal, its density would be less than 19.3 g cm^{-3}.

Solution: Archimedes reasoned that by weighing the crown and then weighing it again submerged in water (Figure 2.11), he could calculate the density with no harm done to the crown. The weight loss when submerged in water would be equal to the weight of the water displaced. Since 1.00 g

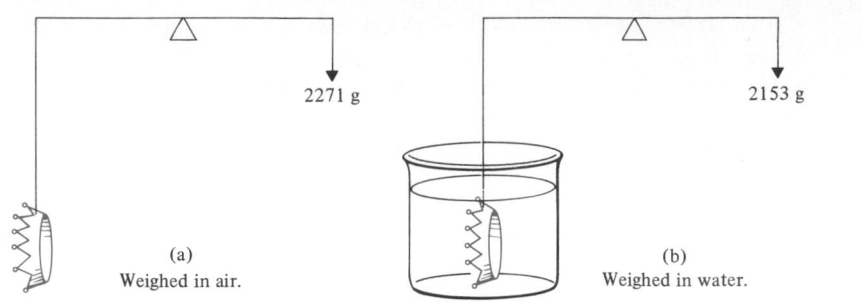

Figure 2.11.   Determination of the purity of gold by Archimedes' method.

of water has a volume of 1.00 cm³, the loss in weight expressed in grams would be equal to the volume of the metal in the crown, expressed in cubic centimeters. Archimedes might have set up his data in the following way.

weight of crown (in air)	2271 g
weight of crown submerged in water	−2153 g
weight of water displaced by crown	118 g
volume of metal in crown	118 cm³

$$\text{density of metal in crown} = \text{mass per unit volume}$$
$$= \frac{2271}{118} \frac{g}{cm^3}$$
$$= \frac{2.271 \times 10^3}{1.18 \times 10^2} \frac{g}{cm^3}$$
$$= 1.92 \times 10^1 \; g \; cm^{-3}$$
$$= 19.2 \; g \; cm^{-3}$$

These results would be comforting to the royal goldsmith although he and the King might quibble over significant figures.

The densities of most of the solid elements can be determined in this way, commonly giving a certainty of four significant figures.

EXAMPLE 3.   *Another example using the one-factor method.* Assuming that Au dust has approximately the same density as bar Au, 19.3 g cm⁻³, determine the weight-lifting prowess of Rip Kirby's antagonist in Figure 2.12. For simplicity you may assume the bucket capacity is 10 qt and that it is filled to the 8-qt level. The final result will be good to one significant figure.

Solution: You are asked to find the weight of the gold and you know that the number of pounds is directly proportional to the number of quarts. Therefore you may begin the problem by writing

$$x \text{ lb} = (8 \text{ qt}) \cdots$$

Furthermore you know that the number of pounds is directly proportional to the density. That is, if the density of the metal were twice as great, the bucket would weigh twice as much. Introducing this factor gives

$$x \text{ lb} = (8 \text{ qt}) \left(\frac{19.3}{1} \frac{\text{g}}{\text{cm}^3} \right) \cdots$$

Alternatively, this last factor could have been obtained from the fact that 19.3 g Au is equivalent to 1 cm³ Au, which can be used to form a one-factor. The cubic centimeter is placed in the denominator because quart is in the numerator, and this will permit cancellation of the two volume units at a later stage.

Since gram is needed in the denominator to cancel gram already in the numerator, use the equality 454 g = 1 lb, to form a one-factor

$$x \text{ lb} = (8 \text{ qt}) \left(\frac{19.3}{1} \frac{\text{g}}{\text{cm}^3} \right) \left(\frac{1}{454} \frac{\text{lb}}{\text{g}} \right) \cdots$$

Since cubic centimeter is needed in the numerator to cancel cubic centimeter already in the denominator, use the equality, 1000 cm³ = 1 ℓ, to form a one-factor

$$x \text{ lb} = (8 \text{ qt}) \left(\frac{19.3}{1} \frac{\text{g}}{\text{cm}^3} \right) \left(\frac{1}{454} \frac{\text{lb}}{\text{g}} \right) \left(\frac{1000}{1} \frac{\text{cm}^3}{\ell} \right) \cdots$$

Now liter is needed in the numerator to cancel liter already in the denominator. The equality 1 ℓ = 1.06 qt forms a one-factor and also provides quart in the denominator to cancel quart already in the numerator.

RIP KIRBY

Figure 2.12. Can you estimate the weight of a bucket of gold? [© *King Features Syndicate, Inc., 1970. World rights reserved.*]

$$x \text{ lb} = (8 \ \cancel{gt}) \left(\frac{19.3}{1} \ \frac{g}{\cancel{cm^3}} \right) \left(\frac{1}{454} \ \frac{\text{lb}}{\cancel{g}} \right) \left(\frac{1000}{1} \ \frac{\cancel{cm^3}}{\cancel{\ell}} \right) \left(\frac{1}{1.06} \ \frac{\cancel{\ell}}{\cancel{gt}} \right)$$

Finally, separate the units from the numbers. The units cancel leaving pounds of gold in the numerator.

$$x \text{ lb} = \frac{(8)(19.3)(1000)}{(454)(1.06)} \text{ lb}$$
$$= 300 \text{ lb!}$$

It is unfortunate that a man like this would turn to a life of crime.

PROBLEMS

1. Apply the rules of exponents to each of the following problems involving powers of ten:
   (a) Write 1,000,000 as a power of ten.
   (b) Write $\dfrac{1}{0.001}$ as a power of ten.
   (c) Calculate $10^3 \times 10^{-23}$.
   (d) Calculate $\dfrac{10^{-6}}{10^{-10}}$.

2. Same as problem 1.
   (a) Write 0.01 as a power of ten.
   (b) Write $\dfrac{1}{(100)^2}$ as a power of ten.
   (c) Calculate $\dfrac{10^3}{10^{10}}$.
   (d) Calculate $10^{-2} \times 10^3 \times 10^6$.

3. (a) Calculate the number of seconds in a week using the one-factor method. Set up all the factors in a chain using each of the following equalities before making the calculation: 1 week = 7 days, 1 day = 24 hr, 1 hr = 60 min, 1 min = 60 sec. (b) Discuss the concept of significant figures as it applies to this calculation.

4. (a) Calculate the number of inches in 1 mile using the one-factor method. Set up all the factors in a chain using each one of the following equalities before making the calculation: 1 mi = 1760 yd, 1 yd = 3 ft, 1 ft = 12 in. (b) Discuss the concept of significant figures as it applies to this calculation.

5. Make the following conversions of units in the metric system. Express your answer in the scientific notation.

   (a) Change 100 cm^3 to cubic meter.

   (b) Change 5000 Å to nanometer.
   (c) Change $10^{-3}\ \mu$m to nanometer.
   (d) Change 500 μg to gram.
   (e) Change 0.01 cm^3 to picoliter.

6. Same as problem 5.

   (a) How many cubic centimeters are in 1 liter?
   (b) Change 10^{-2} nl to picoliter.
   (c) Change 20 km to centimeter.
   (d) Change 500 kg to milligram.
   (e) Change 40 mm to meter.

7. Make the following conversions between English and metric units
   (a) How many centimeters are in 5 ft 11 in.? Compare your answer with the value given on the Australian postage stamp in Figure 2.3.
   (b) Calculate the number of grams in 1 pt of water. ($d = 1.00$ g cm^{-3})

8. (a) If the posted speed limit on a Mexican highway is 80 km hr^{-1}, what should be your maximum speed in miles per hour?

   (b) A professor weighs himself in a Swedish railway station and receives a card stamped "VIKT 76.5" with no units given. If Sweden is on the metric system, how many pounds would you say the professor weighs?

9. Make the following conversion between the Celsius and Fahrenheit temperature scales.
   (a) Show that $0°F = -17.8°C$.

   (b) What is absolute zero expressed in degrees Fahrenheit?

10. Make the following conversions among the Fahrenheit, Celsius, and Kelvin temperature scales.
    (a) Calculate $100°F$ in degrees Celsius. Compare your result with that given on the Australian postage stamp in Figure 2.3.
    (b) What is the absolute temperature in degrees Kelvin corresponding to $273°C$?

11. If the temperature increases by $10°$ on the Celsius scale, how much has it increased on the Fahrenheit scale?

12. If the temperature decreases by $10°$ on the Fahrenheit scale, how much has it decreased on the Celsius scale?

Use the figure shown below to answer problems 13 and 14.

13. Several positions on the 15-cm ruler shown above are marked by lines. Read and report the number of centimeters indicated in (a) through (d) using the maximum number of significant figures. How many significant figures are in each reading?

14. Same as problem 13 but for (w) through (z).

15. (a) If the error in the weight of an 8 lb cat is ± 1 lb, what is the % error? (b) The cat is weighed by a 76 lb boy, first weighing himself holding the cat, and then himself alone. What must be the precision of the scales (i.e., $\pm y$ lb, where y is the error) in order to calculate the weight of the cat with an error of only 1%? (c) Using these same scales what would be the % error in the weight of the boy?

16. In order to determine the weight of a signature, a small piece of paper is weighed before and after writing the signature. The weight is measured on a single pan analytical balance capable of a precision of $\pm 1 \times 10^{-5}$ g. If the weight before signing is 0.84552 g and after signing 0.84559 g, make the following calculations. (a) the weight of the signature (b) the % error in the weight of the piece of paper (c) the % error in the weight of the signature.

17. How many significant figures are in each of the following numbers, assuming that the uncertainty is ± 1 unit in the last digit shown?
    (a) 0.00307   (d) 3.0700×10^4
    (b) 307   (e) 3.070×10^{-3}
    (c) 3.07

18. Same as problem 17.

    (a) 5.302   (d) 5.30×10^6
    (b) 0.053020   (e) 5.302×10^{-6}
    (c) 5.30

19. Express the following numbers using scientific notation, that is, as a number between 1 and 10 times a power of ten.
    (a) $\frac{3}{100}$ (d) 0.6
    (b) 0.0038   (e) 136
    (c) 1,306,247

20. Same as problem 19.
    (a) 0.004   (d) 2001
    (b) $\frac{36}{1000}$ (e) 0.029
    (c) 0.6023×10^{24}

21. Perform the indicated operation and write your answer using the best number of significant figures. Assume an uncertainty of one or two units in the final digit of each number and express your answer in scientific notation.
    (a) $3.8 + 0.496$
    (b) 0.000227×0.003518
    (c) $0.00860/3.02 \times 10^8$
    (d) $7.8 \times 10^8/2.09 \times 10^{-3}$
    (e) $[(4.72 \times 10^{-3}) - (3 \times 10^{-5})]/2.1 \times 10^{-12}$

22. Same as problem 21.
    (a) $0.42 + 273.2$
    (b) 0.0437×0.002156
    (c) $8.76 \times 10^4/0.00318$
    (d) $(6.87 \times 10^6)/(3.0 \times 10^{-4})$
    (e) $[(8.6 \times 10^{-2}) - (4 \times 10^{-3})]/3.2 \times 10^{-8}$

23. The highest-priced* diamond ever sold at auction now belongs to the actress Elizabeth Taylor Burton. Its weight is 69.4 carats (1 carat = 0.200 g). What is its volume in cubic centimeters? The density of diamond is 3.51 g cm^{-3}.

24. Some stars have extremely high densities, which result from a collapse of the atom under pressure. Given that a certain "white dwarf" has a density of 1.5×10^5 g cm^{-3}, calculate the number of tons in 1 in.^{3} of this material.

* 25. In II. Chronicles 4:2, you find "Also he made a molten sea of ten cubits brim to brim, round in compass, . . . and a line of thirty cubits did compass it round about." If a cubit is a unit of length equal to 18 in., calculate the following, in meters: (a) the diameter ("brim to brim"), (b) the circumference ("round about"), (c) calculate π, using the fact that π is the ratio of circumference to diameter of a circle, (d) find the error and per cent error, and give the proper number of significant figures in the calculated value of π, (e) do you think that the diameter and circumference of the "molten sea" were exactly ten and thirty cubits, respectively? Explain.

*26. A tank of liquid mercury (Hg) is used to test Archimedes' law of floating bodies. Which of the following would you expect to sink and which to float? (a) A bar of gold, (b) an iron nail, and (c) a human being (recall that a swimmer can usually float on water by taking a deep breath).

* Cartier's of New York paid $1,050,000 in 1969.

THE GASEOUS STATE: PRESSURE, VOLUME, AND TEMPERATURE

A number of the elements in the periodic table exist as gases under ordinary conditions of temperature and pressure. These elements are all found in the first two rows or last two columns of the table. Specifically, they are the elements that compose the air, namely, nitrogen, oxygen, and argon, which make up 99.96% of dry air, and in addition hydrogen, fluorine, chlorine, helium, neon, krypton, xenon, and radon. All the other elements are either solids or liquids.

The Gaseous Elements

What are the properties of a gas? The most characteristic properties are that a gas fills any container it occupies and exerts a pressure on the walls. **Pressure** is defined as *force per unit area*, such as pounds per square inch.

3.1   THE MEASUREMENT OF PRESSURE

The distinction between the concept of force and that of pressure is subtle but important. Consider in Figure 3.1 the type of footwear used for different kinds of surfaces. In ordinary shoes, you would sink into deep snow; however, snowshoes would decrease the pressure by increasing the area over which the force (weight of the person) is distributed. It is the pressure, not just the force, that determines whether a surface will be penetrated or will collapse. On ice the same force can be distributed over the relatively small area of ice skates without significant penetration.

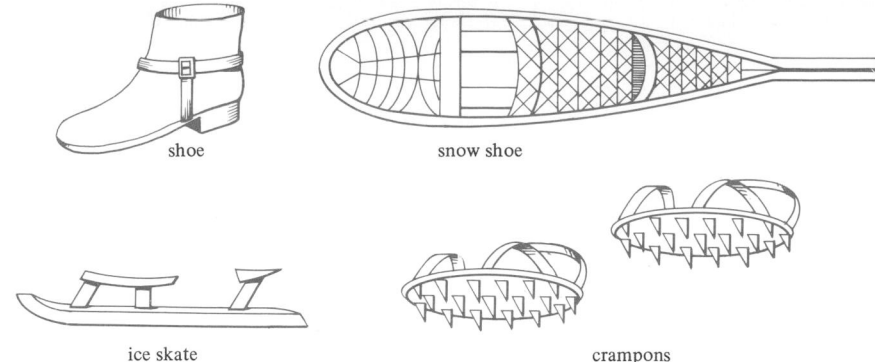

shoe   snow shoe

ice skate   crampons

Figure 3.1. Force *versus* pressure. The pressure a force exerts can be modified by changing the area over which it is applied.

If penetration is desirable, for example when mountain climbers have an ice field to cross, crampons are used to give a small enough contact area to provide pressure to penetrate the ice and prevent slipping.

The brick in Figure 3.2 illustrates the quantitative nature of pressure. In both positions the brick, which weighs 4 lb, exerts a force of 4 lb. However, when it is on end, it exerts a greater *pressure* than when it is on its side.

DEFINITION: *pressure = force/area*

You should become familiar with some of the special features of this definition as applied to liquids. One of the fundamental laws of hydrostatics (liquids at rest) is that the pressures exerted at equal levels of height or depth are the same. To visualize this you may consider the two rectangular vessels in Figure 3.3, one having a square bottom 2 cm on each edge (Figure 3.3a) and the other a square bottom 3 cm on each edge (Figure 3.3b). Both vessels are filled to a height (h) of 10 cm with the dense metallic liquid mercury ($d = 13.6 \, \text{g cm}^{-3}$).

Figure 3.2. Pressure is defined as the ratio of the force to area.

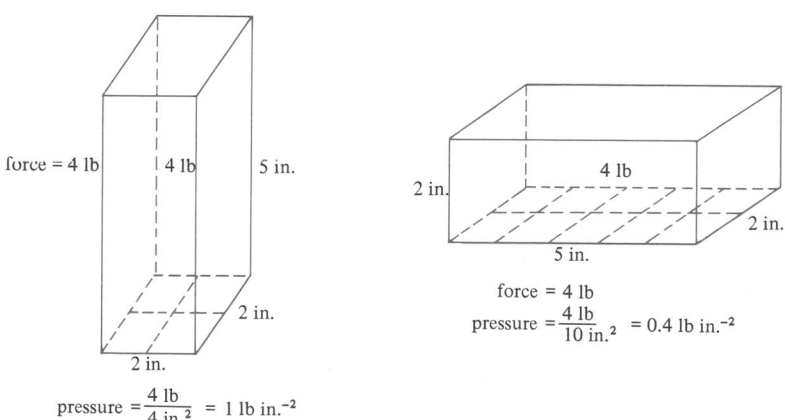

force = 4 lb   4 lb   5 in.

2 in.

2 in.

$$\text{pressure} = \frac{4 \text{ lb}}{4 \text{ in.}^2} = 1 \text{ lb in.}^{-2}$$

2 in.   4 lb   2 in.

5 in.

force = 4 lb

$$\text{pressure} = \frac{4 \text{ lb}}{10 \text{ in.}^2} = 0.4 \text{ lb in.}^{-2}$$

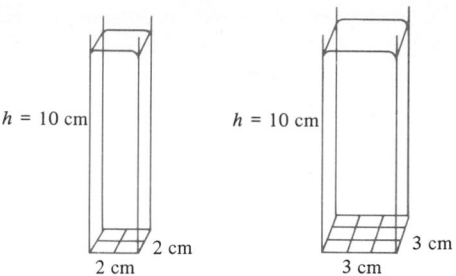

Figure 3.3. Pressures on surfaces having the same height of liquid mercury above them are equal.

(a) Base area = 4 cm²   (b) Base area = 9 cm²

The pressures on the bottoms of both vessels are calculated using the same definition:

$$\text{pressure} = \frac{\text{weight}}{\text{area}}$$

where the weight referred to is that of the liquid supported, x grams. In a physics textbook you would not find mass and weight used interchangeably because you cannot understand *acceleration* of a mass if you confuse mass and weight. Since you must work with acceleration at such an early stage in that discipline, the gram is hardly ever used as a unit of force in physics; it is used only as a unit of mass. However, as long as you do not need to deal with accelerating masses, there is no need to introduce a new unit of force at this point. You may continue to use gram per square centimeter as a unit of force per unit area just as you have been using pound per square inch.

The weight of the mercury, x g, is directly proportional to its volume, $(l)(w)(h)$, in cubic centimeters. A one-factor can be generated from the equivalence 1 cm³ = 13.6 g for mercury, and cubic centimeter is placed in the denominator of the one-factor to cancel with cubic centimeter already in the numerator. The area of the square bottom is obtained by squaring the length of an edge. Use these facts to calculate the pressure in vessels (a) and (b).

(a)
$$\text{pressure} = \frac{\text{weight}}{\text{area}} = \frac{x\,\text{g}}{(2)(2)\ \text{cm}^2}$$

$$= \frac{(10)(2)(2)(\cancel{\text{cm}^3})\left(\dfrac{13.6}{1}\ \dfrac{\text{g}}{\cancel{\text{cm}^3}}\right)}{(2)(2)\ \text{cm}^2} = 136\ \text{g cm}^{-2}$$

(b)
$$\text{pressure} = \frac{(10)(3)(3)(\cancel{\text{cm}^3})\left(\dfrac{13.6}{1}\ \dfrac{\text{g}}{\cancel{\text{cm}^3}}\right)}{(3)(3)\ \text{cm}^2} = 136\ \text{g cm}^{-2}$$

Obviously the pressure does not depend on the area of the bottom of a container if the depth is constant.

The Dependence of Pressure on the Height of a Liquid

Since the pressure does not depend on the area under a column of liquid, you should ask yourself, "How does the pressure of a liquid depend on the height?" To answer this, choose a convenient column base of 1 cm × 1 cm, having an area of 1 cm², such as the column h cm high in Figure 3.4. There are h cubes in the column, each having the volume 1 cm³. The density, d, is the weight of each cube in grams per cubic centimeter. Thus the total weight of cubes is $h \times d$ and, since the base is 1 cm², the unit obtained is grams per square centimeter.

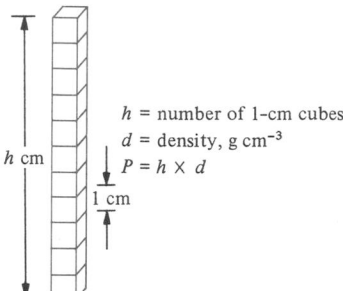

h = number of 1-cm cubes
d = density, g cm⁻³
$P = h \times d$

Figure 3.4. The pressure (P) exerted by a column of liquid h cm high on a base of 1 cm².

$$P = h \,(\text{cm}) \times d \,(\text{g cm}^{-3}) = h \times d \,(\text{g cm}^{-2})$$

Therefore the **pressure under a liquid** is *directly proportional to the height of liquid above the point of measurement.*

EXAMPLE 1. Use the fact that pressure in liquid mercury, at a depth of 10.0 cm, is 136 g cm⁻² to calculate the pressure at a depth of 50.0 cm.

Solution: Since pressure is directly proportional to height, you can say

$$x \,\text{g cm}^{-2} = (50.0 \text{ cm}) \cdots$$

You can make a one-factor from the equivalency, 10.0 cm Hg = 136 g cm⁻² (Figure 3.3), placing grams per square centimeter in the numerator to give appropriate units for pressure in the final result, as follows:

$$x \,\text{g cm}^{-2} = (50.0 \,\cancel{\text{cm}}) \left(\frac{136}{10.0} \, \frac{\text{g cm}^{-2}}{\cancel{\text{cm}}} \right)$$

$$\text{pressure} = 680 \,\text{g cm}^{-2}$$

3.2 THE BEHAVIOR OF GASES

In a liquid the pressure depends on depth; thus, if there are pressure differences at the same depth, a liquid will flow until the pressure differences are equalized. In all the examples considered in this chapter so far, none of the calculated values of pressure has included the pressure due to the atmosphere; however, this quantity must be included if a knowledge of total pressure is required. The average pressure of the atmosphere at sea level is 14.7 lb in.$^{-2}$ (1030 g cm^{-2}). You are most aware of this fact when you try to remove a part of the atmosphere.

Consider the situation in Figure 3.5, for example. Liquid can be drawn through a straw because of the difference in pressure created inside the straw from the air pressure, which is always with us. When air is first withdrawn from the straw by sucking it into the lower pressure region of the lungs, there are momentarily two pressures at the same level in the liquid. On the outside there is the pressure of the atmosphere, and on the inside a reduced pressure due to the removal of some air. Since there is no restriction on flow, the liquid rises in the straw until the pressure exerted by the column of liquid is equal to the pressure of the air removed. Is there a limit to how high the liquid can be drawn by the removal of air (or rather "pushed" by the atmosphere)? The answer is "yes" but, before we discuss this answer, you should consider the experimental method used in most laboratories for measuring the pressure of the atmosphere.

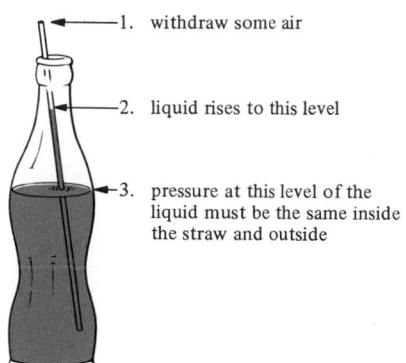

1. withdraw some air

2. liquid rises to this level

3. pressure at this level of the liquid must be the same inside the straw and outside

Figure 3.5. Constant pressure at the same levels in a liquid.

The pressure exerted by air in the atmosphere may be illustrated by constructing a **barometer.** Fill a glass tube about 80 cm long, sealed on one end, with mercury (Hg) to drive out all the air. Place a finger over the open end, invert the tube, and place the open end under the surface of a pool of mercury. When the finger is withdrawn, the mercury level falls, leaving a vacuum above, until the pressure at the level of the pool inside the tube is the same as the pressure on the pool outside (Figure 3.6). Any unequal pressure would cause the mercury to continue flowing. At sea level, on a day of average barometric pressure, the mercury in the tube will stop falling when the height above the mercury pool is 76 cm. This height is independent of the shape of the tubing, as illustrated in Figure 3.6 unless, of course, the tubing is too short. (In very small bore tubing a correction would be necessary because of capillary rise.)

Construction of a Barometer

EXAMPLE 1. The glass tubing with a U bend at the bottom contains liquid mercury. The left leg is open to the atmosphere, and the right leg is closed and has

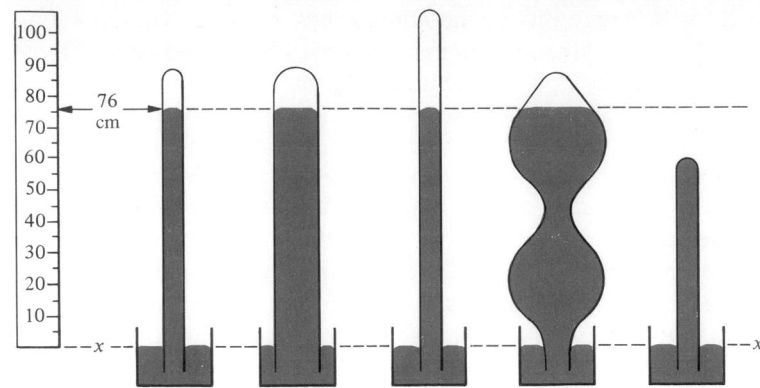

Figure 3.6. The height of the mercury supported in a barometer is independent of its shape and height (unless it is too short). The pressure at level x both inside and outside all the tubes is the same, 76 cm Hg.

a vacuum at the top. (a) Explain why the mercury does not run out. (b) What is the height, h, in cm?

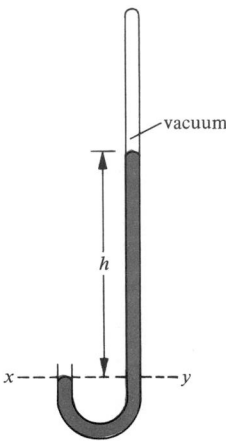

Solution: (a) The mercury does not run out because the pressure at level x in the left leg is equal to the pressure at y in the right leg. If there were some way for air to get into the vacuum space, the mercury would run out, but there isn't.

(b) The pressure at x and y is the same as that of the atmosphere. Since the pressure of the atmosphere supports a column of mercury 76 cm high, the height (h) must be 76 cm.

This experimental value of 76 cm Hg provides a limit to how high water can be drawn by sucking. Since water has a density of 1.00 g cm^{-3} and Hg has a density of 13.6 g cm^{-3}, a barometer using water as liquid would have to be 13.6 times as long or

13.6 × 76.0 cm = 1.03 × 10³ cm (34 ft). Actually the height to which water may be drawn is about 1 ft less than this because water vaporizes under high vacuum. There are two advantages of using mercury as a working fluid in a barometer; it has high density, and it does not vaporize easily.

On a 12,000-ft peak the mercury height would have about $\frac{2}{3}$ the value at sea level, and on the world's highest peak (29,000 ft) about $\frac{1}{3}$ the value at sea level. Thus in a sense the pressure in the atmosphere depends on the "depth" of the air and the earth's gravitational attraction, but this is only because of the tremendous distances involved. For example, it is 50 miles to the limits of the stratosphere which contains 99.9% of the earth's atmosphere. In "ordinary" size containers (as compared with one having a 50 mile dimension!) the gas pressure is the same throughout the container.

3.3   THE KINETIC THEORY OF GASES

If you consider the nature of gases, as opposed to liquids and solids, you will have a basis for the theory of gases.

1. Gases fill the containers they occupy.
2. Gases have the same pressure throughout.
3. Gases are low in density.
4. Gases are easily compressed.

The theory of gases, generally accepted because of the many successful predictions based on it, can be stated as follows: (1) Gases consist of widely separated particles which (2) are in constant linear motion and (3) exert pressure on the walls of a container by colliding with them. The term "kinetic" has to do with motion; thus, the theory is called the **kinetic theory of gases.** The principal features of this theory are illustrated in Figure 3.7, which represents a tiny cubic volume containing a sample of air at sea level.

EXAMPLE 1.   In the weightless condition of a space capsule, properly located between earth and moon, do the occupants exert a force on the "floor" of the

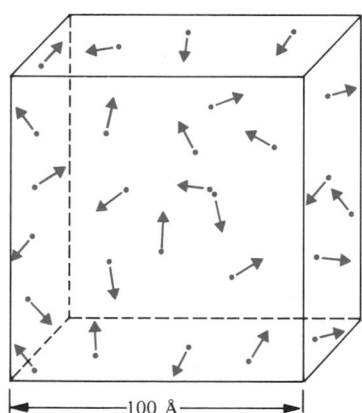

Figure 3.7.   A sample of air at sea level according to the kinetic theory of gases. Note the chaotic motion of the "particles" of air, their large number for such a small volume, and the large fraction of free space. (1 Å = 10^{-8} cm)

—100 Å—

capsule? A pressure? Does the oxygen filling the capsule exert a force on the walls? A pressure?

Solution: The occupants do not exert a force on any of the walls including the "floor" unless they push against it in order to change their position. Therefore, they do not exert a pressure. The oxygen particles filling the capsule are continually colliding with the walls and thus exert both force and pressure on them.

What would be the effect of increasing the number of "particles" of air in Figure 3.7 by a factor of 2? Since the particles would still be widely separated from one another, the increased number would not have any effect on how fast each particle moves. However, the frequency with which they collide with one another and with the walls of the container should increase. In fact, the collision frequency with the wall would be exactly twice as great, causing the pressure to be twice as great.

What would be the effect if, instead of increasing the number of particles, the size of the container were cut in half keeping all of the particles inside? Again, the particles would collide with the walls twice as often. Thus it would be reasonable to expect that the pressure would double. In the next section an experimental law is presented which confirms this expectation.

3.4   BOYLE'S LAW

The study of gases by scientists over the years has revealed many of nature's secrets. One of the earliest discoveries, in the year 1662, was a simple relationship between pressure and volume of an enclosed sample of gas.

Units of Pressure   The centimeter of mercury is a convenient unit of pressure. Another unit of pressure commonly used is called the **atmosphere** (atm); **1 atm** is defined as the pressure that will support 76.0 cm Hg.

In order to derive the law relating pressure to volume, it is necessary to use the *total* pressure of the gas sample. Frequently you may be unaware of the existence of atmospheric pressure. In Figure 3.8 the air pressure in a bike tire is being measured. The gauge has a reading of 75 lb in.$^{-2}$. If the tire were "flat" (that is, with a pressure of 14.7 lb in.$^{-2}$, or 1 atm) the gauge would have a reading of zero. In order to obtain the true pressure or total pressure in the tire, it is necessary to add 14.7 lb in.$^{-2}$ to the gauge pressure.

$$\text{total pressure} = \text{gauge pressure} + 14.7 \text{ lb in.}^{-2}$$
$$= 75 + 14.7$$
$$= 90 \text{ lb in.}^{-2}$$

The following experiments are carried out with a glass syringe that can be closed off with a stopcock, as shown in Figure 3.9. The data from the experiments and

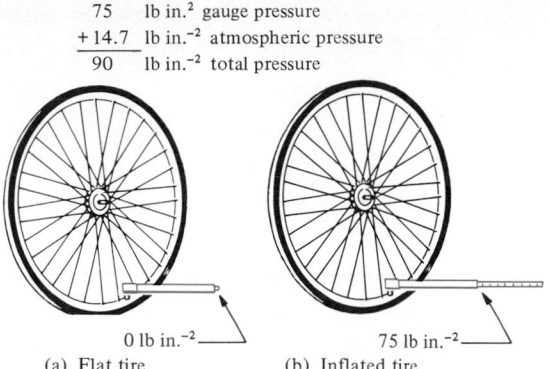

75   lb in.² gauge pressure
+ 14.7 lb in.⁻² atmospheric pressure
90   lb in.⁻² total pressure

Figure 3.8. Total gas pressure *versus* gauge pressure.

0 lb in.⁻²———   75 lb in.⁻²———
(a) Flat tire.   (b) Inflated tire.

calculations based on them are recorded in Table 3.1, where V represents the volume of the trapped air and P represents its pressure.

Fill a syringe with air and close the stopcock so that the air is trapped inside, as in Figure 3.9a. The pressure inside is equal to the pressure outside, since the plunger does not move. If the barometric pressure has the average value at sea level, the pressure of the trapped gas is 1 atm.

Figure 3.9. A sample of air trapped in a syringe under increasing external pressure.

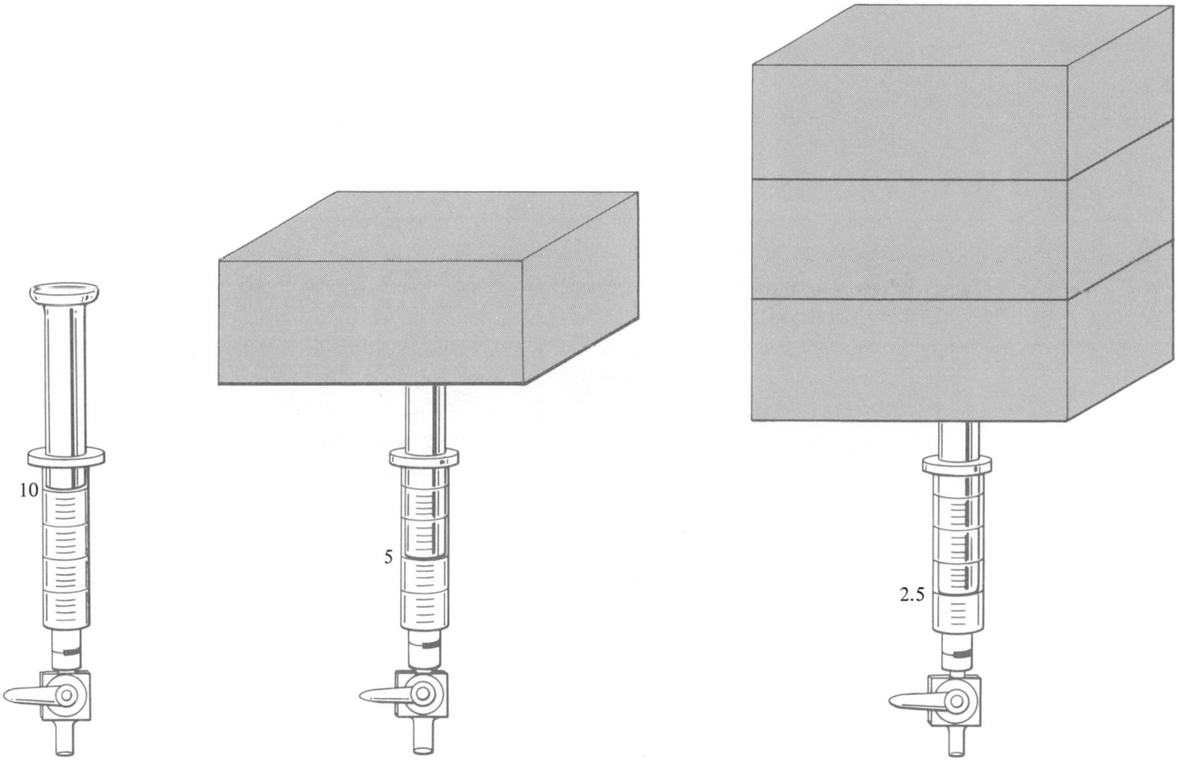

(a) 1 atm   (b) 2 atm   (c) 4 atm

TABLE 3.1 Pressure-volume data for a sample of air at constant temperature

Figure Number	V (cm³)	P (atm)	P × V (atm cm³)	1/P (atm⁻¹)
3.9a	10	1	10	1
3.9b	5	2	10	0.5
3.9c	2.5	4	10	0.25

Now place a brick of such weight on the plunger (Figure 3.9b) that it will exert a force per unit area equal to 1 atm (about 4 lb for a plunger of this 10-cm³ size syringe). The new volume is 5 cm³, and the new pressure inside the syringe is 2 atm. Note that the volume is $\frac{1}{2}$ the original value while the pressure is 2 times the original value. The product of pressure times volume is 10 and remains unchanged. Two more identical bricks placed on the syringe (Figure 3.9c) bring the external pressure to 4 atm (1 atm for each brick and 1 atm for barometric pressure). The new volume is 2.5 cm³, which is $\frac{1}{4}$ the original 10 ml. Once again the product of pressure times volume remains unchanged. It is necessary that the temperature remain constant throughout these experiments.

The measurements and calculations made in this experiment lead us to the discovery that doubling the pressure halves the volume and that the product of the pressure and the volume is always the same. This experiment and its conclusions are a good example of the scientific method. The *organization of observations* is in the form of a mathematical equation

$$P \times V = \text{a constant}$$

and once the product of pressure and volume is known for a sample of gas, the pressure at another volume of the same sample can be *predicted* by calculation. This law was first discovered by Robert Boyle (1627–1691) and it is named after him.

Using the pressure and volume data from Table 3.1 on a graph, with volume data as coordinates (vertical distances) and pressure data as abscissas (horizontal distances), you will obtain the curve shown in Figure 3.10a. It is called a hyperbola. Note that as the volume is decreased toward zero the pressure appears to increase without limit, and as pressure approaches zero the volume appears to increase without limit.

You will obtain a straight line by plotting $\frac{1}{P}$, rather than P, as abscissa. This is illustrated in Figure 3.10b. The line will pass through the point (0,0) if extrapolated. Such a linear relationship indicates that V is proportional to $\frac{1}{P}$, or that V is *inversely* proportional to P.

Examples 1 and 2 show the power of Boyle's law in predicting new volumes under changing conditions of pressure at constant temperature.

EXAMPLE 1. Helium gas fills a weather balloon to a volume of 1000 l at sea level (1 atm). What volume will the gas occupy after the balloon has risen (a) 12,000 ft ($\frac{2}{3}$ atm) and (b) 29,000 ft ($\frac{1}{3}$ atm)? Assume equal pressures inside and outside the balloon.

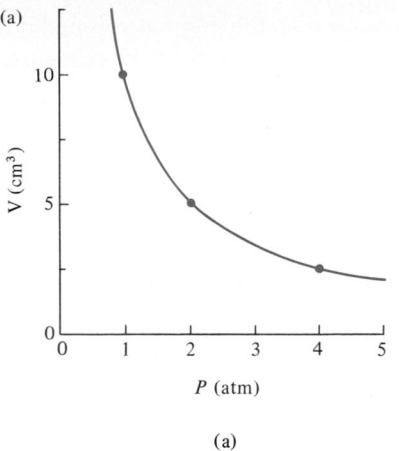

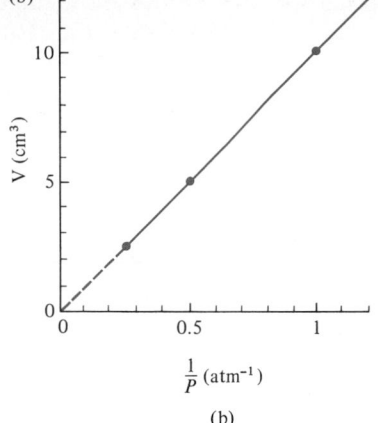

Figure 3.10. Two methods of graphing pressure-volume data for a sample of air trapped in a syringe. All points are taken from Table 3.1. (a) As P increases, V decreases—hyperbola. (b) A linear plot showing V inversely proportional to P.

Solution: Since the new volume is directly proportional to the old volume you may write the following.

$$x \, \ell = (1000 \, \ell) \cdots$$

According to Boyle's law, the new volume is inversely proportional to the new pressure.

(*a*) At 12,000 ft the new pressure is $\frac{2}{3}$ atm, which goes in the denominator.

$$x \, \ell = (1000 \, \ell) \left(\frac{1}{\frac{2}{3} \, \text{atm}} \right) \cdots$$

In order to cancel the units properly, the old pressure, 1 atm, goes in the numerator.

$$x \, \ell = (1000 \, \ell) \left(\frac{1 \, \text{atm}}{\frac{2}{3} \, \text{atm}} \right)$$

$$= (1000 \, \ell) \left(\frac{3}{2} \right)$$

$$= 1500 \, \ell$$

(*b*) At 29,000 ft the new pressure is $\frac{1}{3}$ atm and you obtain

$$x \, \ell = (1000 \, \ell) \left(\frac{1 \, \text{atm}}{\frac{1}{3} \, \text{atm}} \right)$$

$$= (1000)(3) \, \ell$$

$$= 3000 \, \ell$$

Note that this volume is greater than the original volume. It is reasonable that the weather balloon would expand at lower pressure.

EXAMPLE 2. A rubber balloon has a volume of 1500 ml at sea level pressure of 1.0 atm. What volume will it have at a depth of 16 ft underwater where the pressure is $\frac{1}{2}$ atm greater than at the surface?

Solution: The *total* pressures must be used if you wish to use Boyle's law. The new total pressure is $1 + \frac{1}{2} = 1.5$ atm, and the old total pressure was 1.0 atm. Since the new volume is directly proportional to the old volume and inversely proportional to the new pressure, you can write

$$x \text{ ml} = (1500 \text{ ml})\left(\frac{1}{1.5 \text{ atm}}\right) \cdots$$

In order to cancel the units properly introduce the old pressure, 1.0 atm, as a factor in the numerator.

$$x \text{ ml} = (1500 \text{ ml})\left(\frac{1.0 \text{ atm}}{1.5 \text{ atm}}\right)$$

$$= \frac{1500}{1.5} \text{ ml}$$

$$= 1000 \text{ ml}$$

Note that this volume is smaller than the original volume. It is reasonable that the rubber balloon should contract at higher pressure.

In all the examples concerning gases in Section 3.4, it was necessary to assume a constant temperature. In Section 3.5 you will assume that the pressure is constant and observe the effect of changing the temperature.

3.5 CHARLES' AND GAY-LUSSAC'S LAW: THE GAS THERMOMETER

Increasing the temperature of "air particles" makes them move faster. Around 1780 two French scientists, Charles and Gay-Lussac were doing experiments with hot air. These experiments eventually led to a simple law on which the absolute temperature scale is based (see Section 2.3). Actually their "experiments" dealt with launching hot air balloons by replacing cold air in the balloons by a fewer number of moles of hot air. As a result the balloons were lighter. A given balloon could be made to float in the surrounding cold air if the weight of hot air, balloon, and gondola were less than that of an equal volume of cold air. This, of course would be predicted by Archimedes' law (Section 1.2).

If a *fixed* number of moles of gas is cooled at constant pressure, the volume it occupies decreases. The gas-filled rubber balloon in Figure 3.11 approximates these

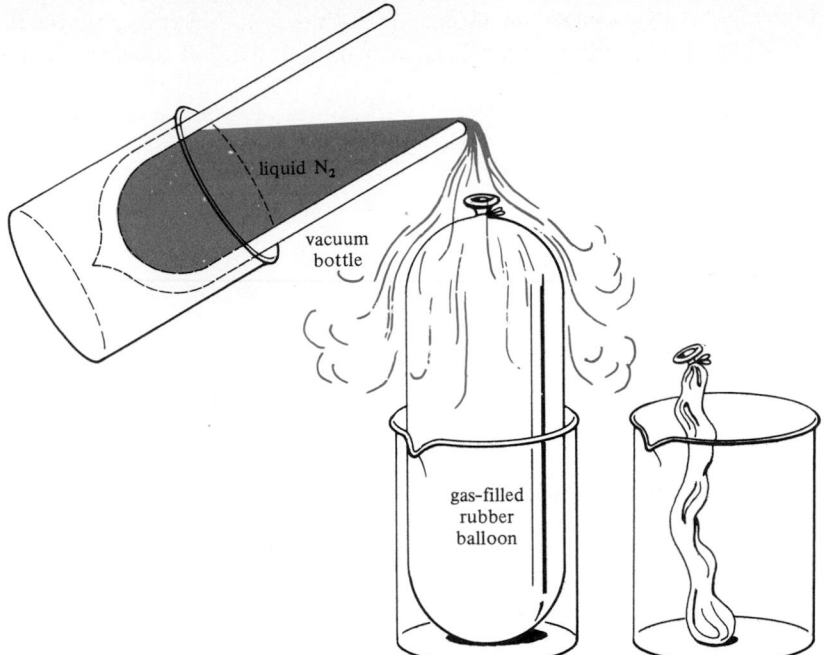

liquid N₂

vacuum
bottle

gas-filled
rubber
balloon

Figure 3.11. The effects of temperature on gas volume at constant pressure.

conditions because the number of moles of gas inside does not change and the pressure inside remains almost the same as the barometric pressure outside. Liquid nitrogen in the thermos bottle has its normal boiling temperature (Section 7.5) of $-196°C$. Pouring the cold liquid over the baloon causes the internal balloon pressure to decrease because the average speed of the gas molecules is decreased and their collisions with the walls of the balloon exert less force. The walls of the balloon contract until the concentration of gas molecules has increased to the point where pressure inside is again equal to the pressure outside.

Exercise 1.   Using the kinetic theory of gases (Section 3.3) and the fact that the average speed of gas molecules increases as temperature increases, explain why a hot air balloon contains fewer moles of air than a cold air balloon. *Hint:* The balloon is open at the bottom.

A gas thermometer can be constructed that depends on the different volumes occupied by a gas at different temperatures. Although a rubber balloon is *not* ordinarily used as a gas thermometer, because of the difficulty of measuring its volume, it could be used in principle, and the results would be like those shown in Figure 3.12.

If the volume of gas at $0°C$ is $1.00 \ \ell$, the volume at $-136°C$ will be half that, or $0.50 \ \ell$; and the volume at $-205°C$ will be $\frac{1}{4}$ that or $0.25 \ \ell$. These data are collected in Table 3.2, which includes the extrapolated values of zero volume at $-273°C$.

TABLE 3.2 The extrapolation of the gas thermometer to absolute zero*

Volume (l)	t (°C)	T (°K)
1.00	0	273
0.50	−136	137
0.25	−205	68
(0.00)	(−273)	(0)

* *The pressure is 1 atm at all temperatures.*

If the number 273 is added to each of the Celsius temperatures, t(°C) and shown in the last column, there is a very interesting result. These numbers, called T (°K or Kelvin), are directly proportional to the volume. In such a proportion the relationship between variables is linear, and when one variable is equal to zero the other is also. Expressed as an equation, the relation is

$$V = \text{a constant} \times T$$

This law is known as Charles' and Gay-Lussac's law and it applies to *all* gases at fixed pressure.

If we start at 273°K and double the temperature to 546°K, the volume is doubled. Decreasing the temperature by half, to 137°K, decreases the volume by half. The extrapolated value of V at $T = 0$°K is 0 l. The extrapolated value is obtained by graphing the data in the manner shown in Figure 3.13. The extrapolation is necessary because no gas can be found that does not liquefy before its temperature reaches absolute zero. There is no problem of estimation because the graph is linear and you need only extend the straight line. To use Charles' and Gay-Lussac's law it is necessary to convert from the Celsius to the Kelvin scale, and this is done by using the equation

$$°K = °C + 273$$

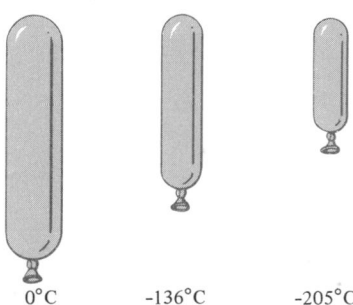

0°C   −136°C   −205°C
273°K   136°K   68°K

Figure 3.12. Gas volume is directly proportional to absolute temperature at constant pressure.

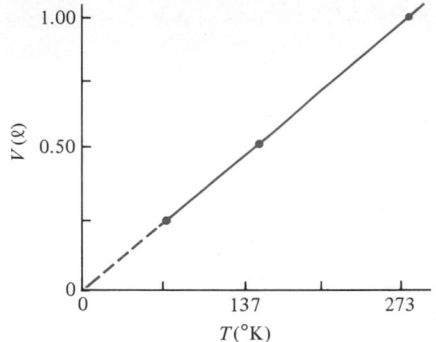

Figure 3.13. The extrapolation of the gas thermometer to absolute zero.

EXAMPLE 1. If the gas-filled balloon in Figure 3.11 has a volume of 850 ml at 25°C, what volume will it have after it is cooled to the temperature of liquid nitrogen, −196°C, at constant pressure?

Solution: In order to use Charles' and Gay-Lussac's law it is necessary to change from Celsius to Kelvin, as follows.

$$\text{old temperature} = 25°C = 25 + 273°K = 298°K$$
$$\text{new temperature} = -196°C = -196 + 273°K = 77°K$$

Since the *new* volume is directly proportional to the *old* volume, and directly proportional to the *new* temperature, you can write

$$\text{new volume} = x \text{ ml} = (850 \text{ ml})(77°K) \cdots$$

In order to cancel the units properly introduce the old temperature as a factor in the denominator.

$$x \text{ ml} = (850 \text{ ml})\left(\frac{77°\text{K}}{298°\text{K}}\right)$$
$$= \frac{(850)(77)}{298} \text{ ml}$$
$$= 220 \text{ ml}$$

Note that this new volume is smaller than the old volume, which correspond to your observation of the experiment.

EXAMPLE 2. The volume occupied by a certain gas sample is 120 ml at −40°C. At what temperature will it have twice this volume if the pressure is unchanged?

Solution: The easy answer to this question is, "At twice the temperature," using Charles' and Gay-Lussac's law. However you must be careful to convert the temperature to Kelvin.

old temperature $= -40°C = -40 + 273°K = 233°K$
new temperature $= 2 \times$ old temperature
$= 2 \times 233°K$
$= 466°K$

To change this temperature to Celsius, subtract $273°K$.

new temperature $= 466 - 273°C = 193°C$

EXAMPLE 3.   On the following graph, which represents the volume of a gas at constant pressure, predict the volume, x ml, at the higher temperature.

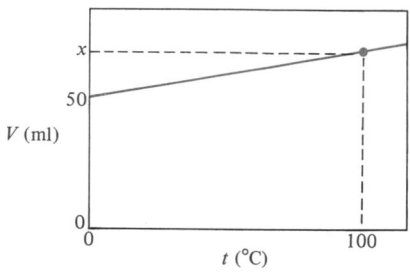

Solution: The relation between volume in milliliters and temperature in degrees Celsius is linear, but in order to have a direct proportion it is necessary to convert temperature to degrees Kelvin.

old temperature $= 0°C = 0 + 273°K = 273°K$
new temperature $= 100°C = 100 + 273°K = 373°K$

The *new* volume is directly proportional to the *old* volume, and directly proportional to the *new* temperature.

new volume $= x$ ml $= (50$ ml$)(373°K) \cdots$

In order to have the units cancel out properly you must insert the old temperature as a factor in the denominator.

$$x \text{ ml} = (50 \text{ ml})\left(\frac{373°K}{273°K}\right)$$

$$x \text{ ml} = \frac{(50)(373)}{273} \text{ ml}$$

$$= 68 \text{ ml}$$

Note that this new volume is greater than the old volume, as it appears to be on the graph.

Up to this point you have learned how the volume of a gas behaves with changes in pressure and temperature provided *one* of them remains *constant*. What if both the temperature *and* the pressure change at the same time? With your knowledge of proportionality, both effects can be calculated in a single chain of factors.

1. Boyle's Law

$$V = \frac{a\ constant}{P} \qquad (\text{at constant temperature})$$

2. Charles' and Gay-Lussac's Law

$$V = a\ constant \times T \qquad (\text{at constant pressure})$$

3. Combined law

$$V = a\ constant \times \frac{T}{P}$$

or

$$\frac{P \times V}{T} = a\ constant$$

If the *old* conditions of pressure, volume, and temperature are indicated by subscript 1 and the new conditions by 2, the meaning of the constant in the combined law is shown below.

COMBINED GAS LAW $\qquad \dfrac{P_1 \times V_1}{T_1} = \dfrac{P_2 \times V_2}{T_2}$

EXAMPLE 4.   A 20.0 l sample of gas at 100°C and 3.00 atm pressure is heated to 200°C and subjected to 6.00 atm pressure at the same time. What will be the new volume?

Solution: In order to use Charles' and Gay-Lussac's law or the combined gas law, it is necessary to use absolute temperatures.

$$\text{old temperature} = 100°C = 100 + 273°K = 373°K$$
$$\text{new temperature} = 200°C = 200 + 273°K = 473°K$$

You should summarize all the old and new conditions and then substitute them into the combined gas law. You must have the same units for the new conditions as for the old.

Old Conditions	New Conditions
$V_1 = 20.0\ l$	$V_2 = x\ l$
$T_1 = 373°K$	$T_2 = 473°K$
$P_1 = 3.00\ atm$	$P_2 = 6.00\ atm$

$$\frac{P_1 \times V_1}{T_1} = \frac{P_2 \times V_2}{T_2}$$

$$\frac{(3.00 \text{ atm})(20.0 \text{ } l)}{373°K} = \frac{(6.00 \text{ atm})(x \text{ } l)}{473°K}$$

Multiply both sides by $\dfrac{473°K}{6.00 \text{ atm}}$ to obtain

$$\frac{(473°\cancel{K})(3.00 \cancel{\text{ atm}})(20.0 \text{ } l)}{(6.00 \cancel{\text{ atm}})(373°\cancel{K})} = x \text{ } l$$

$$x \text{ } l = 12.7 \text{ } l$$

You may also work this problem without using the combined gas law.
The new volume is *directly* proportional to the old volume, *directly* proportional to the new temperature, and *inversely* proportional to the new pressure. You may begin the calculation by writing down the corresponding factors.

$$\text{new volume} = x \text{ } l = (20.0 \text{ } l)(473°K)\left(\frac{1}{6.00 \text{ atm}}\right) \cdots$$

In order to obtain the proper cancellation of units, insert the old temperature as a factor in the *denominator* and the old pressure as a factor in the *numerator*.

$$x \text{ } l = (20.0 \text{ } l)\left(\frac{473 \text{ }\cancel{°K}}{373 \text{ }\cancel{°K}}\right)\left(\frac{3.00 \text{ }\cancel{\text{atm}}}{6.00 \text{ }\cancel{\text{atm}}}\right)$$

$$= \frac{(20.0)(473)(3.00)}{(373)(6.00)} \text{ } l$$

$$= 12.7 \text{ } l$$

Note that this second method is not exactly the same as the one-factor method you worked with earlier, but it has similar features. All the factors can be strung out in a line before any calculations are performed, and it is a necessary (but not sufficient) condition that the unrequired units cancel. In this problem, all the units on the right-hand side cancel except "liter," which is an acceptable unit for the new volume.

EXAMPLE 5. If a balloon of hydrogen gas has a volume of 2.0 l at sea level (760 mm Hg) and 25°C, what volume will it have at the summit of Mt. Everest at a pressure of $\frac{1}{3}$ atm and a temperature of $-10°C$?

Solution: (a) In order to use Boyle's law it is necessary to express both pressures in the same units.

$$\text{old pressure} = 760 \text{ mm Hg} = 1.00 \text{ atm}$$
$$\text{new pressure} = \tfrac{1}{3} \text{ atm}$$

The temperature must be converted to degrees Kelvin in order to be proportional to volume. Then you may use the combined gas law as shown.

Old Conditions	New Conditions
$V_1 = 2.0 \; l$	$V_2 = x \; l$
$P_1 = 1.00$ atm	$P_2 = \tfrac{1}{3}$ atm
$T_1 = 298°$K	$T_2 = 263°$K

$$\frac{P_1 \times V_1}{T_1} = \frac{P_2 \times V_2}{T_2}$$

$$\frac{(1.00 \text{ atm})(2.0 \; l)}{298°\text{K}} = \frac{(\tfrac{1}{3} \text{ atm})(x \; l)}{263°\text{K}}$$

Multiply both sides by $263°\text{K}/\tfrac{1}{3}$ atm to obtain the following expression.

$$\frac{(263°\text{K})(1.00 \text{ atm})(2.0 \; l)}{(\tfrac{1}{3} \text{ atm})(298°\text{K})} = x \; l$$

$$\frac{(3)(263)(1.00)(2.0)}{298} \, l = x \; l$$

$$5.3 \; l = x \; l$$

An alternate way you may work this problem is by proportionalities. The new volume is *directly* proportional to the old volume, *inversely* proportional to the new pressure, and *directly* proportional to the new temperature.

$$\text{new volume} = x \; l = (2.0 \; l) \left(\frac{1}{\tfrac{1}{3} \text{ atm}} \right) (263°\text{K}) \cdots$$

The old pressure and temperature are inserted as factors to cancel the units properly.

$$x \; l = (2.0 \; l) \left(\frac{1.00 \text{ atm}}{\tfrac{1}{3} \text{ atm}} \right) \left(\frac{263 \; °\text{K}}{298 \; °\text{K}} \right)$$

$$= \frac{(2.0)(3)(1.00)(263)}{298} \, l$$

$$= 5.3 \; l$$

Special Forms of Boyle's Law and Charles' and Gay-Lussac's Law

You may use the combined gas law under conditions of constant temperature $(T_1 = T_2)$. In this case a special form of Boyle's Law is obtained.

BOYLE'S LAW: $\qquad\qquad\qquad P_1 \times V_1 = P_2 \times V_2$

Likewise, under conditions of constant pressure $(P_1 = P_2)$, the combined gas law is valid and gives a special form of Charles' and Gay-Lussac's law.

CHARLES' AND GAY-LUSSAC'S LAW:

$$\frac{V_1}{T_1} = \frac{V_2}{T_2}$$

COMBINED GAS LAW:
$$\frac{P_1 V_1}{T_1} = \frac{P_2 V_2}{T_2}$$

---

32 gm CH_4

MW = 16 gm/mole

25° C

360 mm Hg

V = ?

$PV = nRT$

$V = \frac{nRT}{P} = \frac{(2)(0.08205)(298°k)}{\frac{360}{760} ATM}$

1 mole = 22.4 liters at STP   44.8 l

V = 22.4 l/mole

T = 273°k

P_1 = 1 ATM = 760 mm Hg

$$\frac{P_1 V_1}{T_1} = \frac{P_2 V_2}{T_2}$$

$$V_2 = \left(\frac{P_1 V_1}{T_1}\right)\left(\frac{P_2}{T_2}\right) = \frac{(760\ mm)(44.8\ l)}{273°k} \quad \frac{298°k}{360\ mm\ Hg}$$

1. A hammer is used to strike a 150-lb blow on a metal bolt using a chisel. If the chisel and the bolt have an area of contact of 0.00037 in.^{2}, what is the pressure exerted on the bolt?

2. A sharp pencil is pressed on a piece of paper with a force of 3.2 lb. If the size of the pencil dot is 2.1×10^{-4} in.^{2}, calculate the pressure exerted on the paper.

3. A well made barometer, like the one illustrated on the left in Figure 3.6, is pushed down 3.0 cm into the liquid mercury pool of the container.
(a) What happens to the difference in height of the two mercury levels? (b) Is there any change in the mercury level in the lower container? Explain.

4. Consider a barometer like that illustrated on the left in Figure 3.6.

(a) Will there be an error in reading barometric pressure if there is a little air present in the tube space above the mercury? (b) If so, is the column level too high or too low? Explain.

5. A compressed gas cylinder having a volume of 2.2 ft^3 contains oxygen gas at a total pressure of 1800 lb in.$^{-2}$ What volume must a room have to contain this oxygen if it is permitted to expand to a pressure of 14.7 lb in.$^{-2}$?

6. A room having dimensions 6 ft $\times$ 6 ft $\times$ 6 ft contains air at a total pressure of 14.7 lb in.$^{-2}$ If all the air is pumped into a compressed gas cylinder it reaches a total pressure of 2100 lb in.$^{-2}$ What is the volume of the cylinder?

$$V_1 = 6ft^{3} $$
$$P = 14.7 \, lb/in^2$$
$$V_2 = 2100 \, lb/in^2$$
$$P_2 =$$
$$\frac{P_1}{V_1} = \frac{P_2}{V_2}$$
$$V_2 = \frac{P_1 V_1}{P_2} = \frac{(14.7)(6)^{3}}{2100}$$

7. The pressure inside a rubber balloon is very close to that of the barometric pressure around it. What would you expect to happen to a balloon filled with hydrogen gas, released, and allowed to ascend to great altitude? Explain, using Boyle's law.

8. A certain open container of mercury has an air bubble rising in it. If the air bubble has a volume of 0.10 ml at a depth of 38 cm, what size will it be as it breaks through the surface? The barometric pressure is 76 cm Hg.

$$V_1 = 0.10 \, ML$$
$$P_1 = 1 \, ATM + .5 \, ATM$$
$$P_1 = 1.5 \, ATM$$
$$V_2 = ?$$
$$P_2 = 1 \, ATM$$

$$V_2 = \frac{P_1 V_1}{P_2} = \frac{(1.5)(.1)}{1} = .15$$

9. Increased temperature has the effect of making gas particles travel faster. Using the kinetic theory of gases, explain how this has the effect of increasing the gas pressure in *two* ways.

10. According to Avogadro's hypothesis heavy gas particles must exert the same pressure as light particles if they are equal in number (per unit volume) and at the same temperature. This must result from a slower speed of travel for the heavier particles. Give two ways that this slower speed would tend to make the heavier particles exert the same pressure as the light ones.

11. The graphs in Figure 3.10 indicate that at high pressure the volume of a gas sample approaches zero. (a) Make a graphical estimate of the volume corresponding to a pressure of 5 atm. (b) Using Boyle's law, calculate the volume in Table 3.1 corresponding to a pressure of 1000 atm.

12. Same as problem 11 (a) Make a graphical estimate of the volume corresponding to a pressure of 10 atm. (*Hint:* First calculate 1/P.) (b) Using Boyle's law calculate the volume in Table 3.1 corresponding to a pressure of 2000 atm.

13. A rubber balloon is filled to a volume of 1400 ml with air on a cold day at 40°F. If the balloon is taken inside and warmed in an oven to 200°F, calculate the new volume, assuming there has been no change in balloon pressure. (*Note:* It is necessary to use the absolute zero on the Fahrenheit scale, –459°F, or convert these temperatures to degrees Celsius and then to degrees Kelvin in order to use Charles' and Gay-Lussac's law.

14. If the volume occupied by the air in a balloon at 75°F and 1 atm is 100 cm³, what volume would the same amount of air occupy if allowed to expand at the same pressure at a final temperature of 250°F? (See note, Problem 13.)

15. A 25.0-ml sample of dry gaseous oxygen is collected in a syringe similar to that in Figure 3.9, at 23.0°C and 767 mm Hg. If on the next day the temperature has changed to 28.0°C and the barometric pressure to 752 mm Hg, what will be the new volume of oxygen?

16. A weather balloon filled with helium gas has a volume of 1000 m³ at ground level (25°C, 1 atm). If it ascends to an altitude of 115,000 ft where the temperature is −50°C and the pressure is $\frac{1}{250}$ atm, what will be the new volume?

*17. The Hg barometer shown in position I below is properly constructed with a vacuum space above the column of Hg. In position II, show where the Hg level will be if a small amount of air has leaked into the space above the column. Position III represents the improperly made barometer shown in position II, simply pushed down farther into the dish of Hg. Show where the Hg level in this barometer is relative to I and II. Explain any difference in its level from the level in II.

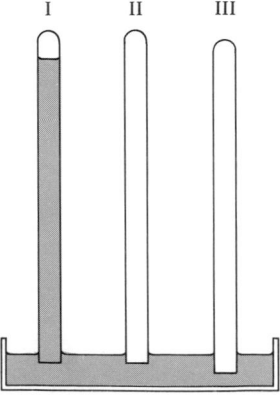

*18. The inverted water bottle shown below is commonly used for young chickens because it furnishes a good water supply at a safe and accessible level. Answer the following questions in terms of barometric and hydrostatic pressures.
(a) What keeps the water in the bottle? (b) Why isn't the bottle full?
(c) How does the bottle release more water?

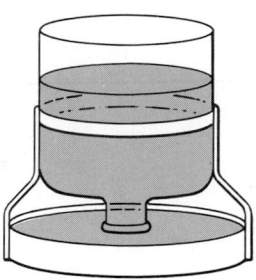

*19. A barrel can be burst by sealing a long thin pipe vertically into the lid and filling the barrel and the pipe with water. A pressure is exerted on the inside of the barrel which is the sum of barometric pressure (the pipe is open at the top) and the pressure of the column of water. Calculate the height in feet necessary to exert a pressure of 2.0 atm greater than barometric pressure.

$$34 \, ft = 1 \, atm \qquad 34 \times 2 = 68$$

$$P_1 = 1 \, atm$$

$$P_2 = 2 \, atm$$

4

MOLECULES, ATOMS, AND MOLES

It was from their study of gases that scientists gained a good understanding of the meaning of such phrases as "chemical combinations" (Figure 1.4), "chemical changes" (page 8), "relative weights of the ultimate particles" (page 9), and the meaning of formulas such as H_2O and H_2S (page 10). You have seen that the pressure, volume, and temperature of a gas sample are all interdependent and simply related to one another. To complete the description of gas behavior one other important quantity is needed—*the number and kind of gas molecules*. You will find that nature has greatly simplified things for us by requiring different kinds of gas molecules to exert the same pressure under equal conditions of volume and temperature.

4.1 MOLECULES

Most of the gases in nature consist of molecules. Historically the concept of the atom (Section 4.3) came before that of the molecule, and a confusing situation arose. The **atom** was accepted as the *smallest particle of an element*, and **molecules** were thought of as *chemical combinations of two or more atoms*, but it was believed that only unlike atoms would combine into molecules.

There was a large body of evidence observed by chemists in reactions involving

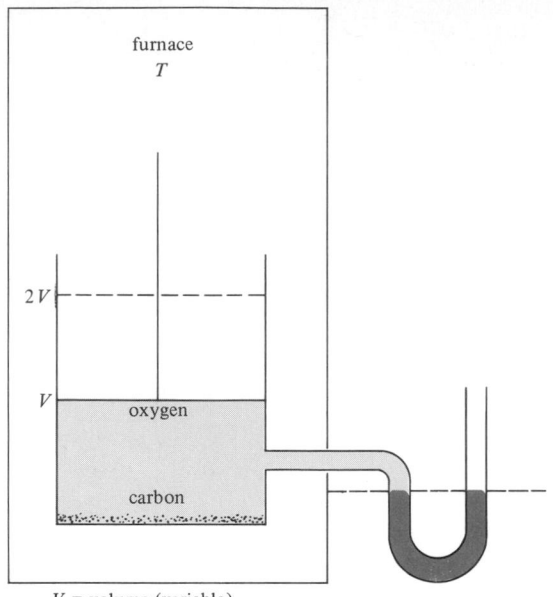

V = volume (variable)
P = pressure (constant)
T = temperature (constant)

Figure 4.1. Volume changes for the reactions between carbon and oxygen.

two or more gases that needed a simple explanation. The ratios of the volumes of gases that reacted or were formed could always be expressed in small integers.

Consider for example two ways of reacting oxygen gas with the black solid, carbon, which behaves like coal (an impure form of carbon), in a furnace of constant temperature. In method 1 an excess of oxygen is provided, and all the carbon will be consumed; in method 2 an excess of carbon is provided, and only part of it will be consumed before the oxygen is depleted. The entire experiment is carried out in a closed container with an adjustable volume (Figure 4.1), like the cylinder-piston arrangement in an automobile engine. A mercury **manometer,** similar in principle to the barometers described in Section 3.2 will provide measurements of the total gas pressure. If the mercury levels in both arms are the same, the pressure inside the closed container is the same as that of the atmosphere. The following results are observed in the two experiments.

Method 1: Although the carbon disappears, there is no change in the pressure at all and no need to change the volume.

Method 2: The pressure begins to rise (the left leg in the manometer begins to fall) but, if the volume is expanded to maintain constant pressure, the final volume of gas is *exactly* twice the original volume.

What is the simplest explanation of these facts? First of all the solid carbon occupies a much smaller volume than a gas and its effect on the volume can be ignored. Assuming that in method 1 the volume remains constant because there is *no change in the total number of molecules of gas* and that in method 2 the volume doubles because the *total number of molecules of gas doubles*, the following molecular formulas will fit the theory, respectively. Note that a numerical **subscript** *indicates the number of atoms*

$$(1) \quad C(s)^* + O_2(g)^* \longrightarrow CO_2(g)^* \qquad (4.1)$$

(in excess)   carbon
dioxide

$$(2) \quad 2\,C(s) + O_2(g) \longrightarrow 2\,CO(g) \qquad (4.2)$$

(in excess)   carbon
monoxide

in a molecule, but a number that appears in front of a molecule as a **coefficient** *represents the number of molecules.* These ideas taken together comprise the **atomic theory.**

In reaction (4.1) every time an O_2 molecule is consumed a CO_2 molecule is produced so there is no change in total number of gas molecules. In reaction (4.2) each O_2 molecule that disappears produces two CO molecules, and the number of gas molecules doubles. Such reactions indicate that certain atoms of the same element react with each other to form elemental molecules. The combination of identical atoms (that is, atoms of the same element) to form molecules was proposed by Avogadro in 1811, based on his observation of gas reactions.

If oxygen did not exist in two-atom (diatomic) molecules but only as single atoms and if a single atom could not be split in a chemical reaction, it would be difficult to explain reactions in which one volume of oxygen gas is consumed to produce two volumes of carbon monoxide.

$$C(s) + O(g) \longrightarrow 2\,CO(g) \qquad \text{(impossible)}$$

Data for reactions (4.1) and (4.2) and for some others like them involving two or more gases are summarized in the following list.

1. solid carbon   + 1 volume oxygen (in excess) $\longrightarrow$
     1 volume carbon dioxide + oxygen mixture

2. solid carbon (in excess) + 1 volume oxygen $\longrightarrow$
     2 volumes carbon monoxide

3. 1 volume hydrogen   + 1 volume chlorine $\longrightarrow$
     2 volumes hydrogen chloride

4. 2 volumes hydrogen   + 1 volume oxygen $\longrightarrow$
     liquid water

---

* (s), (l), or (g) following the symbol of an element or formula of a compound represents the solid, liquid, or gaseous state, respectively, of that substance.

5. 1 volume nitrogen $\quad$ + 3 volumes hydrogen $\quad\longrightarrow$

2 volumes ammonia

6. 1 volume ammonia $\quad$ + 1 volume hydrogen chloride $\longrightarrow$

solid ammonium chloride

In each of these new examples there is a simple "molecular explanation" as shown in equations (4.3) through (4.6).

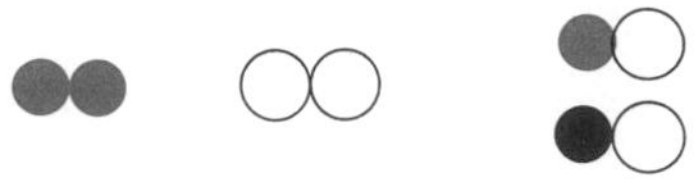

$$H_2 \quad + \quad Cl_2 \quad \longrightarrow \quad 2\,HCl \tag{4.3}$$

hydrogen $\qquad$ chlorine $\qquad\qquad$ hydrogen
chloride

Following Avogadro, chemists say that equal volumes of H_2 and Cl_2 have equal numbers of molecules, and react in a 1:1 ratio (volume) because each *molecule* of H_2 reacts with a single *molecule* of Cl_2.

Similarly hydrogen and oxygen react in a *volume* ratio of 2:1 because of the molecular reaction

$$2\,H_2 \quad + \quad O_2 \quad \longrightarrow \quad 2\,H_2O \tag{4.4}$$

water

that is, two *molecules* of H_2 react with one *molecule* of O_2.

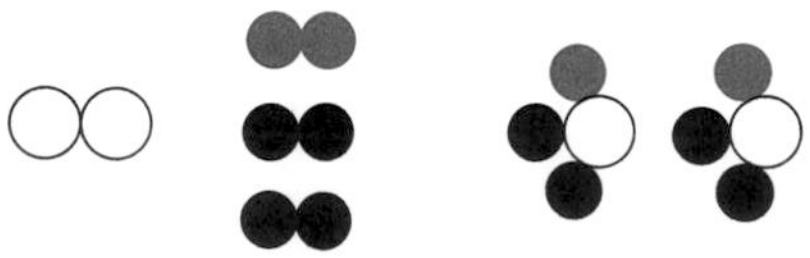

$$N_2 \quad + \quad 3\,H_2 \quad \longrightarrow \quad\quad 2\,NH_3 \tag{4.5}$$

ammonia

$$NH_3 \quad + \quad HCl \quad \longrightarrow \quad NH_4Cl \tag{4.6}$$

ammonia $\qquad$ hydrogen $\qquad\qquad$ ammonium
chloride $\qquad\qquad\qquad$ chloride

The remarkable fact is that equal numbers of *any* gaseous molecules at the same temperature and enclosed in equal volumes exert the *same* pressure. This simple hypothesis, from which the correct formulas for gaseous molecules could be deduced, was not accepted by chemists until 1860, after the death of Avogadro. **Avogadro's hypothesis** was, and is now firmly established, that *equal volumes of different gases at the same pressure and temperature have the same number of molecules.*

4.2 BALANCING CHEMICAL EQUATIONS

Reactions like (4.1) through (4.6) are called equations when written with the proper coefficients on left and right because the number of atoms of each kind on the left is *equal* to that on the right. Equations are then said to be **balanced.** Balancing equations is a skill that you should master because of the amount of useful information that can be obtained from such equations. You will not need to predict what products are formed (the substances on the right) when the reactants are changed (the substances on the left) in this chapter; however, the development of this ability is a major objective of this course. In this chapter you will need to know how to balance the equations only if both reactants and products are given to you. Some typical equations are shown with suggested strategies for balancing. Remember that, once the reactants and products are chosen, *it is not permissible to change subscripts* but only coefficients.

1. CO and O_2 to give CO_2

   (a) Write down reactants and products.

$$CO + O_2 \longrightarrow CO_2$$

   (b) Since any coefficient of CO_2 gives an even number of O atoms on the right, the total number of O atoms on the left must be even. Therefore the coefficient of CO must be even. Try the smallest possible even number, 2.

$$2\,CO + O_2 \longrightarrow CO_2$$

   (c) The number of C atoms on the left is 2. Make it 2 CO_2 on the right.

$$2\,CO + O_2 \longrightarrow 2\,CO_2$$

   (d) Adjust the coefficient of O_2, if necessary, to give the proper number of O atoms. Since there are $2 + 2 = 4$ O atoms on the left and $2 \times 2 = 4$ O atoms on the right, the coefficient should be 1, which is understood if no coefficient is written. Note that the use of a coefficient is like multiplication and every subscript in a formula is multiplied by that coefficient. Therefore,

$$2\,CO + O_2 \longrightarrow 2\,CO_2$$

   is balanced.

Environmental note. This is one of the reactions that might be utilized to decrease air pollution due to the burning of gasoline. Automobiles always produce some CO, a highly poisonous gas. If the CO in the exhaust can be made to react completely with O_2 from the air, the relatively nonpolluting CO_2 is formed. In California there is a law that sets permissible limits for pollutants in the exhaust from automobiles. If an automobile produces more than the limit of pollution, it cannot be sold in the state.

2. Fe and O_2 combine to form Fe_2O_3. This is a simplified version of the rusting of Fe (iron).
   (a) Write reactants and products.

   $$Fe + O_2 \longrightarrow Fe_2O_3$$

   (b) Since any coefficient of O_2 will give an even number of O atoms on the left and the number of O atoms in Fe_2O_3 is odd, multiply Fe_2O_3 by the smallest even number, 2.

   $$Fe + O_2 \longrightarrow 2\,Fe_2O_3$$

   (c) You now have $2 \times 2 = 4$ Fe atoms and $2 \times 3 = 6$ O atoms on the right. Place a 4 in front of Fe on the left and 3 in front of O_2 on the left to give $3 \times 2 = 6$ O atoms and you will have a balanced equation.

   $$4\,Fe + 3\,O_2 \longrightarrow 2\,Fe_2O_3$$

3. If some elements appear in several formulas on left and right, it is usually better to begin by balancing an element that appears in only one formula on left and right. Consider the reaction in which NO_2 combines with H_2O to form HNO_3 and NO.
   (a) Write reactants and products.

   $$NO_2 + H_2O \longrightarrow HNO_3 + NO$$

   (b) The element H appears in a single formula on the left and a single formula on the right. Begin with it.

   $$NO_2 + H_2O \longrightarrow 2\,HNO_3 + NO$$

   (c) There is a 1 understood as coefficient of H_2O. Do not change it without changing the coefficient of HNO_3. At this point you could choose to balance either the N or the O. Choosing N, there are $2 + 1 = 3$ N atoms on the right, which can be balanced by taking $3\,NO_2$. Alternatively, there are $2 \times 3 + 1 = 7$ O atoms on the right, and since the coefficient of H_2O cannot be changed, take $3\,NO_2$ to

give 3 × 2 + 1 = 7 O atoms on the left. Either way the balanced
equation is

77

4.3
Relative Weights
of Atoms:
Atomic Weight

$$3 NO_2 + H_2O \longrightarrow 2 HNO_3 + NO$$

Exercises 1. Balance the following equations using the smallest whole numbers
possible:

(a) $PbO + O_2 \longrightarrow Pb_3O_4$
(b) $Fe_2O_3 + CO \longrightarrow Fe + CO_2$
(c) $HBr + H_2SO_4 \longrightarrow SO_2 + Br_2 + H_2O$

Answers: (a) 6, 1, 2; (b) 1, 3, 2, 3; (c) 2, 1, 1, 1, 2.

As you gain more chemical knowledge you will discover methods of balancing
equations that are more systematic than the trial and error methods used here, but the
fundamental requirement of the indestructibility of atoms will be the same.

4.3   RELATIVE WEIGHTS OF ATOMS: ATOMIC WEIGHT

Using Avogadro's hypothesis, chemists can determine the number of atoms in a
molecule and from that determine the relative weights of different atoms. Take a big
enough volume of H_2 gas (22.4 ℓ at STP)* to weigh 2.0 g; this same volume of O_2 gas
under the same conditions weighs 32.0 g. What is the relative weight of one H atom
to one O atom? The answer must be 2.0 to 32.0 or 1.0 to 16.0 because, they both have
two atoms per molecule and according to Avogadro's law, 22.4 ℓ of $H_2(g)$ and
22.4 ℓ $O_2(g)$ at STP must contain the same number of molecules. This tremendously
large number, which is called **Avogadro's number,** has the value 6.02×10^{23} and is
usually referred to as **1 mole.** It is a counting unit in chemistry such as "dozen" and
"gross" are in other fields.

Since H is the lightest known atom, it is convenient to call 1.0 g mole^{-1} (more
precisely, 1.008 g mole^{-1})† the **atomic weight** of H, which leads to 16.0 g mole^{-1} for
the atomic weight of O.

DEFINITION: *The atomic weight of any element is the weight in grams of 1 mole of atoms of
the element.*

---

*STP, standard temperature and pressure, means that the conditions under which the gas volume is
measured are 0°C and 1 atm pressure.

†This precise value for the atomic weight of H is based on the modern choice of 1 mole carbon-12 having
the exact weight 12.0000 . . . g, which is discussed in Section 5.6.

Historically, the concepts of atoms and atomic weight came from the work of the remarkable English school teacher, John Dalton (1766–1844). **The atomic theory,** which he was the first to state clearly, can be summarized as follows:

1. Atoms exist and are indestructible.
2. Elements consist of identical atoms.
3. Compounds exist of identical combinations of different atoms.

Dalton based the atomic theory on three experimental laws which have come to be described in the following ways:

*Law of conservation of mass (weight): There is no change in mass (weight) during chemical reaction.**

For example: Consider the following chemical reaction.

$$2\,H_2 + O_2 \longrightarrow 2\,H_2O$$

There are 4 H atoms and 2 O atoms before reaction and the same number after; therefore, there is no change in mass (weight).

Law of constant composition: Any sample of a given compound, regardless of its source, has the same composition by weight of the elements in it.

For example: Any visible sample of the compound, water, consists of an extremely large number of identical molecules having the same formula, H_2O. Since the atomic weights of H and O are 1.008 and 16.00 respectively, the proportion by weight in H_2O must have the constant value 2.016/16.00.

Law of multiple proportions: When two elements combine to form two or more compounds, the weights of one element that combine with the same weight of the other element are in the ratios of small whole numbers.

For example: Hydrogen and oxygen form two compounds, water (H_2O) and hydrogen peroxide (H_2O_2). The weights of O that combine with 2.016 g of H are 16.00 g and 32.00 g, respectively. These weights are in the ratio of small whole numbers, $\frac{1}{2}$.

4.4 ATOMIC WEIGHTS AND THE PERIODIC TABLE

If we use Avogadro's law and the law of multiple proportions, we can assign atomic weights to all of the known elements. Atomic weight is a property of elements that has great practical significance. With very few exceptions, the elements in the

---

*Theoretically, there *is* a change in mass, since the energy of a chemical reaction has a mass equivalent, but it is so small that no one has been able to measure it.

TABLE 4.1* Atomic weights of the first 20 elements (Note that K = 39.1 seems to be out of line.)

Periods		Principal Groups							
		I	II	III	IV	V	VI	VII	0
	1	H hydrogen 1.008							He helium 4.00
	2	Li lithium 6.94	Be beryllium 9.01	B boron 10.8	C carbon 12.0	N nitrogen 14.0	O oxygen 16.0	F fluorine 19.0	Ne neon 20.2
	3	Na sodium 23.0	Mg magnesium 24.3	Al aluminum 27.0	Si silicon 28.1	P phosphorus 31.0	S sulfur 32.1	Cl chlorine 35.5	Ar argon 39.9
	4	K potassium *39.1*	Ca calcium 40.1						

Complete tables of atomic weights are found on the inside covers of the textbook.

periodic table are arranged from one end to the other in the order of increasing atomic weights. Indeed Mendeleev insisted that there were *no* exceptions and that some of the experimentally determined relative weights were in error, but in this instance he proved to be mistaken.

The atomic weights of the first 20 elements are given in Table 4.1 Mendeleev would not have accepted the value for K because it does not fit his rule, but today we know that it is correct. A more adequate rule will be presented when you are introduced to the structure of the atom (Section 5.6).

4.5 A MOLE OF MOLECULES: THE MOLECULAR WEIGHT

Since 1 mole of O_2 weighs 32.0 g, the molecular weight of O_2 is 32.0 g mole^{-1}. Likewise, the molecular weights of H_2 and Cl_2 are 2.02 and 71.0 g mole^{-1}, respectively. If the molecule is a compound (composed of two or more atoms of different elements), the molecular weight is the weight of the appropriate number of moles of each atom. For example, the molecular weight of HCl is the weight of 1 mole H plus the weight of 1 mole Cl, or $1.008 + 35.5 = 36.5$ g mole^{-1}.

DEFINITION: *The molecular weight of a compound is the weight of 1 mole of molecules of the compound.**

*Since some solid compounds do not consist of simple molecules, a broader definition of molecular weight is based on the assumption that the simplest *formula* of the compound is a unit.
   For example: Sodium chloride, or table salt, has the formula NaCl. Although it does not consist of NaCl molecules (Section 6.4), its molecular weight is calculated as

$$23.0 + 35.5 = 58.5 \text{ g mole}^{-1}.$$

TABLE 4.2 The relationships between molecules, atoms, moles, grams, atomic weights and molecular weights

H_2	$+$	Cl_2	$\longrightarrow$ 2 HCl
1 molecule H_2	$+$	1 molecule Cl_2	$\longrightarrow$ 2 molecules HCl
12 molecules H_2	$+$	12 molecules Cl_2	$\longrightarrow$ 24 molecules HCl
1 dozen molecules H_2	$+$ 1 dozen molecules Cl_2	$\longrightarrow$ 2 dozen molecules HCl
6×10^{23} molecules H_2	$+$ 6×10^{23} molecules Cl_2	$\longrightarrow$ 12×10^{23} molecules HCl
1 mole H_2	$+$	1 mole Cl_2	$\longrightarrow$ 2 moles HCl
2 moles H_2	$+$	2 moles Cl_2	$\longrightarrow$ 4 moles HCl
2.02 g H_2	$+$	71.0 g Cl_2	$\longrightarrow$ 73.0 g HCl

$2 H_2$	$+$	O_2	$\longrightarrow$ $2 H_2O$
2 molecules H_2	$+$	1 molecule O_2	$\longrightarrow$ 2 molecules H_2O
6×10^{23} molecules H_2	$+$ 3×10^{23} molecules O_2	$\longrightarrow$ 6×10^{23} molecules H_2O
1 mole H_2	$+$	$\frac{1}{2}$ mole O_2	$\longrightarrow$ 1 mole H_2O
2.02 g H_2	$+$	16.0 g O_2	$\longrightarrow$ 18.0 g H_2O
10 mole H_2	$+$	5 mole O_2	$\longrightarrow$ 10 mole H_2O

1 molecule NH_3	contains	1 atom N and 3 atoms H
1 mole NH_3	contains	1 mole N and 3 moles H

molecular weight NH_3 = atomic weight N + 3 × (atomic weight H)

molecular weight NH_3 =   14.0   + 3 × (1.01)

molecular weight NH_3 = 17.0 g mole^{-1}

The mole itself is a number like a dozen or a gross and it should be thought of as such, even though it is tremendously large and it is not an exact number. A mole is, to four significant figures, 6.022×10^{23} of anything. This number is also called Avogadro's number. The examples in Table 4.2 will help you understand the relationship between atoms, molecules, moles, grams, atomic weights, and molecular weights.

Examples 1 through 5 will illustrate how the *one-factor* method may be used to solve problems relating weights of atoms and molecules.

EXAMPLE 1.   How many grams of Cl are in 6 moles of Cl?

Solution: Since the number of grams Cl is directly proportional to the number of moles Cl, begin in the following way.

$$x \text{ g Cl} = (6 \text{ moles Cl}) \cdots$$

A one-factor with "mole Cl" in the denominator comes from the atomic weight table, 1 mole Cl = 35.5 g Cl.

$$x \text{ g Cl} = (6 \text{ moles Cl})\left(\frac{35.5}{1} \frac{\text{g Cl}}{\text{mole Cl}}\right)$$
$$= (6)(35.5 \text{ g Cl})$$
$$= 213 \text{ g Cl}$$

EXAMPLE 2.   How many grams NH_3 are in 3 moles NH_3?

Solution: Since the number of grams NH_3 is directly proportional to the number of moles NH_3, begin in the following way.

$$x \text{ g } NH_3 = (3 \text{ moles } NH_3) \cdots$$

A one-factor with "mole NH_3" in the denominator is needed. The weight of *1 mole NH_3* is the weight of 1 mole N plus the weight of 3 moles H, which with the aid of atomic weight tables you may calculate as follows.

$$14.0 + (3 \times 1.008) = 17.0 \text{ g.}$$

$$x \text{ g } NH_3 = (3 \text{ moles } NH_3)\left(\frac{17.0}{1} \frac{\text{g } NH_3}{\text{mole } NH_3}\right)$$

$$= (3)(17.0 \text{ g } NH_3)$$

$$= 51.0 \text{ g } NH_3$$

EXAMPLE 3. How many moles F_2 are in 30.0 g F_2?

Solution: Since the number of moles F_2 is directly proportional to the number of grams F_2, you may begin as follows.

$$x \text{ mole } F_2 = (30.0 \text{ g } F_2) \cdots$$

A one-factor is needed with "g F_2" in the denominator. From the atomic weight table the equalities 1 mole F = 19.0 g can be used to calculate

$$1 \text{ mole } F_2 = 2 \text{ mole } F = 38.0 \text{ g}$$

$$x \text{ mole } F_2 = (30.0 \text{ g } F_2)\left(\frac{1}{38.0} \frac{\text{mole } F_2}{\text{g } F_2}\right)$$

$$= \frac{30.0}{38.0} \text{ mole } F_2$$

$$= 0.789 \text{ mole } F_2$$

EXAMPLE 4. (*a*) How many O atoms are in 10.0 g O_2 (ordinary oxygen)?
(*b*) How many O atoms are in 10.0 g O_3 (ozone, a form of oxygen present in the stratosphere)?

Solution: (a) Since the number of O atoms is directly proportional to the number of grams O_2 you may begin as follows:

$$x \text{ atoms O} = (10.0 \text{ g } O_2) \cdots$$

A one-factor is needed with "g O_2" in the denominator. Since the number of moles O_2 will lead eventually to the number of O atoms, an appropriate equality can be calculated from the atomic weight table.

$$1 \text{ mole } O_2 = 2 \text{ moles O} = 32.0 \text{ g O}$$

$$x \text{ atoms O} = (10.0 \text{ g } O_2)\left(\frac{1}{32.0} \frac{\text{mole } O_2}{\text{g } O_2}\right) \cdots$$

A one-factor with "mole O_2" in the denominator comes from the equality 1 mole $O_2 = 6.02 \times 10^{23}$ molecules O_2, which will lead to the number of O atoms.

$$x \text{ atoms O} = (10.0 \text{ g } O_2)\left(\frac{1}{32.0} \frac{\text{mole } O_2}{\text{g } O_2}\right)\left(\frac{6.02 \times 10^{23} \text{ molecules } O_2}{1 \text{ mole } O_2}\right) \cdots$$

Finally, since 1 molecule $O_2 = 2$ atoms O, you may complete the chain of one-factors, placing "molecule O_2" in the denominator to cancel with "molecule O_2" already in the numerator.

$$x \text{ atoms O} = (10.0 \text{ g } O_2)\left(\frac{1}{32.0} \frac{\text{mole } O_2}{\text{g } O_2}\right)\left(\frac{6.02 \times 10^{23}}{1} \frac{\text{molecules } O_2}{\text{mole } O_2}\right)$$

$$\left(\frac{2}{1} \frac{\text{atoms O}}{\text{molecule } O_2}\right)$$

$$= \frac{(10.0)(6.02 \times 10^{23})(2)}{32.0} \text{ atoms O}$$

$$= \frac{(10.0)(6.02)(2)}{32.0} \times 10^{23} \text{ atoms O}$$

$$= 3.76 \times 10^{23} \text{ atoms O}$$

(b) The simple answer to this question is that, since the weight of an atom is independent of its molecular form, 10.0 g O_3 contains 3.76×10^{23} atoms O, the same as 10.0 g O_2 in part (a). A detailed procedure, which you may compare with part (a), leads to the same value.

$$x \text{ atoms O} = (10.0 \text{ g } O_3)\left(\frac{1}{48.0} \frac{\text{mole } O_3}{\text{g } O_3}\right)\left(\frac{6.023 \times 10^{23}}{1} \frac{\text{molecules } O_3}{\text{mole } O_3}\right)$$

$$\left(\frac{3}{1} \frac{\text{atoms O}}{\text{molecule } O_3}\right)$$

$$= \frac{(10.0)(6.023 \times 10^{23})(3)}{48.0} \text{ atoms O}$$

$$= 3.76 \times 10^{23} \text{ atoms O}$$

Of course the number of molecules is different:

For O_2:

$$(3.76 \times 10^{23} \text{ atoms O})\left(\frac{1}{2} \frac{\text{molecule } O_2}{\text{atoms O}}\right) = 1.88 \times 10^{23} \text{ molecules } O_2$$

For O_3:

$$(3.76 \times 10^{23} \text{ atoms O})\left(\frac{1}{3} \frac{\text{molecule } O_3}{\text{atoms O}}\right) = 1.25 \times 10^{23} \text{ molecules } O_3$$

EXAMPLE 5. How many grams of H_2O are in 2 molecules of H_2O?

Solution: Since the number of grams of H_2O is directly proportional to the number of molecules of H_2O, you may begin as follows.

$$x \text{ g } H_2O = (2 \text{ molecules } H_2O) \cdots$$

To eliminate "molecules H_2O," use an equality which provides "molecules H_2O" in the denominator, 1 mole $H_2O = 6.02 \times 10^{23}$ molecules H_2O.

$$x \text{ g } H_2O = (2 \text{ molecules } H_2O) \left(\frac{1}{6.02 \times 10^{23}} \frac{\text{mole } H_2O}{\text{molecules } H_2O} \right) \cdots$$

To eliminate "mole H_2O" in the numerator, use an equality which provides "mole H_2O" in the denominator. Since 1 mole H_2O contains 2 mole H and 1 mole O which weigh $2 \times 1.008 + 16.0 = 18.0$ g you may complete the chain as follows:

$$x \text{ g } H_2O = (2 \text{ molecules } H_2O) \left(\frac{1}{6.02 \times 10^{23}} \frac{\text{mole } H_2O}{\text{molecules } H_2O} \right) \left(\frac{18.0}{1} \frac{\text{g } H_2O}{\text{mole } H_2O} \right)$$

$$= \frac{(2)(18.0)}{6.02 \times 10^{23}} \text{ g } H_2O$$

$$= \frac{(2)(18.0)}{6.02} \times \frac{1}{10^{23}} \text{ g } H_2O$$

$$= 5.98 \times 10^{-23} \text{ g } H_2O$$

This is a very small weight, as it should be for only two molecules.

4.6 LIMITING REACTANTS

It is not necessary that all the reactant molecules, those on the left in a chemical equation, combine in a chemical reaction. There may be an excess of one of the reactants. The reaction will be limited by the reactant in short supply, and it is up to the chemist to recognize which one that is. For example in the reaction

$$2 H_2 + O_2 \longrightarrow 2 H_2O$$

suppose there are four molecules H_2 and four molecules O_2. How many molecules H_2O can possibly be formed? Since it takes two molecules H_2 for every O_2, the H_2 must be in short supply; it must be the **limiting reactant.** The reaction could be written,

$$4 H_2 + 2 O_2 \longrightarrow 4 H_2O,$$

and the number of water molecules that can be formed is four. There will be $2 O_2$ molecules that react and $2 O_2$ molecules that do not.

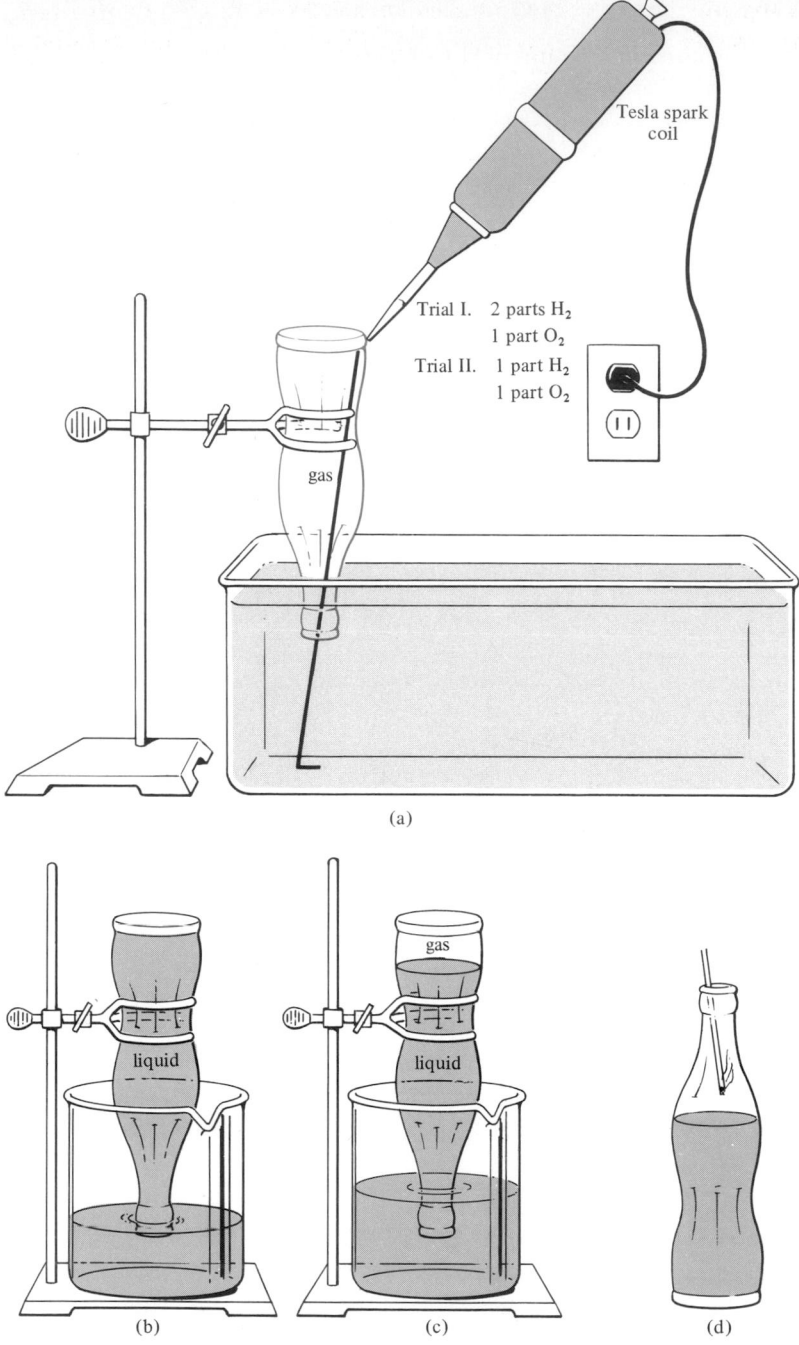

Trial I.   2 parts H_2
            1 part O_2
Trial II. 1 part H_2
            1 part O_2

Tesla spark coil

gas

(a)

liquid

gas

liquid

(b)   (c)   (d)

Figure 4.2. Limiting reactants in the reaction $H_2 + 2\,O_2 \longrightarrow 2\,H_2O$. (a) Bottle filled with gas and immersed in water. A wire is inserted to attract the spark. (*Note:* This equipment should be placed behind an explosion screen before closing the ignition circuit!) (b) Case I. Fills completely with water. (c) Case II. Fills three quarter capacity with water. (d) Identification of the excess gas. It causes a glowing splint to burst into flames.

EXAMPLE 1. In the reaction between 10 moles N_2 and 10 moles H_2 to give NH_3, (a) What is the limiting reactant? (b) What is the maximum amount of NH_3 that could possibly be formed?

Solution: (a) $N_2 + 3 H_2 \longrightarrow 2 NH_3$. Since 3 moles H_2 is consumed for every mole N_2, H_2 must be the limiting reactant. (b) Using the one-factor method we calculate the number of moles NH_3. We begin with the limiting reactant. The appropriate one-factor is (2 moles NH_3)/(3 moles H_2) because 2 moles NH_3 is equivalent to 3 moles H_2, and the units cancel properly.

$$(10 \text{ moles } H_2)\left(\frac{2 \text{ moles } NH_3}{3 \text{ moles } H_2}\right) = \frac{(10)(2)}{3} \text{ moles } NH_3 = 6.7 \text{ mole } NH_3$$

The one-factor method avoids confusion and facilitates rechecking results. As a student of chemistry you should use it as often as possible.

The results for the reaction between H_2 and O_2 may be demonstrated by an apparatus like that shown in Figure 4.2a. In Case I the bottle is filled to $\frac{1}{3}$ capacity with $O_2(g)$ by displacement of water and then filled the rest of the way with $H_2(g)$. The mole ratio of H_2 to O_2 is then $\frac{2}{3}/\frac{1}{3} = 2/1$. A metal wire inserted in the bottle serves to attract a spark from the Tesla spark coil which ignites the mixture. Of course, this must be done behind an explosion screen. The results are shown in Figure 4.2b. The bottle is suddenly and completely filled with water because a vacuum is produced by the disappearance of all gas.

In Case II the bottle is $\frac{1}{2}$ filled with $O_2(g)$ and $\frac{1}{2}$ filled with $H_2(g)$ giving a mole ratio of H_2 to O_2 which is $\frac{1}{2}/\frac{1}{2} = 1/1$. When this mixture is ignited the bottle becomes $\frac{3}{4}$ filled with water (Figure 4.2c). If this bottle is inverted and tested with a glowing wooden splint, the splint bursts into flame indicating the presence of pure $O_2(g)$. A sample of $H_2(g)$ under similar conditions would burn as it combined with air, but the splint would not reignite.

A conceptualized version of this reaction mixture and the change it undergoes is shown in Figure 4.3 with equal numbers of O_2 and H_2 molecules (four of each). The

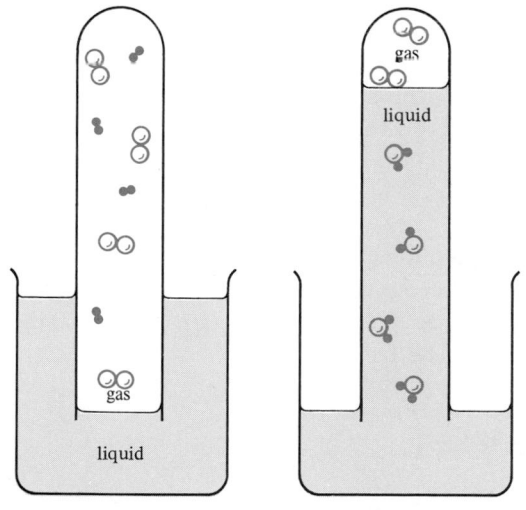

Figure 4.3. Conceptualized model of limiting reactants. 4 $H_2(g)$ + 4 $O_2(g) \longrightarrow$ 4 $H_2O(l)$ + 2 $O_2(g)$

change in volume during reaction is determined by the change in total number of gas molecules. Since the initial number of gas molecules is eight and the final number is two, the ratio of final gas volume to initial gas volume is $\frac{2}{8} = \frac{1}{4}$. This explains why the other $\frac{3}{4}$ of the volume becomes filled with water.

4.7   WEIGHT RELATIONS

Many practical problems in chemistry can be solved using a balanced chemical equation and atomic weights. The fertilizer chemist wants to know how much sulfur he must buy to make sulfuric acid to solubilize a ton of phosphate rock. The farmer wants to know how much lime it will take to neutralize the acidity in his soil. The space engineer wants to know how much hydrogen is needed for a certain amount of oxygen in the fuel cells which will furnish the electrical power during flight. The automotive engineer wants to know how much air the carburetor must furnish for the complete combustion of the amount of gasoline injected into the cylinder. All of these questions and many similar ones can easily be answered if the chemical equation can be written.

The chemistry student must learn to think in moles as easily as in pounds or grams. A problem stated in terms of how much of reactant A reacts with how much of reactant B to give how much of product C is most directly answered in terms of moles. All that is necessary is to write a balanced chemical equation and the coefficients represent the relative number of moles of each substance.

EXAMPLE 1.   A certain fuel cell produces electricity from the reaction between hydrogen and oxygen to form water. How many kilograms hydrogen is necessary to react completely with 10.0 kg oxygen?

Solution: (a) The first step is to write the formulas for each molecule. These formulas have been determined in previous experiments.

$$H_2 \qquad O_2 \qquad H_2O$$

hydrogen   oxygen   water

(b) The second step is to balance the chemical equation, placing reactants on the left and product on the right.

$$2\,H_2 + O_2 \longrightarrow 2\,H_2O$$

(c) The third step is to write down the given quantity and multiply it by appropriate one-factors to convert it to the desired quantity.
(1) The first one-factor converts 10.0 kg O_2 into grams O_2 so that the conversion to moles will be facilitated.

$$(10.0 \text{ kg } O_2)\left(\frac{1000}{1} \frac{\text{g } O_2}{\text{kg } O_2}\right) \cdots$$

(2) The second one-factor converts grams O_2 to moles O_2; 1 mole O_2 weighs 32.0 g because the atomic weight of oxygen is 16.0, and there are two atoms per molecule.

$$(10.0 \text{ kg } O_2)\left(\frac{1000}{1}\frac{\text{g } O_2}{\text{kg } O_2}\right)\left(\frac{1}{32.0}\frac{\text{mole } O_2}{\text{g } O_2}\right)\cdots$$

(3) The third one-factor converts moles O_2 to moles H_2. From the balanced chemical equation 1 mole O_2 is equivalent to 2 moles H_2.

$$(10.0 \text{ kg } O_2)\left(\frac{1000}{1}\frac{\text{g } O_2}{\text{kg } O_2}\right)\left(\frac{1}{32.0}\frac{\text{mole } O_2}{\text{g } O_2}\right)\left(\frac{2}{1}\frac{\text{moles } H_2}{\text{mole } O_2}\right)\cdots$$

(4) The fourth one-factor converts moles H_2 to grams H_2; 1 mole H_2 weighs 2.02 g.

$$(10.0 \text{ kg } O_2)\left(\frac{1000}{1}\frac{\text{g } O_2}{\text{kg } O_2}\right)\left(\frac{1}{32.0}\frac{\text{mole } O_2}{\text{g } O_2}\right)\left(\frac{2}{1}\frac{\text{moles } H_2}{\text{mole } O_2}\right)\left(\frac{2.02}{1}\frac{\text{g } H_2}{\text{mole } H_2}\right)\cdots$$

(5) The fifth one-factor converts grams H_2 into kilograms H_2.

$$(10.0 \text{ kg } O_2)\left(\frac{1000}{1}\frac{\text{g } O_2}{\text{kg } O_2}\right)\left(\frac{1}{32.0}\frac{\text{mole } O_2}{\text{g } O_2}\right)\left(\frac{2}{1}\frac{\text{moles } H_2}{\text{mole } O_2}\right)\left(\frac{2.02}{1}\frac{\text{g } H_2}{\text{mole } H_2}\right)$$
$$\left(\frac{1}{1000}\frac{\text{kg } H_2}{\text{g } H_2}\right)$$

(d) After all the necessary one-factors are strung out in a chain, an operation that can be carried out all on one line, multiply and divide the numbers.

$$\frac{(10.0)(1000)(2)(2.02)}{(32.0)(1000)} = 1.26 \text{ kg } H_2$$

EXAMPLE 2. An early theory of combustion was that when materials burn there is a substance released called "phlogiston" which causes the flame and heat. The idea may have come from the fact that many things such as coal, wood, and candles lose weight when they burn. We know that combustion is actually a combination of a material *with* something, namely the oxygen in the air. Consider a paraffin candle, which has a typical molecular formula of $C_{25}H_{52}$. How many grams of product are formed by the combustion of 1.00 g candle when the candle burns in air?

Solution: (a) Write the balanced equation using the appropriate molecular formulas. First we must know from experiment *all* reactants and products.

$$C_{25}H_{52} + O_2 \longrightarrow CO_2 + H_2O$$

(1) Place a 25 before the CO_2 because there are 25 C atoms on the left and we must have 25 on the right.

$$C_{25}H_{52} + O_2 \longrightarrow 25\ CO_2 + H_2O$$

(2) Place a 26 before the H_2O because there are 52 H atoms on the left and there must be 52 on the right.

$$C_{25}H_{52} + O_2 \longrightarrow 25\ CO_2 + 26\ H_2O$$

(3) The total number of O atoms on the right is 50 in the form of CO_2 and 26 in the form of H_2O. The total is 76. The number of O_2 molecules needed is $\frac{76}{2} = 38$, since each molecule has two atoms. Place a 38 in front of O_2. We now have a balanced chemical equation. In a balanced chemical equation *the number of atoms of every kind is equal on left and right.*

$$C_{25}H_{52} + 38\ O_2 \longrightarrow 25\ CO_2 + 26\ H_2O$$

(b) Use the factor method to convert from 1.00 g $C_{25}H_{52}$ to grams of CO_2 plus grams H_2O. One mole of $C_{25}H_{52}$ weighs $(25 \times 12.0) + (52 \times 1.01) = 300 + 52.5 = 353$ g.
(1) First calculate the weight of CO_2 formed.

$$(1.00\ \text{g}\ C_{25}H_{52})\left(\frac{1\ \text{mole}\ C_{25}H_{52}}{353\ \text{g}\ C_{25}H_{52}}\right)\left(\frac{25\ \text{mole}\ CO_2}{1\ \text{mole}\ C_{25}H_{52}}\right)\left(\frac{44.0\ \text{g}\ CO_2}{1\ \text{mole}\ CO_2}\right)$$

$$= \frac{(25)(44.0)}{353} = \frac{(2.500)(4.40)(10^2)}{(3.53)(10^2)} = 3.12\ \text{g}\ CO_2$$

(2) Then calculate the weight of H_2O formed.

$$(1.00\ \text{g}\ C_{25}H_{52})\left(\frac{1\ \text{mole}\ C_{25}H_{52}}{353\ \text{g}\ C_{25}H_{52}}\right)\left(\frac{26\ \text{mole}\ H_2O}{1\ \text{mole}\ C_{25}H_{52}}\right)\left(\frac{18.0\ \text{g}\ H_2O}{1\ \text{mole}\ H_2O}\right)$$

$$= \frac{(26)(18.0)}{353} = \frac{(2.600)(1.80)(10^2)}{(3.53)(10^2)} = 1.33\ \text{g}\ H_2O$$

(c) The total grams of products per gram of candle is the sum of the two.

$$\text{weight of products} = 3.12 + 1.33 = 4.45\ \text{g}$$

The products weigh over four times as much as the original candle! The proponents of the "phlogiston theory" would have been surprised at the results of these calculations because they believed a candle to be composed of *pure* "phlogiston" and to leave no product at all! Theories which can be used by first year chemistry students to calculate weight relations in chemical reactions did not come easily, and old theories were very slow in being displaced. To the good chemist, however, a bad theory is better than no theory at all, so long as an honest attempt is made to disprove it.

Environmental note. The products of burning gasoline in the internal combustion engine should go unnoticed by the man on the street because the typical gasoline molecule, C_8H_{18}, is very similar to the typical paraffin molecule, $C_{25}H_{52}$. However the short time permitted for the combustion causes some CO to form by incomplete reaction and the high temperatures cause some combination of N_2 and O_2 in the air to form NO. Both CO and NO are highly poisonous. Besides that, more than 3 g of O_2 is consumed for every gram of gasoline.

EXAMPLE 3. Consider the reaction between 5.0 g Cl_2 and 5.0 g Br_2 to form BrCl. What is the maximum weight of product that can be formed?

Solution: Your first step is to write a balanced chemical equation using the reactants and products given.

$$Cl_2 + Br_2 \longrightarrow 2\ BrCl$$

This equation indicates that every molecule of Cl_2 requires 1 molecule of Br_2. Your second step is to determine which reactant is limiting, that is the one with the fewer number of moles. Turning to a table of atomic weights you may obtain the following information using the symbol to represent the atomic weight.

$$Cl = 35.5\ g\ mole^{-1}$$
$$Br = 79.9\ g\ mole^{-1}$$

Since Br atoms weigh more than Cl atoms, there are fewer of them in a 5.0 g sample and Br_2 is the limiting reactant. Your third step is to write a proportionality between the number of grams BrCl and the number of grams Br_2.

$$x\ g\ BrCl = (5.0\ g\ Br_2) \cdots$$

The fourth step introduces a one-factor with "g Br_2" in the denominator. Since 1 mole Br_2 = 2 mole Br = 2 × 79.9 g Br_2 = 160 g Br_2, you may write

$$x\ g\ BrCl = (5.0\ g\ Br_2)\left(\frac{1}{160}\ \frac{mole\ Br_2}{g\ Br_2}\right) \cdots$$

In a fifth step you may introduce a one-factor with "mole Br_2" in the denominator. The balanced chemical equation provides the equivalency of 1 mole Br_2 and 2 mole BrCl.

$$x \text{ g BrCl} = (5.0 \text{ g Br}_2)\left(\frac{1 \text{ mole Br}_2}{160 \text{ g Br}_2}\right)\left(\frac{2 \text{ mole BrCl}}{1 \text{ mole Br}_2}\right) \cdots$$

Finally, you may introduce a one-factor with mole BrCl in the denominator, using the equality 1 mole BrCl = 79.9 + 35.5 = 115.4 g BrCl.

$$x \text{ g BrCl} = (5.0 \text{ g Br}_2)\left(\frac{1 \text{ mole Br}_2}{160 \text{ g Br}_2}\right)\left(\frac{2 \text{ mole BrCl}}{1 \text{ mole Br}_2}\right)\left(\frac{115.4 \text{ g BrCl}}{1 \text{ mole BrCl}}\right)$$

$$= \frac{(5.0)(2)(115.4)}{160} \text{ g BrCl}$$

$$= 7.2 \text{ g BrCl}$$

Per Cent Composition by Weight   The law of constant composition (Section 4.3) says that the composition by weight of any compound is the same for any sample of that compound. You can calculate the weight per cent of any element in a compound by using its formula and a table of atomic weights (see Example 4).

EXAMPLE 4.   Liquid bromine, Br_2, is obtained from magnesium bromide, $MgBr_2$, which is dissolved in the ocean. What is the weight per cent of Br in pure $MgBr_2$?

Solution: The meaning of weight per cent in this example is simply the number of grams Br in 100 g $MgBr_2$. The number of grams Br is directly proportional to the number of grams $MgBr_2$.

$$x \text{ g Br} = (100 \text{ g MgBr}_2) \cdots$$

You will find the following atomic and molecular weights useful.

$$1 \text{ mole Br} = 79.9 \text{ g}$$
$$1 \text{ mole Mg} = 24.3 \text{ g}$$
$$1 \text{ mole MgBr}_2 = (2 \times 79.9) + 24.3 = 184 \text{ g}$$

If you use these equalities and the formula $MgBr_2$, all the necessary one-factors are available.

$$x \text{ g Br} = (100 \text{ g MgBr}_2)\left(\frac{1 \text{ mole MgBr}_2}{184 \text{ g MgBr}_2}\right)\left(\frac{2 \text{ mole Br}}{1 \text{ mole MgBr}_2}\right)\left(\frac{79.9 \text{ g Br}}{1 \text{ mole Br}}\right)$$

$$= 86.8 \text{ g Br} \quad \text{or} \quad 86.8\% \text{ Br}$$

Exercise 1.   The Alcoa company gets most of its aluminum from bauxite ore, which has the formula Al_2O_3 in the dried form. What is the weight per cent of Al in this compound? (Atomic weights: Al = 27.0 g mole^{-1}, O = 16.0 g mole^{-1})

Answer: 52.9% Al

The **empirical formula** of a compound is sometimes called its simplest formula. For most compounds in this book the true formulas and the empirical formulas are the same, for example, CO, CO_2, H_2O, NH_3. Only one could have been written more simply, HO for H_2O_2, hydrogen peroxide. The **true formula** gives the actual number of atoms of each element in a molecule of the compound. How is the empirical formula related to the true formula?

DEFINITION: *The empirical formula is the formula obtained from the true formula by dividing each subscript by the largest common denominator.*

EXAMPLE 1. True formulas are given for the following compounds. What are the corresponding empirical formulas?
(a) Acetylene, C_2H_2.
(b) Hydrazine, N_2H_4.
(c) Benzene, C_6H_6.

Solution: (a) Divide each subscript by 2: CH.
(b) Divide each subscript by 2: NH_2.
(c) Divide each subscript by 6: CH.

Sometimes you may have only enough information about a compound to determine its empirical formula and not its true formula. For example, you can determine the empirical formula of a compound if you know which elements it contains and the number of grams of each. In the first step you should calculate the number of *moles* of each element using its atomic weight. In the second step you find the smallest *whole* (integral) numbers of moles that have the same ratio as those calculated. The *true formula* can be obtained from the *empirical formula* if the approximate molecular weight is known.

EXAMPLE 2. In the furnace shown in Figure 4.1 suppose that 50.0 g carbon and 133 g oxygen react completely with one another to form a single gaseous oxide of carbon. What is the empirical formula of this compound?

Solution: You are looking for the two smallest whole numbers that can serve as subscripts in the formula of the unknown compound, $C_\square O_\square$. These numbers cannot be fractions because they represent the number of atoms in the empirical formula. You can expect both numbers to be fairly small (less than eight) for simple compounds such as gases.

In the first step you should use one-factors to calculate the number of moles of each element in the compound (atomic weights: $C = 12.01$, $O = 16.00$ g mole^{-1}). In calculating empirical formulas you should use as many significant figures as permitted.

$$x \text{ mole C} = (50.0 \text{ g C})\left(\frac{1}{12.01} \frac{\text{mole C}}{\text{g C}}\right) = 4.16 \text{ mole C}$$

$$y \text{ mole O} = (133 \text{ g O})\left(\frac{1}{16.00} \frac{\text{mole O}}{\text{g O}}\right) = 8.31 \text{ mole O}$$

In the second step the mole ratio is changed to small whole numbers by dividing by the smaller number of moles.

4.16 mole C is combined with 8.31 mole O

$\dfrac{4.16}{4.16}$ mole C is combined with $\dfrac{8.31}{4.16}$ mole O

1.00 mole C is combined with 2.00 mole O

The *empirical formula* is CO_2.

If no additional information about the unknown compound is given, the true formula might be CO_2, C_2O_4, C_3O_6, C_4O_8, . . . , all of which have the same approximate ratio of 4.16 mole C to 8.31 mole O.

Suppose that the molecular weight of a sample of gas is measured and that it is found to be approximately 44 g mole^{-1}. Can you then choose the *true formula*? The answer is yes, of course. The only formula with the correct molecular weight is CO_2, which is, therefore, the true formula of the compound, carbon dioxide.

EXAMPLE 3.   The per cent composition by weight of a certain gaseous hydrocarbon found in natural gas is 20.1% H and 79.9% C.
(a) Find the empirical formula.
(b) If the molecular weight of the unknown gas is known to be approximately 30 g mole^{-1}, what is its true formula?

Solution: You will find it convenient to base these calculations on 100.0 g hydrocarbon, which contains 20.1 g H and 79.9 g C.
(a) The empirical formula is $C_\square H_\square$. Find the number of moles of C and H in 100.0 g compound.

$$x \text{ mole C} = (79.9 \text{ g C})\left(\frac{1}{12.01} \frac{\text{mole C}}{\text{g C}}\right) = 6.65 \text{ mole C}$$

$$y \text{ mole H} = (20.1 \text{ g H})\left(\frac{1}{1.008} \frac{\text{mole H}}{\text{g H}}\right) = 19.9 \text{ mole H}$$

6.65 mole C is combined with 19.9 mole H

$\dfrac{6.65}{6.65}$ mole C is combined with $\dfrac{19.9}{6.65}$ mole H

1.00 mole C is combined with 2.99 mole H

The empirical formula is CH_3.

(b) The possible true formulas and their approximate molecular weights are listed below.

$$CH_3, \quad C_2H_6, \quad C_3H_9, \quad \cdots$$
$$15, \quad 30, \quad 45, \quad \cdots \text{ g mole}^{-1}$$

Therefore, the true formula is C_2H_6, ethane.

Exercise 1. One of the liquids used in rocket propellants is an oxide of nitrogen having the following composition by weight: 30.4% N and 69.6% O. (a) What is its empirical formula? (b) If the molecular weight is approximately 92 g mole^{-1}, what is its true formula?

Answer: (a) NO_2; (b) N_2O_4, dinitrogen tetroxide

PROBLEMS

1. Calculate the volume in cubic centimeters of each product in the reactions shown below. Assume that all solids and liquids have negligible (zero) volume and that conditions of all gases are standard (0°C and 1 atm).

   (a) $S(s, \text{ excess}) + O_2(g, \text{ 300 cm}^3) \longrightarrow SO_2(g)$

   (b) $2\,CH_4(g, \text{ 400 cm}^3)$
      $+ \, 3\,O_2(g, \text{ 600 cm}^3) \longrightarrow$
           $2\,CO(g) + 4\,H_2O(l)$

2. Same as Problem 1.

   (a) $2\,NO(g, \text{ 300 cm}^3)$
      $+ \, O_2(g, \text{ 150 cm}^3) \longrightarrow 2\,NO_2(g)$

   (b) $2\,C_2H_2(g, \text{ 100 cm}^3)$
      $+ \, 5\,O_2(g, \text{ 250 cm}^3)$
          $\longrightarrow 4\,CO_2(g) + 2\,H_2O(l)$

3. Balance the following equations using the smallest whole numbers possible:

   (a) $C + O_2 \longrightarrow C_2O_3$
   (b) $Fe_3O_4 + CO \longrightarrow Fe + CO_2$
   (c) $N_2 + H_2 \longrightarrow N_2H_4$
   (d) $N_2H_4 + H_2O_2 \longrightarrow N_2 + H_2O$
   (e) $Al_2O_3 + C \longrightarrow Al_4C_3 + CO$

4. Same as problem 3.

   (a) $NH_3 + O_2 \longrightarrow NO + H_2O$
   (b) $HNO_3 \longrightarrow NO_2 + O_2 + H_2O$
   (c) $PbS + PbO \longrightarrow Pb + SO_2$
   (d) $Fe + H_2O \longrightarrow Fe_3O_4 + H_2$
   (e) $CH_4 + O_2 \longrightarrow CO_2 + H_2O$

5. A shoe box is like a diatomic molecule: each box contains two shoes and each molecule contains two atoms. (There the similarity ends.)

   (a) How many shoes are in a gross of shoe boxes? (1 gross = 144)

   (b) How many H atoms are in 1 mole of H_2 molecules?

6. Same as problem 5.

   (a) How many shoes are in 3 gross of shoe boxes (1 gross = 144)

   (b) How many H atoms are in 3 moles H_2 molecules?

7. The world population is estimated at 4,000,000,000 persons. How many moles of persons populate the earth?

8. There are approximately 5,000,000 bee hives in the United States with a summertime occupancy of 60,000 honeybees per hive (a) What is the total summertime honeybee population? (b) How many moles of honeybees are in this population?

9. An individual sugar crystal is placed on a single pan balance and found to weigh 2.5×10^{-4} g. If the sugar, or sucrose, molecule has the molecular formula $C_{12}H_{22}O_{11}$, how many molecules are in this single crystal?

10. A certain medicine dropper delivers drops of water that weigh an average of 0.052 g. How many water molecules (H_2O) are in a single drop, on the average?

$\#Moles = \dfrac{wt}{molecular\ wt}$ $MW = \dfrac{wt}{\#moles} = g/mole$

11. What is the weight in grams of (a) 3 moles N_2? (b) 3 moles N? (c) 3 molecules N_2? (d) 3 atoms N?

12. How many atoms are in (a) 71.0 g Cl_2? (b) 71.0 g Cl? (c) 35.5 g Cl?

13. How many moles H are in (a) 1 mole H_2O? (b) 3 moles N_2H_4? (c) 2.0 g H_2?

14. How many grams H are in (a) 17.0 g NH_3? (b) 2 moles H_2O? (c) $\frac{1}{2}$ mole HCl?

15. What is the maximum number of molecules or moles of H_2O that can be produced from each of the following reaction mixtures:
(a) 1 molecule H_2 and 1 molecule O_2
(b) 1 mole H_2 and 1 mole O_2
(c) 46 molecules H_2 and 22 molecules O_2
(d) 0.3 mole H_2 and 0.1 mole O_2

16. In the reaction between Al and O_2 to give Al_2O_3
(a) how many moles Al_2O_3 can be produced from 1 mole Al and 1 mole O_2?
(b) how many moles O_2 are needed to produce 0.1 mole Al_2O_3?

17. How many grams H_2O can be produced (a) from 1 mole H_2 and 1 mole O_2? (b) from 0.3 mole H_2 and 0.1 mole O_2?

18. How many moles O_2 would be needed to consume the Hope diamond, which weighs 8.8 g (a) following the reaction $2\,C + O_2 \longrightarrow 2\,CO$, and (b) following the reaction $C + O_2 \longrightarrow CO_2$?

19. Molecules of NO and O_2 combine to give the single product NO_2. How many g NO_2 can be produced from 100 g NO and 100 g O_2?

20. Assume that you are going on a space flight and will use fuel cells to supply your electrical power. If the overall reaction in the fuel cell is $2\,H_2 + O_2 \longrightarrow 2\,H_2O$, how many pounds O_2 will you carry for every pound H_2?

21. Calculate the per cent by weight of the following elements in the pure compounds indicated.
(a) Pb in PbO_2
(b) Mg in Mg_3N_2

22. Same as problem 21.
(a) Ca in CaC_2
(b) Fe in Fe_3O_4

23. A 50.0 g sample of a certain sulfide of copper is found to contain 10.1 g S and 39.9 g Cu. What is the empirical formula of this compound?

24. A 60.0 g sample of a certain oxide of tin is found to contain 12.7 g O and 47.3 g Sn. What is the empirical formula of this compound?

25. A certain hydride of boron consists of 21.9% H and 78.1% B by weight.
(a) What is the empirical formula of this compound?
(b) If the compound has an approximate molecular weight of 28 g mole^{-1}, what is its true formula?

26. A certain hydride of carbon consists of 7.7% H and 92.3% C by weight.
(a) What is the empirical formula of this compound?
(b) If the compound has an approximate molecular weight of 78 g mole^{-1}, what is its true formula?

*27. A 215-ml sample of a certain gaseous hydride of nitrogen at STP (0°C and 1 atm) weighs 0.308 g. The composition of the compound is 12.6% H and 87.4% N by weight. What is its true formula?

*28. The number of moles of a substance contained in a volume of 1.00 l is an important unit for expressing concentration. What is the number of moles of H_2O in 1.00 l $H_2O(l)$?

5

ELECTRICAL NATURE
OF MATTER AND
ATOMIC STRUCTURE

In the earlier chapters you have seen how the atoms of elements combine to form different molecules and compounds. In this chapter you will see that atoms themselves are composed of **fundamental particles** called protons, electrons, and neutrons, which are simpler than the atom. Some of these particles are electrically charged. You will learn to build an atom up by proper combination and placement of the fundamental particles, and you will begin to learn how this arrangement, or structure, is related to the chemical and physical properties of the atom.

5.1   ELECTRICAL CHARGE

Electricity is a common part of our everyday existence. It starts our cars in the mornings, streaks across the skies in afternoon thunderstorms, and plays images on our TV screens in the evenings. The atoms of the elements are electrically neutral only because they contain equal numbers of positively and negatively charged particles. It is when the two charges become separated that we have the best evidence of their existence.

Charges can be separated as easily as passing a comb through your hair on a dry day. The crackling sound and the attractive force that holds bits of paper to the comb are evidences of electricity. By experimenting with the forces between electrical charges, it can be shown that there are only two types. One type attracts the other but repels its own kind. One charge is arbitrarily called positive and the other negative. The **proton** is a single particle with a positive charge and the **electron** is a single particle with a negative charge.

The experiments by which the forces of attraction and repulsion between charges are determined in terms of the size of the charges and the distance of separation between the charges are described in the next two sections. After a long discussion of the behavior of electricity Bertrand Russell, the English mathematician and philosopher, said, "Some readers may expect me at this stage to tell them what electricity 'really is.' The fact is that I have already said what it is. It is not like St. Paul's Cathedral, it is a way in which things behave."* This same irreverent attitude about all models in chemistry, including the model of the atom itself, is characteristic of the good scientist.

5.2   EFFECT OF CHARGE SIZE ON FORCE

How *do* electrons and protons behave? They attract one another and the combination of an electron with a proton is neutral, so they must have equal charge. Two protons, or two electrons, repel one another with the same force if they are the same distance apart.

In Figure 5.1 the length units in terms of d (d, $2d$, and $d/2$) represent distances of separation, from center to center, between charged particles. The symbol is a number having a unit of length such as centimeters or Ångstroms. The arrows with terms in F written over them represent the forces in units, such as pounds or dynes (metric system), that are exerted on each particle and the directions in which they are exerted. If the arrows point toward one another, the force is one of attraction; if they point away from one another, the force is one of repulsion. The length of the arrow represents the magnitude of the force; an arrow that is twice as long ($2F$) as another arrow (F) represents twice as great a force.

In Figure 5.1a all the force arrows are the same length, but the direction for unlike charges (attractive, inward) is opposite to the direction for like charges (repulsive, outward).

All particles in Figure 5.1 are assumed to be point charges, that is, the charge is concentrated at one point. Single electrons and single protons can be considered to be point charges because they are so small, and simple particles composed of unequal numbers of electrons and protons often behave as if they were point charges. If a particle has two more protons than electrons, it is charged ⊕⊕; if it has two more electrons than protons, it is charged ⊖⊖. The effect of increasing the charge and keeping the distance the same is shown in Figures 5.1b and 5.1c. A particle with a net charge of two protons attracts an electron with twice the force shown in Figure 5.1a, and a similar attraction occurs between a ⊕ particle and ⊖ particle, if the distances are the same (Figure 5.1b). If both particles are doubly charged, the force of attraction is four times as great, at the same distance of separation (Figure 5.1c). These observations are expressed mathematically by saying that the force of attraction

*Bertrand Russell: *ABC of Atoms*, E. P. Dutton, New York, 1923.

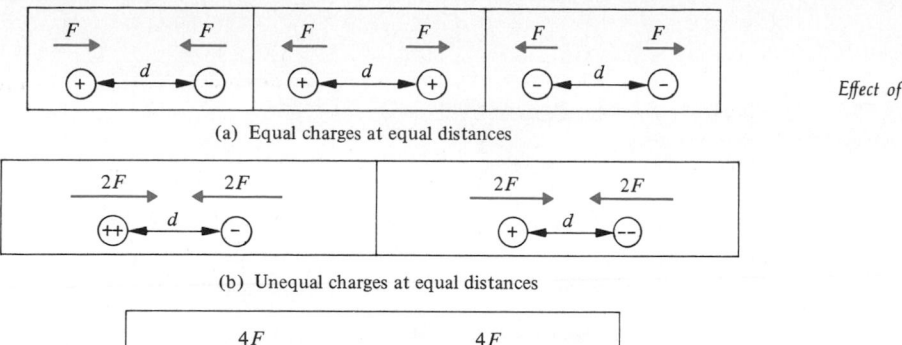

(a) Equal charges at equal distances

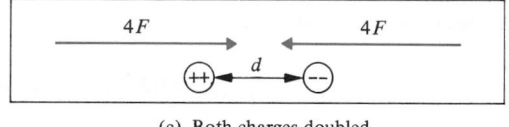

(b) Unequal charges at equal distances

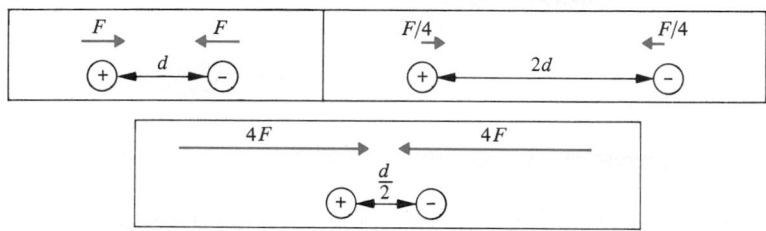

(c) Both charges doubled

(d) Equal charges at unequal distances

Figure 5.1. Forces between charged particles.

between two unlike charges at a given distance of separation is directly proportional to the product of the two charges.

force = a constant × (charge 1) × (charge 2), no change in distance   (5.1)

5.3   EFFECT OF DISTANCE ON FORCE

If the distance of separation changes, it will also affect the force of attraction. For unit charges ($+1$ or -1) at various distances, the effects are shown in Figure 5.1d. Apparently the forces become very great at short distances and decrease rapidly as the distance increases. Mathematically, for given charges, the changes of force with distance are expressed by an inverse square proportionality.

$$\text{force} = \frac{\text{a constant}}{(\text{distance})^2} \quad \text{(no change in charge)} \qquad (5.2)$$

The relation is said to be inverse because as the distance *increases* the force *decreases*. The force is said to be inversely proportional to the *square* of the distance because the distance must be multiplied by itself, or squared, in order to give the correct relationship.

5.4 COULOMB'S LAW: EFFECT OF CHARGE AND DISTANCE ON FORCE

If you wish to compare forces of electrical attraction between pairs of charges in which both the charge and the distance change, you must combine the proportionalities of equation (5.3) with those of equation (5.4).

(a) Constant distance

$$\text{force} = \text{a constant} \times (\text{charge 1}) \times (\text{charge 2}) \qquad (5.3)$$

(b) Constant charge

$$\text{force} = \frac{\text{a constant}}{(\text{distance})^2} \qquad (5.4)$$

(c) Changing charge and distance

$$\text{force} = \text{a constant} \times \frac{(\text{charge 1}) \times (\text{charge 2})}{(\text{distance})^2} \qquad (5.5)$$

This general equation is known as *Coulomb's law*.

EXAMPLE 1.   Use Coulomb's law (equation 5.5) to calculate the unknown forces between the charges shown in the adjoining figure.

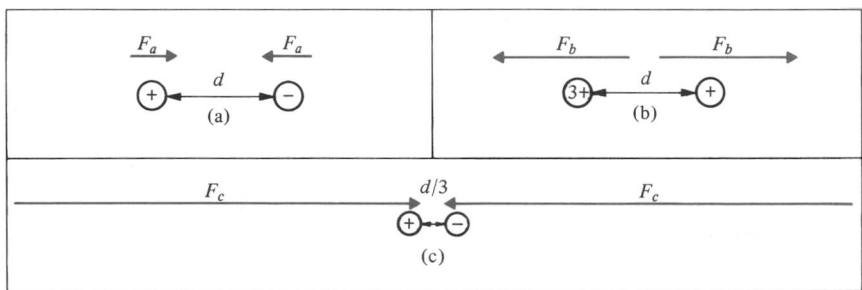

Solution: In working this problem you may compare the unknown forces F_b and F_c with the given force F_a. The answer will be expressed in terms of F_a. It would be necessary to know the numerical value of F_a in pounds, dynes, or some other unit of force to calculate the numerical values of F_b and F_c.

(a) Since the charges have opposite sign, you know that the force is attractive. You may give each charge the numerical value +1 in equation (5.5)

$$\text{force} = \text{a constant} \times \frac{(\text{charge 1}) \times (\text{charge 2})}{(\text{distance})^2} \qquad (5.5)$$

$$F_a = \text{a constant} \times \frac{(1) \times (1)}{d^2}$$

$$= \frac{\text{a constant}}{d^2}$$

(b) The charges in diagram (b) are both positive, therefore the force is repulsive. Use $+3$ and $+1$ in equation (5.5) as in paragraph (a).

$$F_b = \text{a constant} \times \frac{(3) \times (1)}{d^2}$$

$$= 3 \times \frac{\text{a constant}}{d^2}$$

If you compare this result with that of paragraph (a), you can express F_b in terms of F_a. The constant in Coulomb's law cancels out.

$$F_b = 3 \times F_a$$

The force arrow in diagram (b) should be exactly three times longer than that in diagram (a).

(c) Since the charges in diagram (c) are of opposite sign, you know that the force is attractive. Call the charges both $+1$ and substitute into equation (5.5) as in paragraph (a).

$$F_c = \text{a constant} \times \frac{(1) \times (1)}{(d/3)^2}$$

$$= \text{a constant} \times \frac{(1) \times (1)}{d^2/9}$$

$$= 9 \times \frac{\text{a constant}}{d^2}$$

Compare this result with that of paragraph (a) so that you can express F_c in terms of F_a. The constant in Coulomb's law cancels out.

$$F_c = 9 \times F_a$$

5.5 HYDROGEN, THE SIMPLEST ATOM

The simplest atom of all, the H atom, consists of 1 p^+ (proton) and 1 e^- (electron). It is electrically neutral but, by adding enough energy, the electron can be separated from the proton to form an H$^+$ **ion.**

DEFINITION: *An ion is an electrically charged atom or chemical combination of atoms.*

All three equations that follow are equivalent ways of representing the ionization of an H atom by addition of enough energy to overcome the attraction between electron and proton.

$$
\begin{array}{rcl}
\text{hydrogen atom} & + \text{ energy} \longrightarrow & \text{hydrogen ion} + \text{ electron} \\
\text{(1 proton, 1 electron)} & + \text{ energy} \longrightarrow & \text{1 proton} \quad + \text{1 electron} \\
\text{H} & + \text{ energy} \longrightarrow & \text{H}^+ \quad + \quad e^-
\end{array}
$$

You can see that an H^+ ion is nothing more than a bare proton.

It is interesting to consider the size and shape of an H atom, how much it weighs, and whether the weight is distributed evenly between proton and electron like the charge.

EXAMPLE 1. Assuming that all the H atoms in 1 mole H have the same mass, calculate the mass of a single H atom.

Solution: Use the equalities 6.02×10^{23} atoms H $= 1$ mole H and 1 mole H $= 1.01$ g H to set up the following one-factors:

$$
x \text{ g H} = (1 \text{ atom H})\left(\frac{1}{6.02 \times 10^{23}} \frac{\text{mole H}}{\text{atoms H}}\right)\left(\frac{1.01}{1} \frac{\text{g H}}{\text{mole H}}\right)
$$

$$
= \frac{1.01}{6.02 \times 10^{23}} \text{ g}
$$

$$
= 1.67 \times 10^{-24} \text{ g}
$$

This is a very small mass indeed, but the remarkable fact is that nearly all of it is due to the proton. The mass of the electron his been measured and found to be only 1/1840 that of a proton.

Chemists believe in a **nuclear model** of the atom, which means that practically all the mass is located in a very tiny nucleus at the center of the atom. The nucleus of an H atom consists of a proton. The electron, which is in constant motion around it, gives the atom a spherical shape. The average distance of separation between the proton and the electron is called the **radius** of the H atom. This distance is about 0.5 Å, giving the H atom a diameter of approximately 1 Å. Certain scattering experiments, based on firing nuclear particles through very thin samples of gold foil, have indicated that the nuclei of atoms have diameters on the order of 10^{-13} cm, or 10^{-5} Å. The illustration in Figure 5.2 summarizes the information on size and mass of the H atom. The diameter of the atom is 100,000 times the diameter of the nucleus, but the nucleus contains 99.95% of the atomic mass. The proportions are approximately the same for all the atoms in the periodic table.

You may think of the space in an atom as being "occupied" in the same way that space in our solar system is "occupied." However, there is a rule in nature (Section 13.3) which requires that, in atoms, only so many electrons can "occupy" the same space at the same time (two electrons per "orbital") and extra electrons are rigorously excluded.

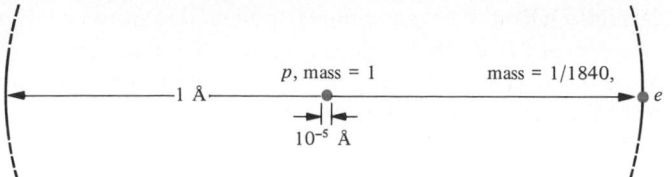

Figure 5.2. The size and mass of an H atom compared with its nucleus (not to scale).

Chemists have abandoned the orbit model of the atom, which meant that electrons moved in certain specified paths or orbits around the nucleus. Although an H atom is spherical in shape, it is not necessary to think of the electron as moving in a circular path. The modern view, shown in Figure 5.3, comes from quantum mechanics, a mathematical approach to space and energy on the atomic scale. Quantum mechanics describes the electron in the atom in terms of its probability of location rather than its orbit. The position of the electron is represented by a probability cloud, or orbital, which is simply a collection of points showing where the electron spends its time. Regions where the points are closest together represent higher probabilities of location. Figure 5.3a shows the probability cloud of a normal H atom, where p represents the proton and each dot the position of the electron at a particular time. A spherical shell having radius, r, contains a certain number of points proportional to the probability, P, of the electron being that distance from the nucleus. For shells of given thickness, this probability goes through a maximum at a certain distance from the nucleus, as shown in Figure 5.3b. For shells that are too small, the volume of the shell is too small to contain more than a few points; for shells that are too large, the points are too far apart to give more than a few points in the shell. The distance or radius corresponding to the maximum probability is 0.5 Å for the H atom, which is its most probable radius.

Clearly, you cannot choose the exact radius of an H atom with great precision. The electron cloud is too diffuse for that. However, you can say with considerable

Figure 5.3. Probability model of the H atom. (a) Cloud, or orbital, with a spherical shell superimposed. (b) Probability of location within shells having different radii. (c) Sphere of maximum probability, $r = 0.5$ Å.

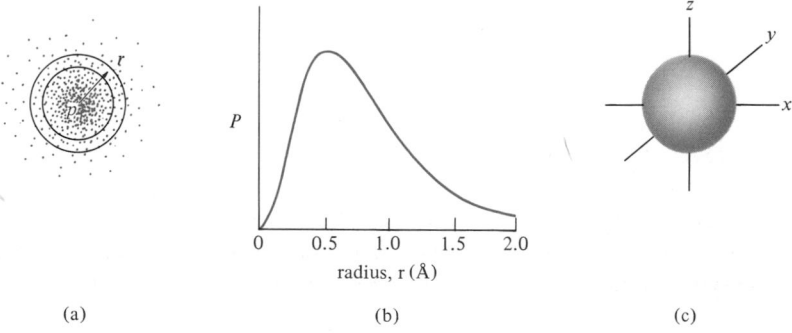

(a)　　　　　　　　　　　　(b)　　　　　　　　　　　　(c)

certainty what the shape of the H atom is. It is spherical. No matter in which direction you proceed from the nucleus (Figure 5.3c), whether it is along the x, y, or z axis, or in between, the probability of finding the electron depends only on the distance, r, from the nucleus. All positions on the sphere, therefore, are equally probable, and in this sense the H atom is spherical in shape. The sphere having radius 0.5 Å is the surface of maximum probability and can be said to represent the most probable size and shape of the atom.

The electrons in atoms are held in place by the attraction between their negative charges and their positive nuclei. All atoms can gain one additional electron and produce energy **(exothermic process),** but the tendency is very slight for most atoms. Hydrogen for example forms a **hydride ion** (H⁻) when it picks up an electron.

$$\text{H} \quad + \quad e^- \quad \longrightarrow \quad \text{H}^- \quad + \text{ energy}$$

(1 proton, 1 electron) + 1 electron $\longrightarrow$ (1 proton, 2 electrons) + energy

Additional electrons are possible, but such additions depend on the absorption of energy **(endothermic process).**

5.6 ATOMIC STRUCTURE AND THE PERIODIC TABLE

The He atom contains 2 protons and 2 electrons. If He gained an electron, it would be He⁻; if it lost an electron, it would be He⁺. In both cases it would still be a form of the element helium. In other words, the number of protons determines what an element is, not the number of electrons. The number of protons in the nucleus is called the **atomic number.**

DEFINITION: *The atomic number of an atom is the number of protons in its nucleus. Atomic number = Z.*

The atomic number of He is 2. The atomic number of H is 1. The elements in the periodic table are arranged in the order of their increasing atomic number, *without exception,* as shown in Table 5.1.

Thus the elements succeeding H and He are Li, Be, B, C, N, O, F, Ne, and so on, having atomic numbers 3, 4, 5, 6, 7, 8, 9, 10, and so on. In Chapter 13 you will be given the explanation for the large gaps that occur, such as that between Ca ($Z = 20$) and Br ($Z = 35$). Chemists have been able to discover new elements to fill all gaps from one end of the periodic table to the other, and there are no spaces left for yet undiscovered elements except at the lower end. This is an important feature of the arrangement.

A second feature of the arrangement in Table 5.1 is the fact that the number of electrons in each atom is equal to the number of protons. This is a requirement for an electrically neutral element, which is the normal form of the pure substance.

TABLE 5.1 Arrangement of the periodic table by atomic number and electron shells

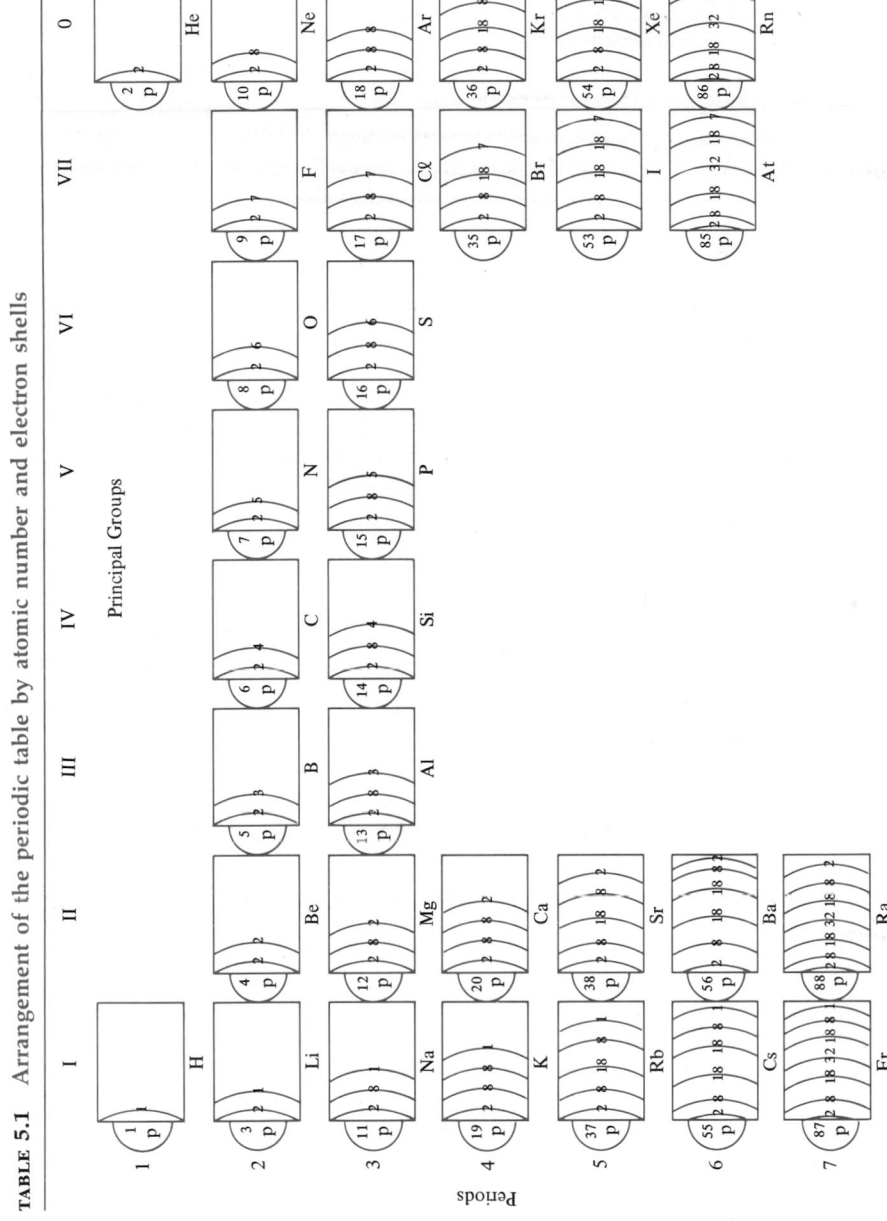

EXAMPLE 1.   Show that the number of protons in elementary bromine is equal to its number of electrons.

Solution: Bromine, Br $(Z = 35)$, is located in principal group VII and period 4 of Table 5.1. It has 35 protons. Starting with the innermost shell and moving outwards, the total number of electrons is given by the sum, $2 + 8 + 18 + 7 = 35$.

A third important feature of Table 5.1 is that the electrons are arranged in shells, and the number of electrons in the outermost shell is the same as the principal group number, except in the case of the inert gases, group 0. Among the inert gases, the first, He, has two electrons in its only shell, and the rest (Ne, Ar, Kr, Xe, and Rn) have eight electrons in their outermost shells.

EXAMPLE 2.   Show that the principal group number and the number of outermost shell electrons is the same for the halogens.

Solution: The halogens (F, Cl, Br, I, and At) are in principal group VII. Atomic fluorine has two electrons in its first shell and seven in its outermost, second, shell; atomic chlorine has two electrons in its first shell, eight electrons in its second shell, and seven electrons in its outermost, third, shell; and so on.

It is this fact, that atoms in the same principal group have the same number of outermost shell electrons, that gives them similar physical and especially similar chemical properties.

The atomic number is the most important characteristic of an element. For example, consider the process

$$F \quad + e^- \longrightarrow \quad F^- \quad + \text{ energy.}$$
$$\text{(9p2e7e)} \qquad\qquad \text{(9p2e8e)}$$

Since the product still his 9p, it is still a form of the element fluorine with atomic number 9.

The Neutron   As you saw in Table 3.2, the regularity of increasing atomic *weight* in the first series of twenty elements in the periodic table is not quite so perfect as that of increasing atomic *number*. The complete periodic table, found on the front end paper of the text, shows other examples of this irregularity in atomic weight. If atoms with more protons are heavier, why is it that there are exceptions in the order of increasing atomic weights, for example at potassium, K $(Z = 19)$, and at tellerium, Te $(Z = 52)$? The fact is that another kind of particle having almost the same mass as the proton but without electrical charge, exists in the nuclei of atoms. This particle is called a **neutron** (n). This particle permits an explanation of why He, which has only twice as

TABLE 5.2 The fundamental particles that compose atoms

Name	Symbol	Charge	Approximate Mass
proton	p	$+1$	1^*
electron	e	-1	$\frac{1}{1840}$
neutron	n	0	1

$^*1 \ unit = 1.67 \times 10^{-24} \ g$

many protons as H, has an atomic weight four times as great. The typical He nucleus contains $2p$ and $2n$.

All the atoms of all the elements can be constructed from these three fundamental particles, p, n, and e. Their most important properties are summarized in Table 5.2. The neutron is found to be slightly heavier than the proton, if the precision in its mass is carried out to the fourth significant figure.

The number of neutrons in the nucleus of an atom of a given element is not invariable like the number of protons. For example there are two kinds of hydrogen in nature, hydrogen-1, $(1p)$ and hydrogen-2, $(1p1n)$. A third kind, hydrogen-3, $(1p2n)$, can be made artificially but disintegrates within a short time. In each of the three kinds, the neutral atom contains one electron external to the nucleus. There are two kinds of helium, helum-3, $(2p1n)$, and helium-4, $(2p2n)$, each containing two electrons externally. Lithium exists in two forms, lithium-6, $(3p3n)$ and lithium-7, $(3p4n)$, with three external electrons. On the other hand, beryllium exists in only one form, beryllium-9, $(4p5n)$, with four external electrons.

ISOTOPES

The principal contribution to the mass of an atom is, of course, from its protons and neutrons. Their total number is called the **mass number** of the particular atom.

DEFINITION: *mass number = Z + the number of neutrons*

Each atom of a given element having a characteristic mass number is called an **isotope** of the element.

EXAMPLE 3.   Give the number of isotopes of each of the first four elements in the periodic table. (Consult the preceding paragraph.)

Solution:

$Z = 1$	hydrogen-1, hydrogen-2, hydrogen-3	three
$Z = 2$	helium-3, helium-4	two
$Z = 3$	lithium-6, lithium-7	two
$Z = 4$	beryllium-9	one

DEFINITION: *Isotopes of an element have the same atomic number but differ in mass number.**

*Some elements have only a single isotope, for example, beryllium-9 and fluorine-19

EXAMPLE 4. What are the mass numbers of the following isotopes? (a) $3p3n$, (b) $4p5n$, (c) H with two neutrons.

Solution: mass number = Z + the number of neutrons.
(a) mass number = 3 + 3 = 6
(b) mass number = 4 + 5 = 9
(c) mass number = 1 + 2 = 3 (H has atomic number 1.)

Calculation
of the
Atomic Weight

The existence of isotopes provides an explanation for atomic weights that are not whole numbers, such as chlorine (Cl = 35.5). Chlorine exists in two isotopic forms in nature, chlorine-35 and chlorine-37, with mass numbers 35 and 37, respectively. The isotopes are well mixed together throughout the world so that no matter where the chlorine sample comes from, 75% of the atoms are the chlorine-35 isotope and 25% are chlorine-37. The total weight per mole of atoms is determined as

Cl = weight of 1 mole of naturally occurring Cl atoms (the "atomic weight")
= (weight of 0.75 mole chlorine-35) + (weight of 0.25 mole chlorine-37)

$$= (0.75 \text{ mole chlorine-35}) \left(\frac{35}{1} \frac{\text{g chlorine-35}}{\text{mole chlorine-35}} \right) +$$

$$(0.25 \text{ mole chlorine-37}) \left(\frac{37}{1} \frac{\text{g chlorine-37}}{\text{mole chlorine-37}} \right)$$

= 26.2 g chlorine-35 + 9.3 g chlorine-37

Cl = 35.5 g mole^{-1}, since the original basis was 1 mole.

This is only an approximation (good to three significant figures) because a unit mass number is only approximately equal to 1 g mole^{-1}.

EXAMPLE 5. Boron consists of two isotopes in nature, 19% boron-10 and 81% boron-11. Assuming that the mass numbers are the actual isotopic weights in grams per mole, calculate the atomic weight of B.

Solution: B = weight of 1 mole of naturally occurring B atoms (the "atomic weight").

B = (weight of 0.19 mole boron-10) + (weight of 0.81 mole boron-11)

$$= (0.19 \text{ mole boron-10}) \left(\frac{10}{1} \frac{\text{g boron-10}}{\text{mole boron-10}} \right) +$$

$$(0.81 \text{ mole boron-11}) \left(\frac{11}{1} \frac{\text{g boron-11}}{\text{mole boron-11}} \right)$$

= 1.9 g boron-10 + 8.9 g boron-11

B = 10.8 g mole^{-1}, since the original basis was 1 mole.

Nearly all the elements have more than one isotope, but usually one isotope predominates in nature. For example H is 99.984% hydrogen-1, C is 98.892% carbon-12, and oxygen is 99.76% oxygen-16. Therefore the atomic weights for most elements are very nearly whole numbers.

EXAMPLE 6. Consult the periodic table located on the front end paper of the text and identify all the elements in the first three periods having integral, or whole number, atomic weights, to the first decimal place.

Solution: H = 1.0, He = 4.0, Be = 9.0, C = 12.0, N = 14.0, O = 16.0, F = 19.0, Na = 23.0, Al = 27.0, P = 31.0

Chemists have chosen a particular isotope for their definition of the mole and consequently their basis for atomic weight.

DEFINITION: *The mole is the number of atoms in 12.0000 . . . g carbon-12.*

This is the most precise definition of the mole. Since the atomic weight of every element is defined as the weight in grams of 1 mole of atoms of that element, every atomic weight is based on carbon-12 as the standard. Fortunately, the isotopic mixtures of most elements are nearly enough constant throughout the world to possess the same atomic weights at every locality.

A simple ion consists of a single atom. Examples 7 and 8 give the atomic number, mass number, number of protons, number of electrons, number of neutrons, and net charge in each of the isotopes or isotopic ions.

Examples of Isotopes in Atoms and Simple Ions

EXAMPLE 7. Li^+ (lithium-7).

Solution: Lithium is the third element in the periodic table. Therefore it has atomic number 3. As given above the isotope has mass number 7. Since lithium has atomic number 3, it has 3 protons. It is positively charged; thus it has one more proton than electrons. Since it has 3 protons it must have 2 electrons. The mass number is 7; therefore, the number of neutrons plus the number of protons is 7, and 7 minus 3 leaves 4 neutrons. The net charge is +1, as given.

EXAMPLE 8. Uranium with 89 electrons and 143 neutrons.

Solution: Uranium is the ninety-second element in the periodic table; therefore it has atomic number 92. Since the mass number is the number of neutrons plus the atomic number, the mass number is 143 plus 92 equals 235. Uranium has atomic number 92; thus, it has 92 protons. It has 89 electrons and 143 neutrons as given. With 89 electrons and 92 protons, it has net charge of 92 minus 89 equals +3. The ion is U^{3+}.

Exercise 1. Fill in the missing information in the table below.

Symbol	Atomic Number	Mass Number	Number of Protons	Number of Electrons	Number of Neutrons
Na^+	11	23	11	10	12
Cl^-		37			
B					6
Au	79	197			
O^{2-}				10	8
Al^{3+}		27			
		24	12	10	

Answer:

(Na^+)	11	23	11	10	12
(Cl^-)	17	37	17	18	20
(B)	5	11	5	5	6
(Au)	79	197	79	79	118
(O^{2-})	8	16	8	10	8
(Al^{3+})	13	27	13	10	14
(Mg^{2+})	12	24	12	10	12

Exercise 2. Find the neutral atom in the periodic table which has the same number of electrons as O^{2-}, F^-, Na^+, Mg^{2+}, and Al^{3+}. What is this number?

Answer: (Ne, 10)

1. If the force of attraction between two particles with charges $+1$ and -1, which are separated by a distance d, is called F (Figure 5.1a), calculate the force for each of the following charge combinations. Specify whether the force is one of attraction (A) or repulsion (R).
   (a) $2+$ and $2+$. (c) $4-$ and $1-$.
   (b) $1+$ and $3+$. (d) $2+$ and $3-$.

2. Same as problem 1.
   (a) $2-$ and $2-$. (c) $1+$ and $3-$.
   (b) $1+$ and $4+$. (d) $3+$ and $2-$.

3. Using the quantity F for the force between $1+$ and $1-$ at a separation d (Figure 5.1a), calcuate the force for each of the following distances of separation:
   (a) $d/4$;   (b) $4d$.

4. Same as problem 3.
   (a) $0.1d$;   (b) $10d$.

5. If the force of attraction between $1+$ and $1-$ charges, which are 1 Å apart, is called F, calculate the force in each of the following combinations in terms of F. Is the force attractive (A) or repulsive (R)?

   (a) $2+$ and $2-$ at 2 Å
   (b) $3+$ and $1-$ at 2 Å
   (c) $3+$ and $1-$ at 0.5 Å
   (d) $2+$ and $3+$ at 2 Å
   (e) $1+$ and $1-$ at 0.1 Å

6. A charge of $1+$ and a charge of $1-$ separated by a distance of 1 Å have a force of attraction, F. In each of the following combinations give the force in terms of F and indicate either attraction (A) or repulsion (R).
   (a) $1+$ and $2+$ at 2 Å
   (b) $2+$ and $2+$ at 2 Å
   (c) $1+$ and $2-$ at 2 Å
   (d) $1+$ and $2-$ at 3 Å
   (e) $2+$ and $2-$ at 0.5 Å

7. Calculate the average weight of a single He atom (a) from the atomic weight of He, and (b) from the approximate weight of its protons and neutrons (1.67×10^{-24} g each, Table 5.2), assuming the He atom contains two of each. Compare the results in parts (a) and (b).

8. Calculate the average weight of a single O atom (a) from the atomic weight of O, and (b) from the approximate weight of its protons and neutrons (1.67×10^{-24} g each, Table 5.2), assuming the O atom contains eight of each. Compare the results in parts (a) and (b).

9. If the complete H atom had the diameter of the Houston Astrodome, about 642 ft, what would be the diameter of its nucleus? Use the proportions illustrated in Figure 5.2 and express your answer in inches.

10. If the nucleus of a hydrogen atom had a diameter large enough to be seen under a good optical microscope, about 2000 Å, what would be the diameter of the complete atom? Use the proportions illustrated in Figure 5.2 and express your answer in inches.

11. Identify the following processes as being either exothermic (energy producing) or endothermic (energy absorbing) and explain. You may assume that all substances are in their gaseous states.
    (a) $Na \longrightarrow Na^+ + e^-$
    (b) $Cl + e^- \longrightarrow Cl^-$
    (c) $Na^+ + Cl^- \longrightarrow NaCl$
    (d) $Cl^- + e^- \longrightarrow Cl^{2-}$
    (e) $He \longrightarrow He^+ + e^-$

12. Same as problem 11.

    (a) $H \longrightarrow H^+ + e^-$
    (b) $H + e^- \longrightarrow H^-$
    (c) $O \longrightarrow O^+ + e^-$
    (d) $O + e^- \longrightarrow O^-$
    (e) $O^- + e^- \longrightarrow O^{2-}$

13. Using the electron shell structures of the atoms in Table 5.1, show by addition that the number of electrons in each of the atoms of the following elements is equal to its atomic number.
    (a) Ra $(Z = 88)$ (c) K $(Z = 19)$
    (b) Kr $(Z = 36)$ (d) Br $(Z = 35)$

14. Using the electron shell structures of the atoms in Table 5.1, show by example that the atoms in a given group have the same number of shells of electrons as the period number in which they are located.

15. Naturally occurring copper contains two principal isotopes. The per cent abundance of each on a mole basis is 70% copper-63 and 30% copper-65. Calculate the atomic weight of this element and compare it with the table value.

16. There are two isotopes of lithium in nature. Lithium-6 has an abundance of 7.4% and lithium-7 has an abundance of 92.6%, on a mole basis. Show that this leads to an atomic weight of 6.9 g mole^{-1} for the element. Compare this value with the table value.

17. Fill in the missing information in the following table.

18. Fill in the missing information in the following table.

Symbol	Atomic Number	Mass Number	No. of p	No. of e	No. of n
H^+		1			
	1			2	0
Ra^{2+}		226			
			10	10	10
S^{2-}					16

Symbol	Atomic Number	Mass Number	No. of p	No. of e	No. of n
Cs^+		133			
Xe					78
			2	0	1
I^-		127			
	3			2	3

*19. Which diagram best represents the changing force of repulsion, F, between two positively charged particles as they are separated by greater distances, d? Explain.

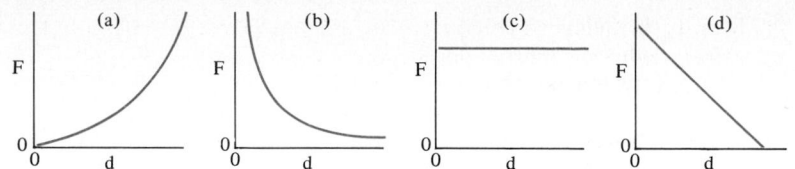

*20. Use the data in Table 5.2 to make the following calculations.
  (a) Express the weight of a single electron in grams.
  (b) How many electrons are in 1/1840 g of electrons?

*21. If an H atom is considered to have a diameter of 1 Å and its nucleus, the proton, a diameter of 10^{-5} Å, how many times denser than the atom is the nucleus?
  (*Hint:* Make use of the volume of a sphere formula, $V = 4\pi r^3/3$, where $r =$ the radius.)

6

CHEMICAL BONDS:
COVALENT AND IONIC

In previous chapters you have encountered a variety of molecules and compounds. Molecules of a substance consist of identical chemical combinations of at least two atoms, such as O_2 and HCl, and they sometimes consist of several thousand atoms, as in protein molecules. The atoms are held together by chemical bonds. If the elements comprising the molecule are dissimilar, such as in HCl and NH_3, the molecules are further designated as compounds. Some substances have the capability of forming three-dimensional crystalline particles with repeating bonds terminating at the surfaces of the crystal. Examples of these solids are diamonds, consisting of carbon, and crystals of table salt, consisting of sodium and chlorine. In these cases the combination is not called a molecule, even though the bonds holding the atoms together are correctly referred to as chemical bonds.

In this chapter you will first learn some of the characteristics of covalent bonds and how the electron distribution of the atom, as shown in Table 5.1, determines the kind, number, and location of the bonds it forms. A simple and yet profound rule explains many of these facts. It is known as the octet rule. The same rule will be used later in the chapter to explain another kind of chemical bond found in nature: the ionic bond.

6.1   THE VALENCE ELECTRONS

The valence electrons are the electrons of an atom that are involved in chemical reactions. The valence electrons are the ones that are located in the outermost shell of a given atom (Table 5.1). Among the principal group elements, electrons in inner shells are never involved in chemical reactions. The single electron that an H atom has must be a valence electron. The next element in the same group, Li, has a total of three electrons, but two of these are in the first shell and are not valence electrons. Only the electron in the outermost (second) shell is a valence electron. The two inner electrons are never involved in the chemical reactions of Li. Likewise the next element in the group, Na, has a single valence electron located in the third shell. The other ten electrons, which occupy inner shells, are never involved in the chemical reactions of Na.

EXAMPLE. Use the distribution of electrons in Table 5.1 to give the number of valence electrons in each of the following atoms: K, P, Br, Ra, C, and S.

Solution: The numbers of valence electrons are 1, 5, 7, 2, 4, and 6, respectively.

Electron Dot Representations A convenient method for symbolizing and locating the valence electrons in isolated atoms is to use dots, or points, to represent them. Since the inner electrons are not involved in chemical reactions, they are not shown. The electron dot representations of all the elements in Table 5.1 are shown in Table 6.1. Note that the number of valence electrons in an atom is the same as the number of the principal group to which it belongs with the exception of group 0.

TABLE 6.1 Electron dot representations of selected principal-group atoms in the periodic table

		Principal Groups							
		I	II	III	IV	V	VI	VII	0
Periods	1	H·							He:
	2	Li·	Be:	·B:	·C̈:	·N̈:	·Ö:	·F̈:	:N̈e:
	3	Na·	Mg:	·Al:	·Si:	·P̈:	·S̈:	·C̈l:	:Är:
	4	K·	Ca:					·B̈r:	:K̈r:
	5	Rb·	Sr:					·Ï:	:Ẍe:
	6	Cs·	Ba:					·Ät:	:R̈n:
	7	Fr·	Ra:						

6.2 SHARED ELECTRON BONDS

Atoms in molecules stay together because of **bond energy.** This is the amount of energy required to break the attraction between the atoms and force them apart. When isolated atoms come together in chemical reaction to form molecules, the reaction is always exothermic or heat producing. The greater the amount of heat given off, the more stable is the resulting molecule with respect to dissociation.

There is a type of bond for which the attraction of positive and negative charges does not offer an explanation. Imagine, for example, the kind of force that holds the atoms in H_2, O_2, and Cl_2 molecules together. Clearly there is no tendency for one side of the molecule to have a positive and the other a negative charge. The dilemma which faced scientists in the 1800's and prevented them from accepting Avogadro's hypothesis about gases (Section 4.1) was the belief that only *unlike* atoms could attract one another and form bonds. However, many molecules composed of identical atoms are known to exist. Even among molecules having unlike atoms the bond energies are never entirely a result of the attraction of positive and negative charges.

The explanation is given to us in terms of quantum mechanics and, although the mathematics is too advanced for an introductory course, you can easily understand the results. These results tell us that an electron bound to an atom has a certain amount of freedom to move about. When the electron can move about an additional atom, it gains freedom of movement. This additional freedom of movement has an energy equivalent that makes the process of sharing electrons exothermic.

The electrons are always shared in pairs. Each pair of shared electrons is called a **single covalent bond.** If two pairs are shared, they are called a **double bond,** and if three pairs are shared, they are called a **triple bond.** A single bond has a *multiplicity* of one, a double bond a multiplicity of two, and a triple bond a multiplicity of three.

DEFINITION: *A covalent bond is a pair of electrons shared between two atoms.*

The manner in which electrons of a given atom are shared depends on the electronic structure of the group 0 element nearest it in the periodic table.

6.3   THE OCTET RULE FOR COVALENT BONDS

We wish next to establish a rule to predict the arrangement of electrons in covalent bonds. It will help to begin with a look at the arrangement of electrons in the valence shells of the atoms in Table 5.1. The inert gases in group 0 provide a model of electronic stability. Atoms of these gases do not combine to form diatomic molecules, as most other elemental gases do. In addition, they do not form compounds with the other elements, with very few exceptions. We must conclude that the electronic structure of the inert gases must be particularly stable.

G. N. Lewis, an American chemist, saw the parallel between the electronic structures in chemically combined atoms and in inert gas atoms. His rule for covalent bonds follows.

> **The octet rule for covalent bonds:** *Covalent bonds are formed by sharing pairs of electrons between atoms in such a way that every atom in the compound attains the electronic structure of the group 0 element nearest it in the periodic table. In all cases except helium, this structure has an **octet** of electrons in its outermost shell. Helium has a **pair** of electrons.*

EXAMPLE 1. Apply the octet rule to the Cl_2 molecule, one of the gaseous halogens. Which inert gas atom do the Cl atoms in Cl_2 resemble?

Solution: Since Cl is in principal group VII, it has seven electrons in its outermost shell (Table 6.1).

$$\cdot \ddot{\underset{\cdot \cdot}{Cl}} :$$

When two Cl atoms combine to form Cl_2, they each contribute an electron to a shared pair, or single bond.

$$: \ddot{\underset{\cdot \cdot}{Cl}} : \ddot{\underset{\cdot \cdot}{Cl}} :$$

As shown, each atom has an octet of electrons, resembling Cl's nearest inert gas neighbor, Ar (argon).

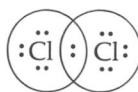

EXAMPLE 2. Show how the electrons in the N_2 molecule, the principal constituent of air, satisfy the octet rule. What is the bond multiplicity? Which inert gas do the N atoms resemble?

Solution: Since N is in principal group V, each atom contains five valence shell electrons.

$$\text{total number of outermost electrons} = 2 \times 5e = 10e$$

There must be at least a single bond holding the two N atoms together.

$$N : N$$

This leaves eight electrons. Use them to satisfy the octet rule for the atom on the left.

$$: \ddot{\underset{\cdot \cdot}{N}} : N$$

This leaves two electrons. Place them on the right-hand atom.

$$: \ddot{N} : N :$$

Since this structure represents all the valence shell electrons available, you must now *rearrange* them to satisfy the octet rule for *both* atoms. Move a pair of electrons from the left hand atom to the shared position.

$$: N :: N :$$

The left hand atom still has eight electrons including those shared in the double bond, but the right hand atom now has six electrons. Move two more electrons from the left-hand atom to the shared position.

$$:N::N:$$

The octet rule is now satisfied for both N atoms.

The bond holding the atoms together is a triple bond. Because of this high multiplicity the N_2 molecule is extremely stable, unlike the O_2 molecule, which is also a constituent of air. Each atom of nitrogen resembles its nearest inert gas neighbor, Ne (neon).

EXAMPLE 3. Which inert gases do the H atoms and N atom in NH_3 resemble? NH_3 is the formula for ammonia, a covalent gaseous molecule with a pungent odor.

Solution: To apply the octet rule, count the number of available valence shell electrons.

3 H atoms, group I	$3 \times 1e = 3e$
1 N atom, group V	$1 \times 5e = 5e$
total valence shell electrons	$8e$

Each H atom must be bound by at least one electron pair. Since this gives it the electronic structure of a He atom, its nearest inert gas neighbor, the H atom never has more than a single bond to anything. Therefore the N atom must be the central atom of the molecule.

$$H:N:H$$
$$\overset{..}{H}$$

This leaves $8 - 6 = 2$ electrons. Use them in an unshared position on the N atom to satisfy its octet and to give it the electron configuration of Ne, its nearest inert gas neighbor.

$$H:\overset{..}{N}:H$$
$$\overset{..}{H}$$

Thus every atom satisfies the octet rule (or the special octet rule for He atoms).

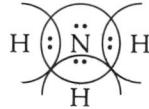

EXAMPLE 4. Show how the octet rule is satisfied for the covalent molecule sulfuric acid, H_2SO_4, whose atoms are arranged as shown.

This dense oily liquid is the most widely used industrial acid in the world.

Solution: Use the number of valence shell electrons determined by the principal group number of each element to calculate the total number of available electrons.

2 H atoms, group I	$2 \times 1e = 2e$
4 O atoms, group VI	$4 \times 6e = 24e$
1 S atom, group VI	$1 \times 6e = 6e$
total valence shell electrons	$32e$

Place two electrons between each H atom and O atom, since this is the only possibility for the H atom.

This leaves 28 electrons. Complete the octet of electrons around every atom.

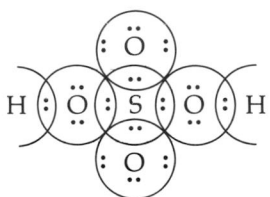

There are exactly 32 electrons accounted for, and the octet rule is satisfied for every atom.

EXAMPLE 5. Show how the octet rule is satisfied for ethylene, C_2H_4. This covalent gaseous molecule is the starting material for one of the principal methods of synthesizing ethyl alcohol.

Solution: First calculate the number of available electrons in valence shells.

2 C atoms, group IV	$2 \times 4e = 8e$
4 H atoms, group I	$4 \times 1e = 4e$
total valence shell electrons	$12e$

The carbon atoms must be bound to each other, since a hydrogen atom can have only a single bond. The use of two electrons in this bond will leave $12 - 2 = 10$ electrons to bind the H atoms and satisfy the octet rule.

$$C:C$$

Each H atom must be bound by a single bond. Trial and error will show that two H atoms must be bound to each C atom.

$$\begin{array}{cc} H & H \\ \cdot\cdot & \cdot\cdot \\ H:C:C:H \end{array}$$

The number now remaining is $12 - 10 = 2$ electrons. If they are placed between the C atoms to form a double bond, the octet rule will be satisfied. An electron pair can be represented by a single line.

$$\begin{array}{cc} H & H \\ \cdot\cdot & \cdot\cdot \\ H:C::C:H \end{array} \quad \text{or} \quad \begin{array}{cc} H & H \\ | & | \\ H-C=C-H \end{array}$$

6.4 IONIC COMPOUNDS AND THE OCTET RULE

There is another way in which the valence shell electrons of atoms may be rearranged to satisfy the octet rule; electrons can be gained or lost rather than shared. If two atoms interact with one another in such a way that one atom loses electrons (becoming positive) and the other atom gains the electrons (becoming negative), the two ions thus formed will attract one another. Such an attraction, or bond, is called **ionic.** In order to predict whether the bonds between particular atoms will be covalent or ionic, you will find the periodic table very helpful. In Section 6.2, we pointed out that it is difficult to imagine anything other than covalent bonds when atoms are identical, such as in H_2, O_2 and Cl_2. On the other hand, you might expect ionic bonds when the atoms are enough "different."

Metals, Nonmetals, and Metalloids

In the discussion of the periodic table in Section 1.4 you saw that metals and nonmetals tend to be grouped separately. The reason for this is that metals tend to lose electrons and nonmetals tend to gain electrons when they interact with one another. Table 6.2 shows a diagonal of elements that cuts completely across the periodic table from the upper left to the lower right hand sides. These elements, called **metalloids,** separate the entire periodic table into metals on the left and nonmetals on the right. The names of these metalloids, which are found in the complete periodic table inside the front cover of the text, are as follows: boron (B), silicon (Si), germanium (Ge), arsenic (As), antimony (Sb), tellurium (Te), and polonium (Po).

TABLE 6.2 Separation of the periodic table into metals and nonmetals by the diagonal of metalloids

	Principal Groups					
	II	III	IV	V	VI	VII
2	4 Be	5 B	6 C			
3		13 Al	14 Si	15 P	nonmetals	
4		31 Ga	32 Ge	33 As	34 Se	
5	metals		50 Sn	51 Sb	52 Te	53 I
6				83 Bi	84 Po	85 At

Periods

Exercise 1. Locate the following elements in the periodic table and identify them as either metals or nonmetals: (a) Li ($Z = 3$), (b) F ($Z = 9$), (c) Fe ($Z = 26$), (d) I ($Z = 53$), (e) Ag ($Z = 47$).

Answer: (a) metal, (b) nonmetal, (c) metal, (d) nonmetal, (e) metal

Ionic Bonds When nonmetals combine with one another, they satisfy the octet rule by forming covalent bonds. When metals and nonmetals combine with one another, they satisfy the octet rule by forming ionic bonds.

> *The octet rule for ionic bonds:* *Ionic bonds are formed by metal atoms losing electrons and nonmetal atoms gaining those same electrons in such a way that every atom in the compound attains the structure of the group 0 element nearest it in the periodic table. In all cases this number of electrons is **eight**, except in the case of He, for which it is **two.***

EXAMPLE 1. Show how the ionic compound NaCl (sodium chloride) satisfies the octet rule.

Solution: The Na Atom ($Z = 11$) has the electronic shell structure, Na ($2e, 8e, 1e$), as shown in Table 5.1. The Cl atom ($Z = 17$) has the structure, Cl ($2e, 8e, 7e$). If the Na atom loses one electron, it attains the structure of Ne ($Z = 10$), its nearest group 0 neighbor; if the Cl atom gains that electron, it attains the structure of Ar ($Z = 18$). Both the sodium ion, Na$^+$, and the chloride ion, Cl$^-$, satisfy the octet rule.

By convention formulas of ionic compounds are written with the positive ion first. It is the attraction between positive and negative ions that holds the compound together, and there is no covalent bond involved. This is illustrated for NaCl in Figure

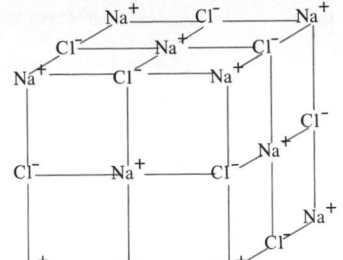

Figure 6.1. Ionic structure of NaCl(s). It is cubic with six
nearest neighbors.

6.1, where the solid crystal has the structure of alternating Na$^+$ and Cl$^-$ ions at the corners of small cubes.

In the ionic solid there is a tendency for as many negative ions as possible to cluster around a positive ion (and vice versa) in order to maximize the attractive energy. On the other hand, there is a limit to how closely the ions can approach because of the size and shape (spherical) of the ions themselves. In NaCl(s) it is the size of the Cl$^-$ ion that determines the number of ions in the cluster because it is the larger of the two ions. Each Na$^+$ ion is surrounded by six nearest-neighbor Cl$^-$ ions, as shown in Figure 6.1. If there were any more than six, not all of the Cl$^-$ ions could be in contact with the Na$^+$ ions. As you have seen earlier, the maximum attractive force between two charged particles occurs when they are in contact with one another, in order to provide the minimum distance of charge separation (Figure 5.1d).

EXAMPLE 2. Compare the ionic compound cesium chloride, CsCl, with NaCl.

Solution: By losing an electron, Cs ($Z = 55$) attains the electronic configuration of the group 0 element Xe ($Z = 54$). The Cl atom gains an electron, as in Example 1. CsCl is an ionic solid, like NaCl, but its crystal structure is such that eight Cl$^-$ ions surround each positive ion. This is possible without bringing the Cl$^-$ ions in contact with one another because the diameter of the Cs$^+$ ion is larger than that of the Na$^+$ ion. In Figure 6.2, each Cs$^+$ ion is at the center of a small cube which has a Cl$^-$ ion at each of its eight corners. The macroscopic crystal is generated by replicating this unit in all three perpendicular directions, and this provides equal numbers of Cs$^+$ and Cl$^-$ ions as required by the formula, CsCl.

The existence of ions in NaCl and CsCl can be demonstrated by their ability to conduct electricity, either in the melt at high temperatures or in aqueous solution.

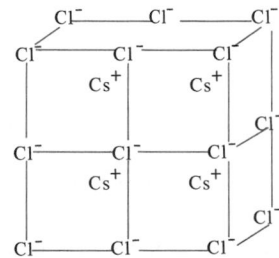

Figure 6.2. Ionic structure of CsCl(s). It is cubic with eight
nearest neighbors.

EXAMPLE 3.   Predict the formula of the compound between Ca and O.

Solution: Since Ca is a metal and O a nonmetal the compound is ionic. Ca ($Z = 20$) can attain the configuration of Ar ($Z = 18$) by losing two electrons and becoming a Ca^{2+} ion; O ($Z = 8$) gains the two electrons to form an oxide ion, O^{2-}, which has the electronic structure of Ne ($Z = 10$). The resulting ionic compound is CaO (calcium oxide).

EXAMPLE 4.   Predict the formula of the ionic compound between H and Na.

Solution: The electronic structure of Na is shown in Table 5.1, as $2e, 8e, 1e$. The outermost electron is lost to produce Na $(2e, 8e)^+$, like Ne ($Z = 10$). Hydrogen is a nonmetal (Section 1.4 and 6.8). In ionic compounds each H ($1e$) gains a single electron to form the hydride ion H $(2e)^-$, which has the same electronic structure as He ($Z = 2$). The formula for the ionic compound sodium hydride is NaH.

There is very little tendency of H to form H^+ in ionic compounds because of the extremely small size of the bare proton (Figure 5.2). The H^+ ion, if formed, would attach itself to almost any unshared pair of electrons in order to attain the He ($Z = 2$) electronic structure, covalently.

Often the charge on a stable ion consisting of a single atom can be predicted by the position of the atom in the periodic table. The principal groups of elements and their simple ionic forms are shown in Table 6.3. The group 0 elements ordinarily do not form ionic compounds. The H^+ ion could only exist in the gas phase.

These ions never exist alone in nature but always have enough ions of the opposite sign to provide **electroneutrality.** For example, if there is 1 mole Na^+ ions, there must be 1 mole Cl^- ions or 0.5 mole O^{2-} ions, or some other combination that provides equal numbers of positive and negative charges.

The formulas for many chemical compounds can be predicted using the ions in Table 6.3. Every positive ion is capable of forming a compound with every negative ion in the table. If we use the smallest whole number of atoms that produces electroneutrality and combine the 13 positive ions with the 8 negative ions, the existence of $8 \times 13 = 104$ different stable binary compounds can be predicted. A **binary compound** is one consisting of only two elements.

TABLE 6.3   Ions composed of single atoms

		I	II	III	IV	V	VI	VII	0
					Principal Groups				
Periods	1	H^+ or H^-							He
	2	Li^+	Be^{2+}	B	C	N	O^{2-}	F^-	Ne
	3	Na^+	Mg^{2+}	Al^{3+}	Si	P	S^{2-}	Cl^-	Ar
	4	K^+	Ca^{2+}					Br^-	Kr
	5	Rb^+	Sr^{2+}					I^-	Xe
	6	Cs^+	Ba^{2+}					At^-	Rn
	7	Fr^+	Ra^{2+}						

EXAMPLE 5. What is the formula of the binary ionic compound produced from Ca^{2+} and Cl^-?

Solution: Since the charge on each Ca^{2+} ion is $+2$, and the charge on each Cl^- ion is -1, it will take one Ca^{2+} to neutralize two Cl^-.

$$1(+2) + 2(-1) = 0$$

Therefore the formula must be symbolized as

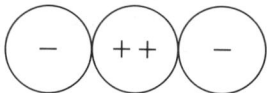

$CaCl_2$, calcium chloride.

EXAMPLE 6. What is the formula of the ionic compound produced from Al^{3+} and S^{2-}?

Solution: Since the charge on each Al^{3+} ion is $+3$ and the charge on each S^{2-} ion is -2, it will take two Al^{3+} to neutralize three S^{2-}.

$$2(+3) + 3(-2) = 0$$

Therefore the formula must be symbolized as

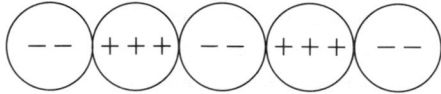

Al_2S_3, aluminum sulfide.

The examples shown in Table 6.4 represent some of the 104 possibilities predicted.

The names of the ionic compounds in Table 6.4 all follow the rules of nomenclature for binary compounds given in Section 1.5. These compounds and all 104 possible combinations from Table 6.3 are unique because of their conformity to the octet rule for ions. For this reason it is not necessary to use the Greek prefixes mono-, di-, tri-, . . . to designate number of atoms.

Nomenclature of Binary Compounds of Metals with Nonmetals

TABLE 6.4 Other examples of ionic compounds

$$Li^+ \ + Cl^- \longrightarrow LiCl \quad \text{(lithium chloride)}$$
$$2Li^+ \ + O^{2-} \longrightarrow Li_2O \quad \text{(lithium oxide)}$$
$$Mg^{2+} + 2F^- \longrightarrow MgF_2 \quad \text{(magnesium fluoride)}$$
$$Mg^{2+} + S^{2-} \longrightarrow MgS \quad \text{(magnesium sulfide)}$$
$$Al^{3+} \ + 3Br^- \longrightarrow AlBr_3 \quad \text{(aluminum bromide)}$$
$$2Al^{3+} + 3O^{2-} \longrightarrow Al_2O_3 \quad \text{(aluminum oxide)}$$
$$K^+ \ + H^- \longrightarrow KH \quad \text{(potassium hydride)}$$
$$Ca^{2+} + 2H^- \longrightarrow CaH_2 \quad \text{(calcium hydride)}$$

A large number of binary compounds of metals with nonmetals, especially those in which the metals are not members of the principal groups, have more than one formula. The Greek prefixes are not used in naming these ionic compounds. The preferred method of nomenclature is to represent the positive charge on the metal by Roman numeral and to use the "-ide" ending on the nonmetal.

EXAMPLE 7. Name the following compounds, using Roman numerals to designate positive charge. In each case assume that the nonmetal has the negative charge given in Table 6.3, and calculate the positive charge on *each* metal ion necessary to provide electroneutrality.

Formula	Name
$FeBr_3$	iron(III) bromide
$FeBr_2$	iron(II) bromide
CuO	copper(II) oxide
Cu_2O	copper(I) oxide
HgI_2	mercury(II) iodide
Hg_2I_2	mercury(I) iodide

Combination Ionic and Covalent Compounds

It is possible to have ionic compounds in which some of the atoms are held together by covalent bonds. A good example is NaOH (sodium hydroxide). This compound, also called lye, dissolves easily in water, breaking down into Na^+ and OH^- (hydroxide) ions. The Na^+ ion satisfies the octet rule as shown in Example 4. For the O atom and the H atom both to satisfy the octet rule, it is necessary to have an extra electron (donated by sodium) and to share electrons covalently as follows:

The arrangement of atoms is

O H

The number of valence electrons is determined from principal group numbers and the net charge on the ion.

1 O atom, group VI	$1 \times 6e = 6e$
1 H atom, group I.	$1 \times 1e = 1e$
net charge on the ion, -1	$\underline{1e}$
total valence shell electrons	$8e$

Use one pair of electrons to form a bond between the atoms, and use the remaining six electrons to satisfy the octet rule for the O atom. The correct configuration is

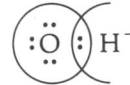

It is necessary to include the single negative charge to indicate that the structure contains one more electron than required for neutrality.

Try the following exercises on your own. Show that the number of valence shell electrons used is correct and that all valence shells satisfy the octet rule.

Exercise 1. Nitric acid

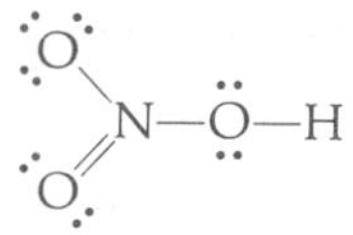

Exercise 2. Peroxide ion

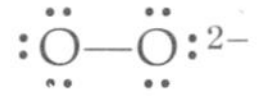

$$:\overset{..}{O}-\overset{..}{O}:^{2-}$$

Exercise 3. Hydronium ion

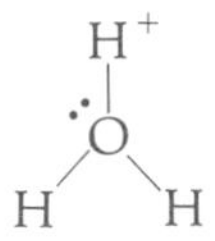

Exercise 4. Phosphorus acid

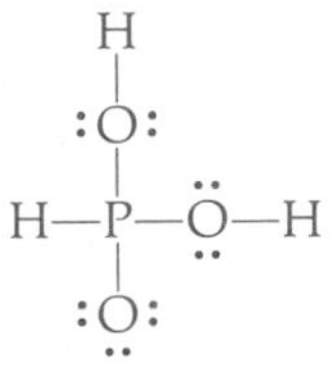

6.5   EXCEPTIONS TO THE OCTET RULE

There are a number of exceptions to the octet rule, among which are molecules having an odd number of valence electrons, such as in NO_2 (nitrogen dioxide). Other exceptions occur among molecules in which more than four single bonds are attached to the central atom, such as in PCl_5 (phosphorus pentachloride).

EXAMPLE 1.   In concentrated HNO_3 (nitric acid), there is always a red gas present filling the bottle. This gas is NO_2 (nitrogen dioxide). If a sample of this gas is cooled somewhat below room temperature, it becomes a yellow liquid, N_2O_4 (nitrogen tetroxide).

Show that NO_2 molecules cannot satisfy the octet rule separately but that the chemical combination of two molecules, to form N_2O_4, can satisfy the rule.

Solution: The atoms in the NO_2 molecule are oriented as shown.

$$O$$
$$O \quad N$$

Calculate the number of available valence electrons.

1 N atom, group V	$1 \times 5e = 5e$
2 O atoms, group VI	$2 \times 6e = 12e$
total valence shell electrons	$\overline{17e}$

An octet of electrons around every atom would necessarily lead to an even number of electrons, but the number of available electrons calculated is odd. Thus, it would be impossible to satisfy the octet rule for every atom in this molecule. You may arbitrarily choose to satisfy the oxygen octets as shown below.

$$:\overset{\cdot\cdot}{\underset{}{O}}:$$
$$|$$
$$:O\!=\!\overset{\cdot\cdot}{\underset{}{N}}\cdot$$

Since the octet around the N atom is not complete, you would expect that chemical combination of two NO_2 molecules would be possible, providing a total of 34 electrons to be shared.

$$:\overset{\cdot\cdot}{O}::\overset{\cdot\cdot}{O}:$$
$$| \quad |$$
$$:\underset{\cdot\cdot}{O}\!=\!N\!-\!N\!=\!\underset{\cdot\cdot}{O}:$$

This molecule, N_2O_4, has all its octets filled. It should be the stable form of the compound at energy levels low enough to prevent dissociation.

EXAMPLE 2. A more subtle and important exception to the octet rule is that of molecular O_2. Show how the octet rule may be satisfied in this molecule.

Solution: The arrangement of atoms is

$$O \quad O$$

The number of valence electrons is calculated as follows.

$$2 \times 6e = 12e$$

Placing eight electrons around the O atom on the left leaves four electrons for the one on the right.

$$:\overset{\cdot\cdot}{\underset{\cdot\cdot}{O}}:\overset{\cdot\cdot}{O}:$$

The obvious solution is to move two unshared electrons from the left atom to the shared position.

$$:\ddot{O}::\ddot{O}:$$

The octet is now complete for both, but we know from the behavior of molecular O_2 that the structure cannot be correct. For example the fact that O_2 molecules are attracted to a magnet is evidence that they have unpaired electrons. It is believed that the true electronic structure of O_2 is

$$\cdot\ddot{O}:\ddot{O}\cdot$$

Molecular O_2, unlike molecular N_2, is very reactive. This is partly a result of the O_2 molecule having only a single bond holding the atoms together. It is also due to the presence of unpaired electrons in O_2 which tend to become paired by chemical combination with other molecules.

6.6 GEOMETRY OF COVALENT BONDS

There is a general tendency for the unshared pairs of electrons and the electron pairs in separate bonds around a given atom in a molecule or ion to move as far apart from one another as possible. This is a result of the repulsion that like charges have for one another (Section 5.1).

The effect of this repulsion is that some molecules and ions are linear, some are bent, some are planar, and some are pyramidal or tetrahedral. Each case will be taken up individually.

Linear Molecules

There is no choice in the geometry of diatomic, or "two-atom," molecules; they must be linear. This means simply that the line connecting the centers of the two atoms is a straight line. Several examples are indicated.

$$:N\equiv N: \qquad H-H \qquad H-\ddot{\underset{\cdot\cdot}{Cl}}:$$

Some *ions* consist of two atoms and they also must be linear.

$$:\ddot{\underset{\cdot\cdot}{O}}-\ddot{\underset{\cdot\cdot}{O}}:^{2-} \qquad :\ddot{\underset{\cdot\cdot}{O}}-H^{-}$$

A three-atom molecule however can be either linear or bent. It is linear if all the electrons on the central atom are used in forming bonds.

EXAMPLE 1. Hydrogen cyanide, HCN, a very poisonous gas, has a linear shape; that is, the line connecting the centers of its three atoms is a straight line.

Show how this would be predicted from the repulsion of the bonding electrons.

Solution: The atoms contribute valence shell electrons as follows: H (1e), C (4e), and N (5e). The total number of valence electrons is 10. They are distributed as follows to satisfy the octet rule.

$$H-C\equiv N:$$

Treat the six electrons in the triple bond *as a unit*. This bond will be as far away as possible from the electron pair bond between the H atom and C atom. This leads to a linear molecule.

EXAMPLE 2. Show that CO_2 (carbon dioxide), one of the gaseous products of burning any carbonaceous material, is linear.

Solution: Use the octet rule to obtain the following electronic structure of CO_2. The C atom is known to be the central atom.

$$:\ddot{O}=C=\ddot{O}:$$

Treat each double bond as a unit; one double bond will move as far away from the other as possible. This leads to a linear structure, as in Example 1.

Exercise 1. Show that OCN^- (the cyanate ion) is linear.

Bent Molecules Another three-atom molecule provides an example of a **bent** structure. That molecule is O_3 (ozone), which is found naturally in the stratosphere where it is produced from ordinary O_2 by the absorption of ultraviolet light. The manner in which O_3 satisfies the octet rule is

$$\ddot{O}$$
$$:\ddot{O} \quad \ddot{O}:$$

The unshared pair of electrons on the central O atom is repelled by the single bond pair as well as the four electrons in the double bond, which is treated as a unit. Ideally, the molecule would be bent at an angle of $360°/3 = 120°$ in order for the electron units to be as far away from each other as possible. Actually the 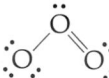 angle has been measured experimentally and found to be approximately 125°.

EXAMPLE 3. One of the most far reaching and meaningful facts of nature is that the H_2O molecule is bent and not linear. The implications of this geometric shape will be discussed in Section 6.7, Polar Molecules. Show how this fact would be predicted from the repulsion of the electron pairs on O.

Solution: There are $6e + 1e + 1e = 8e$ in the valence shells of H and O available to satisfy the octet rule in the H_2O molecule. The resulting electronic structure is

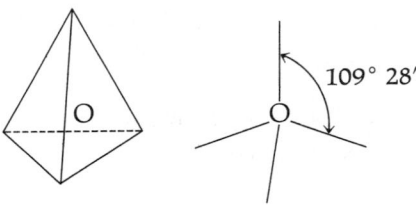

Since the O atom is surrounded by four pairs of electrons (two pairs bonded to H atoms and two unshared pairs) they should repel one another to the greatest angle of separation possible. This angle is $109° \ 28'$, which you obtain by placing the O atom at the center of a regular tetrahedron and drawing lines from the O atom to each of the four corners.

In this three-dimensional figure, each of the four lines makes an angle of $109° \ 28'$ with every other line. This provides the maximum angle of separation for all of the lines.

However, the three atoms in H_2O must all lie in the same plane; therefore, the structure is simply that of a bent molecule and not a three-dimensional structure. The lines connecting the centers of the two H atoms with the center of the O atom form an angle. Ideally the angle would be $109 \ 28'$, but from experiment it is found to be about $105°$.

Exercise 2. Show that the air pollutant SO_2 (sulfur dioxide) is a bent molecule.

All three-atom bent molecules are planar because any three points, not in a straight line, describe a plane. When molecules have more than three atoms, they can be either planar or three-dimensional. Two examples of planar structures will be given. The first is a molecule containing four atoms and the second is an ion with four atoms.   **Planar Molecules**

EXAMPLE 4. Show that SO_3 (sulfur trioxide), a gas which produces sulfuric acid when added to water, is a planar molecule.

Solution: Both the O atoms and the S atom are in principal group VI. Therefore the number of valence electrons is $4 \times 6e = 24e$, and these may be distributed in the following manner to satisfy the octet rule.

$$\ddot{\text{:O:}}$$
$$\|$$
$$\text{S}$$
$$\ddot{\text{:O}}\cdot \quad \cdot\ddot{\text{O:}}$$

The four electrons in the double bond are treated as a unit. They repel the other two bonds, causing the three units to move as far apart as possible. The resulting angle is $\frac{1}{3} \times 360°$ or $120°$ and the molecule is planar. This angle is the same as that obtained experimentally.

EXAMPLE 5. Show that NO_3^- (nitrate ion) is planar.

Solution: The number of valence electrons is calculated as follows.

1 N atom, group V	$1 \times 5e = 5e$
3 O atoms, group VI	$3 \times 6e = 18e$
net charge on the ion, -1	$1e$
total valence shell electrons	$24e$

This is the same number of valence shell electrons calculated in Example 4, and the same results are obtained for the structure; the nitrate ion is planar.

$$O^-$$
$$\|$$
$$N$$
$$O \quad\quad O$$

Pyramidal and Tetrahedral Molecules

Many molecules and ions in nature are three-dimensional. The NH_3 molecule in Example 3 of Section 6.3 is pyramidal in shape.

$$\overset{\cdot\cdot}{N}$$
$$H \quad | \quad H$$
$$H$$

The lone pair electrons repel the three pairs of bonding electrons, and the angles tend toward that of the tetrahedral angle, $109° \, 28'$. The experimental value of the angle between the N—H bonds is $107°$. The structure is called pyramidal because the N atom is located at the top of a pyramid.

There are a number of examples in nature of perfect tetrahedral structures. One of these is exhibited by the methane gas molecule, the principal constituent of natural gas.

EXAMPLE 6. Show that CH_4 (methane) should have a tetrahedral structure.

Solution: With four electrons from the four H atoms and four electrons from the single C atom, there are eight electrons available to satisfy the octet rule. This can be accomplished by the arrangement

$$
\begin{array}{c}
H \\
| \\
H-C-H \\
| \\
H
\end{array}
$$

This planar structure is the simplest way to show the octet rule but, in order for the four electron pairs to be as far apart as possible, the angles between C—H bonds must be 109° 28′, and the molecule is tetrahedral in shape. This matches the angle that is determined experimentally.

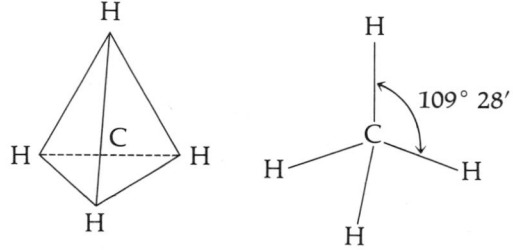

EXAMPLE 7. Show that SO_4^{2-} (sulfate ion) should be tetrahedral in shape.

Solution: The total number of valence electrons is $5 \times 6e + 2e = 32e$, including the two extra electrons that give the ion a net charge of $2-$. The octet rule is satisfied by the structure

$$
\begin{array}{c}
:\ddot{O}:^{2-} \\
| \\
:\ddot{O}-S-\ddot{O}: \\
| \\
:\ddot{O}:
\end{array}
$$

The four pairs of bonded electrons repel one another and the optimum tetrahedral angle is produced.

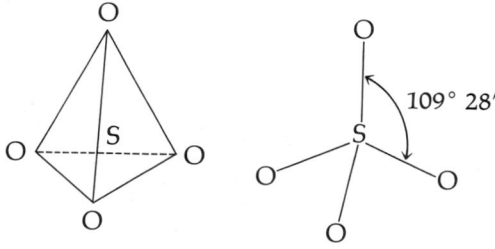

6.7 POLAR MOLECULES

For a molecule to be polar means that, as a result of covalent bond formation, one side of the molecule has accumulated more negative charge than the other side. This leads to the existence of what is called an **electric dipole.** The electric dipole is illustrated in Figure 6.3 as a dumbbell-type structure, but this does not imply that the molecule it represents must be diatomic. We mean that there is some point in every molecule that can be considered the center of the positive charge and another point that can be considered the center of the negative charge and that they are separated by distance, ℓ. If the separated positive charge is $q+$, not necessarily the same as that of a proton or any whole multiple of it, then the separated negative charge must be $q-$, in order to maintain electroneutrality. A quantity which is more important than q and ℓ separately, because it can be measured experimentally whereas q and ℓ cannot, is their product $q \times \ell$ which is called the **dipole moment.** The length of the arrow in Figure 6.3 is proportional to the size of the dipole moment, and it points from the positive side of the dipole toward the negative side.

In order for a molecule to be polar, there are two requirements. First, it must have polar covalent bonds; second, the geometry of these covalent bonds must be such that the center of the positive charge is not located at the same point as the center of the negative charge. In the following subsections you will learn how to predict the existence of polar bonds.

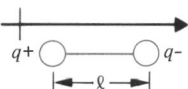

Figure 6.3.   An electric dipole: dipole moment $= q \times \ell$

Polar Bonds   Different elements have different tendencies to attract the electron pairs in the covalent bond between two atoms. The diagonal of metalloids in Table 6.2, which divides the periodic table into metals in the lower left area and nonmetals in the upper right area serves as a reference line for comparison. The farther away a nonmetal is from the diagonal of metalloids, the greater is its tendency to attract electrons.

Hydrogen and the Periodic Table   There is one element which is noticeably out of place with respect to the other elements in the periodic table and with respect to its tendency to gain electrons. That element is hydrogen, which should be located slightly to the right of boron and to the left of carbon.

EXAMPLE 1. Determine which atom in each of the following pairs of elements has the stronger attraction for the electrons in a covalent bond from its position in Table 6.2. From these data predict whether the bond is polar or nonpolar.

(a) C and O   (d) Cl and Cl
(b) O and F   (e) Cl and Br
(c) H and O   (f) C and P

Solution: The first five examples involve little uncertainty with regard to polarity. The last example requires another more quantitative approach.

(a) C and O are in the same period but O is farther to the right. Thus O has the stronger attraction for the electrons. The C—O bond is polar.

(b) Using the same reasoning as in (a), F has a stronger attraction for the electrons than O. The O—F bond is polar.

(c) Since H is to the left of C and since O is well to the right of C, H should be closer to the diagonal. Therefore O has a stronger attraction for the electrons and the H—O bond is polar.

(d) Two Cl atoms obviously have the same attraction for the pair of electrons holding them together and the Cl—Cl bond is nonpolar.

(e) Cl and Br are in the same group but Cl is higher. Cl has a stronger attraction for electrons and the Cl—Br bond is polar.

(f) C and P are both the same distance from the diagonal, and a prediction cannot be made. It is found that C is more electron attracting than P or even S.*

From a knowledge of bond polarity, as predicted in this section, and from a knowledge of covalent bond geometry found in Section 6.6, you will now learn how to predict whether the molecule is polar.

The existence of a polar bond indicates that the electron pairs in that bond are displaced toward one of the two atoms sharing the electrons. In Example 1 a number of polar bonds were predicted. In diatomic molecules the linear geometry is such that, if the bond is polar, the molecule must also be polar.

EXAMPLE 2. Predict whether the following covalent molecules are polar or nonpolar.
(a) Cl_2, (b) BrCl, (c) HCl.

Solution: Refer to Example 1 for bond polarities.

(a) Since the Cl—Cl bond is nonpolar, the Cl_2 molecule is nonpolar. The center of the negative charge and the center of the positive charge are located at the same point in the center of the molecule.

(b) Since the Br—Cl bond is polar the BrCl molecule is polar. Because the shared electron pair is displaced toward the Cl atom, the

*See Chapter 9 in Linus Pauling: *College Chemistry.* W. H. Freeman and Company, San Francisco, 1964.

center of the negative charge is located to the right of the center of the positive charge. This separation of charge tends to hold the BrCl molecule together because of the force of attraction between unlike charges, although it is held together mainly by the covalent bond. The dipole moment points from left to right.

$$Br—Cl$$

The usual way of writing formulas of compounds containing elements in which one is more positive than the others, is to write the most positive element first. Thus the formula BrCl would be preferred to that of ClBr.

(c) If we use the same logic as in Example 1(c), the H—Cl bond must be polar because the shared electrons are displaced toward the Cl atom. Thus the center of the negative charge lies to the right of the center of the positive charge, and the HCl molecule is polar. The dipole moment is directed from left to right.

$$H—Cl$$

Note that, as in the formula for BrCl, the HCl formula is written with the more positive atom first.

When three-atom and four-atom molecules are considered, the geometry may be just as important as the bond polarity. The geometries of the molecules have already been determined in Example 3 (Section 6.6).

EXAMPLE 3. Predict whether the following covalent molecules are polar or nonpolar. In case the molecule is polar, indicate which side is positive. (a) CO_2, (b) H_2O, (c) SO_3, (d) NH_3, (e) CH_4

Solution: Refer to the specific example in Section 6.6.

(a) In Example 2 (of Section 6.6) the CO_2 molecule was found to be linear with the C atom in the middle. Although the C=O bond is polar, the symmetry of the molecule is such that the center of the positive charge and the center of the negative charge coincide. Therefore the molecule, as a whole, is nonpolar.

(b) In Example 3 (Section 6.6) the H_2O molecule was found to be bent. Since both of the O—H bonds are polar and the O atom is more negative, the H side of the molecule is more positive than the O side and the molecule is polar. This is illustrated in Figure 6.4.

(c) In Example 4 (Section 6.6) the SO_3 molecule was shown to be planar with exactly 120° between adjacent S—O bonds. Although the S—O bonds are polar, the SO_3 molecule is nonpolar because of the symmetry.

(d) The ammonia molecule, NH_3, is polar for the same reason that

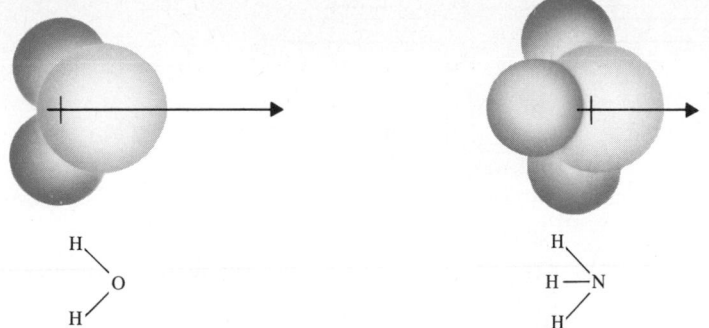

Figure 6.4. Molecular structures and dipole moments of H_2O (water) and NH_3 (ammonia).

H_2O is polar. The N side of the molecule is negative and the H side is positive, as illustrated in Figure 6.4.

(e) In Example 6 (Section 6.6) the CH_4 molecule was shown to be perfectly symmetrical with exactly 109° 28′ between every pair of C—H bonds. Although the C—H bonds are slightly polar, the CH_4 molecule is nonpolar due to its symmetry.

In Section 6.8 you will see how bond polarity can be increased to the point where ionic compounds are formed.

6.8 IONIC CONDUCTION OF ELECTRICITY

The best evidence for the existence of ions is their ability to conduct electricity. For example, NaCl will conduct electricity under the right conditions. If it is heated to its melting point, 808°C, the rigid structure of the solid is broken down and the ions that compose it are able to move in the direction of an electric current. When the two test probes in Figure 6.5 are dipped into molten NaCl, the light bulb indicates if the

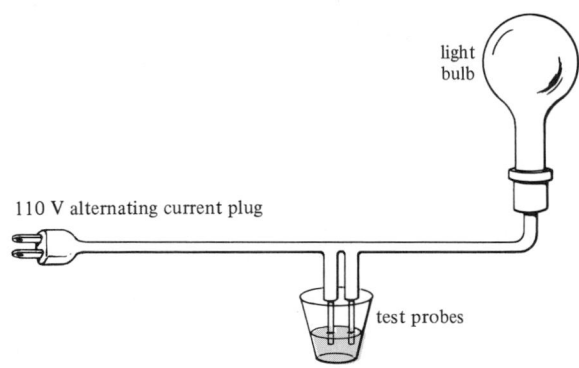

light
bulb

110 V alternating current plug

test probes

Figure 6.5. Conductivity tester.

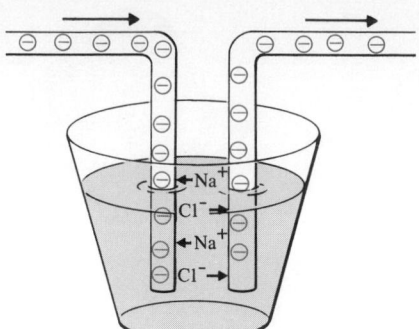

Figure 6.6. Instantaneous conduction of an alternating current of electricity in molten NaCl.

circuit is complete. The Na⁺ ions and Cl⁻ ions move in opposite directions and constitute an electric current. The metal probes conduct electricity by the flow of electrons in the metal while the positive metal ions remain fixed in place. Since the current alternates rapidly in direction, there is no decomposition of the NaCl. Figure 6.6 represents the direction of ionic flow of current at a given instant.

The same effect is obtained if the NaCl is dissolved in water and the test probes are dipped into the aqueous solution. If a direct current is used, chemical reactions occur as well, which will be discussed in Section 12.4.

1. From a knowledge of the structure of substances and the arrangement of their chemical bonds you will recognize that (1) some compounds are *not* molecules and (2) some molecules are *not* compounds, and, of course, (3) some compounds *are* molecules. Choose the best description, either (1), (2) or (3) for each of the substances listed below.
   (a) NaCl
   (b) N_2
   (c) HCN (hydrogen cyanide)
   (d) CO_2 (carbon dioxide)
   (e) SiO_2 (silicon dioxide). One form of this substance is called quartz. It is a three-dimensional crystalline solid in which each Si atom is at the center of a tetrahedron surrounded by four O atoms, each of which is bound to another Si atom, and so on.

2. Same as problem 1.

   (a) H_2
   (b) O_3 (ozone)
   (c) CsCl (cesium chloride)
   (d) H_2O
   (e) BN (boron nitride). This solid substance is just as hard as diamond and has a similar structure (See Figure 9.3), consisting of alternating B atoms and N atoms.

3. Predict the number of valence electrons in single atoms of each of the following elements. You may refer to a periodic table to determine the position of each element.
   (a) Na;   (d) Si;
   (b) Ba;   (e) Al.
   (c) N;

4. Same as problem 3.

   (a) Cl;   (d) O;
   (b) Cs;   (e) B.
   (c) Ca;

5. Draw electron dot structures in which the octet rule is satisfied by every atom in the following covalent molecules and ions. Beneath one atom of each element indicate which inert gas element it resembles.
   (a) F_2;
   (b) CO (carbon monoxide);
   (c) N_3^- (azide ion);
   (d) H_2S (hydrogen sulfide);
   (e) C_2H_2 (acetylene).

6. Same as problem 5.

   (a) PH_3 (phosphine);
   (b) Br_2;
   (c) ClO^- (hypochlorite ion);
   (d) CN^- (cyanide ion);
   (e) H_2O_2 (hydrogen peroxide).

7. The following molecules and ions are held together by covalent bonds. Draw an electron dot structure for each in which the octet rule is satisfied. The appropriate arrangement of atoms is indicated in each case.

(a) H_2SO_3 (sulfurous acid)

$$\begin{array}{ccc} & O & \\ O & S & O \\ H & & H \end{array}$$

(b) ClO_4^- (perchlorate ion)

$$\begin{array}{ccc} & O & \\ O & Cl & O \\ & O & \end{array}$$

(c) CH_3COOH (acetic acid)

$$\begin{array}{cccc} & H & O & \\ H & C & C & O \\ & H & & H \end{array}$$

(d) N_2O (nitrous oxide)

$$N \quad N \quad O$$

8. Same as problem 7.

(a) HNO_2 (nitrous acid)

$$\begin{array}{cc} & N \quad O \\ H & O \end{array}$$

(b) CO_3^{2-} (carbonate ion)

$$\begin{array}{cc} & O \\ & C \\ O & \quad O \end{array}$$

(c) HSO_4^- (hydrogen sulfate ion)

$$\begin{array}{ccc} & O & \\ O & S & O \\ & O & H \end{array}$$

(d) SO_2 (sulfur dioxide)

$$\begin{array}{cc} & O \\ O & S \end{array}$$

9. Compounds of metals with nonmetals are generally ionic, whereas compounds of nonmetals with other nonmetals are generally covalent. Predict whether each of the following compounds is ionic or covalent.
(a) MgO (magnesium oxide);
(b) Na_2SO_4 (sodium sulfate);
(c) $FeCl_2$ (iron(II) chloride);
(d) CCl_4 (carbon tetrachloride).

10. Same as problem 9.

(a) PCl_3 (phosphorus trichloride);
(b) OF_2 (oxygen fluoride);
(c) KNO_3 (potassium nitrate);
(d) $CrCl_3$ (chromium(III) chloride).

11. Show how the octet rule is satisfied in each of the following ionic compounds. Below one ion of each element indicate which inert gas element it resembles. Name each compound.
(a) SrO;   (b) AlH_3;   (c) Na_3N.

12. Same as problem 11.

(a) RaS;   (b) Li_2S;   (c) $MgCl_2$.

13. Write the correct formula and name of every possible binary ionic compound in a table like the one following.

14. Write the correct formula and name of every possible binary ionic compound in a table like the one following.

	S^{2-}	N^{3-}	I^-			P^{3-}	O^{2-}	H^-
K^+		K_3N potassium nitride		Mg^{2+}		Mg_3P_2 magnesium phosphide		
Al^{3+}				Na^+				
Ba^{2+}				Al^{3+}				

15. The following compounds consist of metals that are not in principal groups and nonmetals that are. Assuming the normal ionic charge for the nonmetals, give names for the following compounds.
    (a) $CrCl_2$ and $CrCl_3$.
    (b) HgO and Hg_2O.

16. Same as problem 15.

    (a) $SnCl_2$ and $SnCl_4$
    (b) FeO and Fe_2O_3

17. The following compounds consist of metals that are not in principal groups and nonmetals that are. Assuming the normal ionic charge for the nonmetals give formulas for the following compounds.
    (a) nickel(II) oxide and nickel(III) oxide.
    (b) vanadium(III) bromide and vanadium(IV) bromide.

18. Same as problem 17.

    (a) mangenese(II) oxide and manganese(IV) oxide.
    (b) cobalt(II) iodide and cobalt(III) iodide.

19. Give correct formulas for the following simple ionic compounds. (*Note: Simple ions are ions of single atoms.*)
    (a) potassium oxide;
    (b) lithium hydride;
    (c) aluminum bromide.

20. Same as problem 19.

    (a) barium chloride;
    (b) cesium oxide;
    (c) aluminum fluoride.

21. Assuming that metals react with nonmetals to form ionic compounds and assuming that the nonmetal ions are simple (that is, not covalently bound), complete and balance the following chemical equations.
    (a) $Mg + Cl_2 \longrightarrow$
    (b) $Al + O_2 \longrightarrow$
    (c) $Na + S_8 \longrightarrow$

22. Same as problem 21.

    (a) $Na + Br_2 \longrightarrow$
    (b) $Ca + P_4 \longrightarrow$
    (c) $Al + I_2 \longrightarrow$

23. Make a sketch showing the geometry (linear, bent, planar, pyramidal, or tetrahedral) for each of the following molecules or ions. Include in your sketch all important unshared electron pairs and indicate the theoretical angles between bonds.
    (a) NO_2^- (nitrite ion; N atom central);

    (b) N_2O (see problem 7);
    (c) CCl_4 (carbon tetrachloride; C atom central);
    (d) ClO_3^- (chlorate ion; Cl atom central);
    (e) C_2H_4 (ethylene; Note: Although not predicted by the rule of electron repulsion, this entire molecule lies in a single plane.)

24. Same as problem 23.

    (a) PO_4^{3-} (phosphate ion; P atom central);
    (b) H_2S (hydrogen sulfide);
    (c) I_2;
    (d) CO_3^{2-} (carbonate ion; see problem 8);
    (e) PH_3 (phosphine).

25. Predict whether or not the following bonds are polar and, if so, which atom is the more positive.
    (a) C≡N;   (c) S=O;
    (b) I—Cl;   (d) H—H.

26. Same as problem 25.

    (a) N≡N;   (c) S—Cl;
    (b) B—N;   (d) P=O.

27. Predict which of the following molecules would be polar and explain each in terms of polar and nonpolar covalent bonds and the geometry of those bonds. If the molecule is polar, give the direction of the dipole moment from the positive to negative side.
    (a) CF_4 (carbon tetrafluoride; C atom central);
    (b) P_4;
    (c) SO_2 (sulfur dioxide; S atom central);

    (d) O_2.

28. Same as problem 27.

    (a) HF (hydrogen fluoride);
    (b) $CHCl_3$ (chloroform; C atom central);
    (c) C_2H_2 (acetylene);

    (d) Br_2.

*29. The gray solid, CaC_2 (calcium carbide), is manufactured from limestone rock ($CaCO_3$), found abundantly in the earth, and coke (C), which comes from coal. Adding the carbide to water rapidly produces C_2H_2 (acetylene) gas, as shown in the equation.

$$CaC_2 + 2 H_2O \longrightarrow Ca(OH)_2 + C_2H_2$$

calcium   calcium   acetylene
carbide   hydroxide

Illustrate by sketch how the octet rule is satisfied for every atom involved in this reaction.

*30. There are two different compounds known to have the molecular formula C_2H_6O. One of them, ethyl alcohol or drinking alcohol, has a C atom located between the O atom and the other C atom. The other compound, dimethyl ether, which is not the anesthetic, has the O atom located between the two C atoms. Show how the octet rule is satisfied for every atom in both of these compounds.

*31. Formaldehyde, a liquid which is commonly used as a preservative of animal specimens, has the formula CH_2O. It is called the simplest carbohydrate. How many ways can you arrange the atoms to satisfy the octet rule? Sketch them. If you are restricted to structures in which the C atom has four bonds and the O atom has two bonds, select the correct structure.

*32. In Figure 6.2 the Cs^+ ions are located at the centers of four small cubes with Cl^- ions at all the corners. Show that the number of Cs^+ ions and Cl^- ions are equal by considering the fraction of each Cl^- ion shared by other units (before, behind, above, below, left, and right). *Hint:* Make a new sketch of the figure showing only the Cl^- ions, including those "not visible" in the figure given.

*33. Is NaCl(g) polar? Is NaCl(s) polar? Explain.

7

HEAT AND ENERGY, PHASE CHANGE, AND VAPOR PRESSURE

In 1973–1974 the Organization of Petroleum Exporting Countries (OPEC) quadrupled the price of oil. This action dramatized for the world the predictions of scientists that all reserves of oil and gas would be used up within about 30 years.

Many chemical reactions produce heat, which can be converted into other forms of energy. Combustion reactions in particular produce large amounts of heat. High temperatures and high pressures are produced by these reactions, and industrialized societies have learned how to use them to generate power. In its 1975 report to the government, the Energy Research and Development Administration (ERDA) pointed out that the USA has changed its main source of energy twice before in history. In 1850 this source was *wood;* by 1910, it was *coal;* and by 1970, it was *oil and gas.* Each time the complete changeover took about 60 years; this time, however, it must be accomplished in 30 years. This accelerated pace of change is a good example of the causes of "future shock" described by Alvin Toffler.* Our nation should plan carefully and choose the right alternatives to make technology work in our best interests. It may mean the development of substitutes for gas and oil, but it may just as well mean a change in life style through which we better conserve the resources we have.

Temperature and heat are closely related. You could say that temperature is the

*Alvin Toffler: *Future Shock.* Random House, Inc., New York, 1970.

"intensity of heat" because heat flows naturally from high temperature to low. In this chapter you will see the relationships of heat with changes in temperature, with chemical reactions, and with phase changes of all kinds.

7.1 MEASUREMENT OF HEAT: CALORIMETRY

Heat is a form of energy. It can be measured by the extent to which it changes the temperature of a substance. Different substances respond with different temperature changes for the same amount of heat. The amount of heat required to raise the temperature of 1 g of a substance by 1°C is called its **specific heat.** It is a very characteristic property of that substance. Liquid water, for example, requires 1 **calorie** (cal) to produce this change. This is a convenient way to define the calorie of heat, although scientists actually use electrical methods today because of their greater precision. The amount of heat required to raise the temperature of 1 g liquid Hg by 1°C is only 0.033 cal. The specific heat of Hg(l) is 0.033 cal g^{-1} $°C^{-1}$ and that of H_2O(l) is 1.00 cal g^{-1} $°C^{-1}$ at ordinary temperatures.

EXAMPLE 1. How many calories of heat are required to heat (a) 1 kg H_2O(l) from room temperature (25°C) to the boiling point of water (100°C)? (b) How many calories for 1 kg Hg(l)?

Solution: The amount of heat required is directly proportional to the specific heat. This is indicated by the fact that the unit of specific heat, calories per gram per degree Celsius (cal g^{-1} $°C^{-1}$), has calorie in the numerator. Also the amount of heat required is directly proportional to the increase in temperature, $(100 - 25)$ °C, and to the mass, 1 kg.

(a) $$x \text{ cal} = \left(1.00 \frac{\text{cal}}{\text{g }°\text{C}}\right)([100 - 25]\ °\text{C})(1 \text{ kg}) \cdots$$

The only one-factor needed can be derived from the equality, 1 kg = 1000 g.

$$x \text{ cal} = \left(1.00 \frac{\text{cal}}{\text{g }°\text{C}}\right)([100 - 25]\ °\text{C})(1 \text{ kg})\left(\frac{1000}{1} \frac{\text{g}}{\text{kg}}\right)$$
$$= (1.00)(100 - 25)(1000)$$
$$= (1.00)(75)(1000)$$
$$= 7.5 \times 10^4 \text{ cal}$$

(b) For Hg(l) the only change in calculation is in the specific heat.

$$x \text{ cal} = \left(0.033 \frac{\text{cal}}{\text{g }°\text{C}}\right)([100 - 25]\ °\text{C})(1 \text{ kg})\left(\frac{1000}{1} \frac{\text{g}}{\text{kg}}\right)$$
$$= (0.033)(100 - 25)(1000)$$
$$= 2.5 \times 10^3 \text{ cal}$$

EXAMPLE 2. A refrigerator removes enough heat from 600 g of water in a tray to cool it to 0°C. If the original temperature is 25°C, how much heat must be removed?

Solution: The amount of heat that must be removed is directly proportional to the specific heat of the water, $1.00 \text{ cal g}^{-1} \, ^\circ\text{C}^{-1}$, the change in temperature $(25 - 0) \, ^\circ\text{C}$, and the weight of the water, 600 g. No other factors are needed. The calculation is

$$x \text{ cal} = \left(1.00 \, \frac{\text{cal}}{\text{g} \, ^\circ\cancel{C}}\right)([25 - 0] \, ^\circ\cancel{C})(600 \, \cancel{g})$$
$$= (1.00)(25)(600) \text{ cal}$$
$$= 1.5 \times 10^4 \text{ cal}$$

If you wish to freeze the water after it has reached 0°C, it will be necessary to remove even more heat. This process will be discussed in Section 7.4.

Other forms of energy besides heat can be used to raise the temperature of water. Turning a paddle wheel in water will increase the temperature. It is possible to express the amount of work necessary to turn the paddle wheel in terms of calories by measuring the weight and change in temperature of the water. The work is calculated from the laws of physics. The form of energy that can be most precisely measured is electrical energy, and it is used to define the calorie of heat. Electrical energy can be converted to heat just as it is in a burning light bulb or a coffee percolator. The effect is used to standardize various kinds of heat-measuring equipment, called **calorimeters.**

7.2 HEAT OF CHEMICAL REACTION

You will frequently want to know the amount of heat produced in a chemical reaction. One type of calorimeter used to measure heats of reaction is shown in Figure 7.1. Two substances that will react rapidly and completely with one another (until one is consumed) are placed in the bomb in Figure 7.1. Everything is assembled inside the

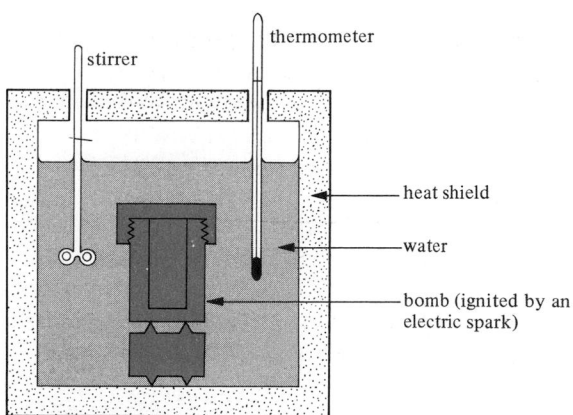

Figure 7.1. A bomb calorimeter used to measure heats of combustion.

heat shield and thermal equilibrium is attained by stirring the water. After carefully measuring the temperature, the reaction is initiated, typically by electrical heating or a spark, and after a short wait a new thermal equilibrium is attained. If no heat is exchanged with the surroundings, the temperature change should be a measure of the heat produced by the reaction.

One substance commonly used to produce a known amount of heat for purposes of calibration is the white crystalline solid benzoic acid, which contains only carbon, hydrogen, and oxygen. It is burned rapidly and completely under pressure in pure gaseous oxygen. This heat of reaction has been accurately determined by the National Bureau of Standards by comparing it with heat produced by electrical energy. Calorimetrists may obtain samples of the reference material from NBS to calibrate their own calorimeters.

Heats of chemical reactions are usually compared on a mole basis. The molar heat of combustion of benzoic acid would be expressed in units of kilocalories per mole (kcal mole^{-1}).

EXAMPLE 1. If exactly 1000 cal of electrical heat is added to a certain calorimeter, the temperature is found to rise 0.408°C. (a) What is the heat of combustion of 1.000 g benzoic acid ($C_7H_6O_2$) if benzoic acid reacts with excess $O_2(g)$ under pressure to produce a temperature rise of 2.58°C? The same calorimeter is used in both experiments. (b) What is the *molar* heat of combustion of benzoic acid?

Solution: You may assume that all the benzoic acid is consumed in this reaction. Since the final temperature is only slightly above the initial temperature, the H_2O produced will be liquid. The balanced equation is

$$2\ C_7H_6O_2(s) + 15\ O_2(g) \longrightarrow 14\ CO_2(g) + 6\ H_2O(l)$$

(a) The unknown value is x kcal. This heat is directly proportional to the increase in temperature during combustion.

$$x\ \text{kcal} = (2.58°C) \cdots$$

You may obtain a one-factor from the electrical calibration, which showed the equivalence of 1000 cal and 0.408°C. You may obtain another one-factor from the equality, 1000 cal = 1 kcal.

$$x\ \text{kcal} = (2.58°C)\left(\frac{1000\ \text{cal}}{0.408\ °C}\right)\left(\frac{1}{1000}\frac{\text{kcal}}{\text{cal}}\right)$$

$$= 6.32\ \text{kcal}$$

This is the heat of combustion of benzoic acid on a weight basis, 6.32 kcal g^{-1}.

(b) The *molar* heat of combustion is the heat of combustion of 1 mole of benzoic acid. It is proportional to the number of grams in 1 mole $C_7H_6O_2$(BA), or 122 g.

$$x\ \text{kcal} = (122\ \text{g BA}) \cdots$$

You may use a one-factor based on the answer in part (a).

149

7.3
Addition and
Subtraction
of Heats of
Chemical Reaction

$$x \text{ kcal} = (122 \text{ g BA})\left(\frac{6.32 \text{ kcal}}{1.000 \text{ g BA}}\right)$$

$$= 771 \text{ kcal}$$

This is the *molar* heat of combustion of benzoic acid, 771 kcal mole^{-1}.

When a person exercises, the source of energy is chemical reaction. For example, carbohydrates are metabolized in the human body ultimately producing $CO_2(g)$ and $H_2O(l)$. The amount of heat in the nutritionist's "Calorie" (spelled with a capital "C") is the same as the "kcal", or 1000 calories. (We will always use kilocalories, as many nutritionists are now doing.) The number of kilocalories in a certain sample of food is the amount of heat produced by burning that sample in excess $O_2(g)$.

EXAMPLE 2.   The heat of combustion of table sugar (sucrose, $C_{12}H_{22}O_{11}$) is 1350 kcal mole^{-1}. How much heat is produced in the metabolism of 1 teaspoon (4.8 g) of sugar according to the following equation?

$$C_{12}H_{22}O_{11}(s) + 12\ O_2(g) \longrightarrow 12\ CO_2(g) + 11\ H_2O(l) + 1350 \text{ kcal mole}^{-1}$$

Solution: The amount of heat in kilocalories is directly proportional to the number of grams sucrose and the molar heat in kilocalories per mole.

$$x \text{ kcal} = (4.8 \text{ g sucrose})\left(1350\ \frac{\text{kcal}}{\text{mole sucrose}}\right)\cdots$$

A single one-factor based on the equality, 342 g sucrose = 1 mole sucrose, is sufficient to complete the calculation.

$$x \text{ kcal} = (4.8 \text{ g sucrose})\left(1350\ \frac{\text{kcal}}{\text{mole sucrose}}\right)\left(\frac{1}{342}\ \frac{\text{mole sucrose}}{\text{g sucrose}}\right)$$

$$= \frac{(4.8)(1350)}{342} \text{ kcal}$$

$$= 19 \text{ kcal}$$

Some nutritionists would say that a teaspoon of granulated sugar contains 19 Calories.

7.3   ADDITION AND SUBTRACTION OF HEATS OF CHEMICAL REACTION

Chemical reactions in nature can go in both directions. If heat is given off in one direction, the same amount is absorbed in the opposite direction. A process that produces heat is called **exothermic.** If the heat is written as part of a chemical equation, the coefficients of each molecule are assumed to represent moles. An example is shown below.

$$2 H_2(g) + O_2(g) \longrightarrow 2 H_2O(l) + 136.6 \text{ kcal} \tag{7.1}$$

The heat of combustion of 2 moles $H_2(g)$, which is exothermic, is 136.6 kcal [or 68.3 kcal (mole $H_2)^{-1}$]. If you want twice as much reaction, which is equivalent to adding two equations together, there is twice as much heat.

$$4 H_2(g) + 2 O_2(g) \longrightarrow 4 H_2O(l) + 273.2 \text{ kcal} \tag{7.2}$$

The reverse reaction requires the absorption of heat. One way to accomplish this is to pass a direct current of electricity through the water (see Section 12.4). The reaction in equation (7.3) is the negative of equation (7.1) for 2 moles $H_2(g)$.

$$136.6 \text{ kcal} + 2 H_2O(l) \longrightarrow 2 H_2(g) + O_2(g) \tag{7.3}$$

This process, which absorbs heat, is called **endothermic.** These experimental results and many others like them show that there is a law of conservation of energy in nature as well as conservation of mass.

> **Law of conservation of energy:** *Energy can only be changed in form; it cannot be created or destroyed.*

Most exothermic reactions are spontaneous. A **spontaneous** reaction is one that proceeds on its own (once it has been initiated) without help from any external source. Thus, the combustion reactions of benzoic acid, sucrose, and gaseous hydrogen are all exothermic and all spontaneous. The decomposition of water to form hydrogen and oxygen is *impossible* on its own, like most endothermic reactions; however, it can be made to occur by applying energy.

Heats of Processes That Are Difficult to Accomplish

Some changes have interesting heats even though they are difficult to accomplish. For example, it is of interest to know the heat that is naturally associated with the conversion of graphite to diamond.

$$C(\text{graphite}) \longrightarrow C(\text{diamond}) + \text{heat} \tag{7.4}$$

Although this conversion has actually been accomplished under certain extreme conditions of pressure and temperature, it is practically impossible to measure the heat directly. There is a much simpler approach to its measurement, however, based on the principle of the conservation of energy.

It is a simple matter to measure the heats of the following reactions in an ordinary bomb calorimeter (provided that you are willing to burn a little diamond).

$$C(\text{graphite}) + O_2(g) \longrightarrow CO_2(g) + 94.05 \text{ kcal} \tag{7.5}$$

$$C(\text{diamond}) + O_2(g) \longrightarrow CO_2(g) + 94.50 \text{ kcal} \tag{7.6}$$

By the conservation of energy principle you can predict the result of taking the negative of reaction (7.6). The heat would have the same value, but the reaction would be endothermic rather than exothermic.

151

7.3
Addition and
Subtraction
of Heats of
Chemical Reaction

$$CO_2 + 94.50 \text{ kcal} \longrightarrow C(\text{diamond}) + O_2(g) \qquad -(7.6)$$

If equation (7.5) is performed as written and is then followed by the negative of equation (7.6), this corresponds, in theory, to converting 1 mole C(graphite) and 1 mole $O_2(g)$ into 1 mole $CO_2(g)$ followed by taking 1 mole $CO_2(g)$ and converting it into 1 mole $O_2(g)$ and 1 mole C(diamond). The net reaction would be as follows:

$$C(\text{graphite}) + 94.50 \text{ kcal} \longrightarrow C(\text{diamond}) + 94.05 \text{ kcal} \qquad (7.5) - (7.6)$$

$$C(\text{graphite}) \longrightarrow C(\text{diamond}) \qquad\qquad + 94.05 - 94.50 \text{ kcal}$$

$$C(\text{graphite}) \longrightarrow C(\text{diamond}) \qquad\qquad - 0.45 \text{ kcal} \qquad (7.4)$$

This result may be surprising. It tells you that the transformation of graphite to diamond is endothermic, and that the reverse is exothermic. Like most exothermic reactions this one is spontaneous, which means that graphite is a more stable form of C than diamond. Diamond should disintegrate into graphite given enough time; however, there is little to worry about if you own diamonds because the reaction is too slow to detect.

The heat of the reaction in the following example is also difficult to measure directly, but for a different reason.

EXAMPLE 1.   It is not possible to measure the heat of the reaction

$$2 \, C(\text{graphite}) + O_2(g) \longrightarrow 2 \, CO(g) \qquad (7.7)$$

under high pressures of $O_2(g)$ because the reaction goes completely to $CO_2(g)$. Show how the calculation of this heat would be possible from the easily measured reactions given in equations (7.5) and (7.8).

$$C(\text{graphite}) + O_2(g) \longrightarrow CO_2(g) + 94.1 \text{ kcal} \quad (7.5)$$

$$2 \, CO(g) + O_2(g) \longrightarrow 2 \, CO_2(g) + 135.2 \text{ kcal} \qquad (7.8)$$

Solution: Multiply equation (7.5) by two because 2 C(graphite) is needed on the left in equation (7.7).

$$2 \, C(\text{graphite}) + 2 \, O_2(g) \longrightarrow 2 \, CO_2(g) + 188.2 \text{ kcal}$$

Take the negative of equation (7.8) because 2 CO(g) is needed on the right in equation (7.7).

$$2 \, CO_2(g) \longrightarrow 2 \, CO(g) + O_2(g) - 135.2 \text{ kcal} \qquad -(7.8)$$

Add the last two equations and cancel moles that appear on both left and right to obtain the desired equation.

$$2\,C(graphite) + 2\,O_2(g) + 2\,CO_2(g) \longrightarrow$$
$$2\,CO_2(g) + 2\,CO(g) + O_2(g) + 188.2\,kcal - 135.2\,kcal$$

$$2\,C(graphite) + O_2(g) \longrightarrow 2\,CO(g) + 53.0\,kcal \quad (7.7)$$

Since heat appears on the right hand side of equation (7.7), the net reaction is exothermic.

7.4 HEAT AND PHASE CHANGE

Water exists in three different phases; namely, solid, liquid, and gas. Heat is absorbed in the phase changes, $H_2O(s) \longrightarrow H_2O(l) \longrightarrow H_2O(g)$. Likewise heat is given off when these phase changes occur in the opposite direction. Therefore, it is necessary to stipulate in which phase water exists, if there is any doubt, when it occurs as a reactant or product in a reaction such as equation (7.1). Similar changes can be observed for all stable substances that do not undergo chemical decomposition when heated.

Phase changes occur abruptly as the temperature is increased. As a matter of fact, once the phase change has begun the temperature will not increase at all as the heat is slowly added until the phase change is complete. A graphical illustration of this fact is shown in Figure 7.2. We start with $1\,g\,H_2O(s)$, or ice, at 0°. The horizontal axis indicates the amount of heat absorbed by the H_2O; the vertical axis indicates the temperature of the H_2O resulting from the corresponding amount of heat.

The heat absorbed at 0°C while the $H_2O(s)$ is melting is $80\,cal\,g^{-1}$ or $1.44\,kcal$ mole^{-1}. It is called the **heat of fusion.** The heat absorbed at 100 C while $H_2O(l)$ is boiling is $540\,cal\,g^{-1}$ or $9.7\,kcal$ mole^{-1}. It is called the **heat of vaporization.** In balanced equations it is always the *molar* heats that are used.
Melting (fusion):

$$H_2O(s) + 1.44\,kcal \longrightarrow H_2O(l) \quad at\ 0°C.$$

Boiling (vaporization):

$$H_2O(l) + 9.7\,kcal \longrightarrow H_2O(g) \quad at\ 100°C.$$

Reversals of the phase changes result in reversals of heats.
Freezing:

$$H_2O(l) \longrightarrow H_2O(s) + 1.44\,kcal \quad at\ 0°C.$$

Condensation:

$$H_2O(g) \longrightarrow H_2O(l) + 9.7\,kcal \quad at\ 100°C.$$

Thus, melting and boiling are endothermic, whereas freezing and condensation are exothermic.

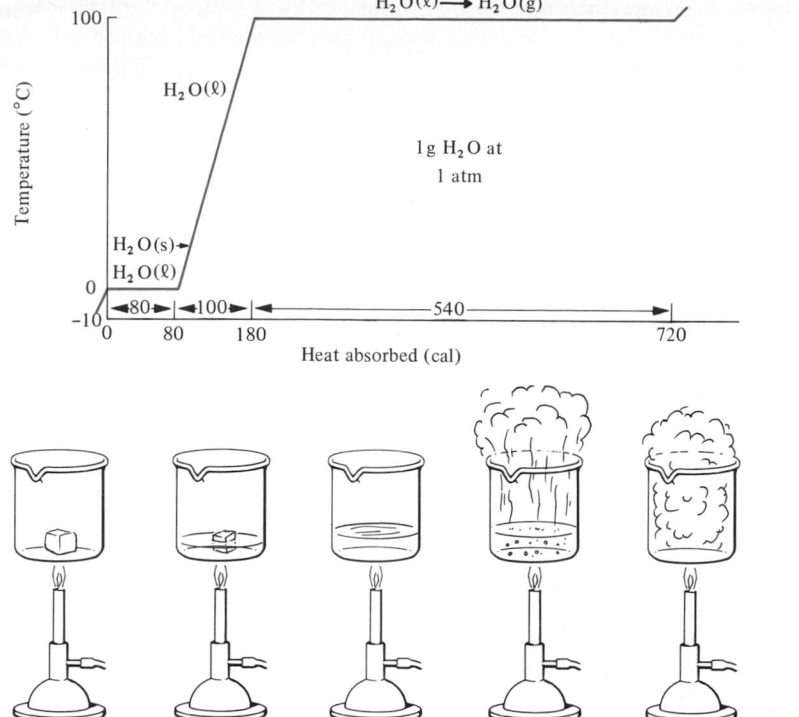

Figure 7.2. Steps in heating H_2O from below its freezing point (ice) to above its boiling point (steam).

EXAMPLE 1. Which of the following heats of combustion is more exothermic?

$$2\,H_2(g) + O_2(g) \longrightarrow 2\,H_2O(l) + heat \qquad (a)$$
$$2\,H_2(g) + O_2(g) \longrightarrow 2\,H_2O(g) + heat \qquad (b)$$

Solution: You can consider equation (a) as a two-step process. The first step is the exothermic reaction in equation (b) in which $2\,H_2O(g)$ is the product. The second step is the condensation of $2\,H_2O(g)$ into $2\,H_2O(l)$, which gives off another 2×9.7 kcal, or 19.4 kcal (see above). Therefore, equation (a) is 19.4 kcal *more* exothermic than equation (b).

7.5 PROPERTIES OF SOLIDS, LIQUIDS, AND VAPORS

The volumes of liquids and solids can be decreased very little by increasing the pressure even to very high values. The molecules of a substance in its liquid or solid phase are so close together that they are in contact with one another. The existence of liquids is somewhat of a paradox. In the liquid phase molecules have enough energy to move around, which gives the liquid its **fluid** property, but not enough to escape one another. Thus a liquid does not have a shape of its own but fills the lower portion of its container. Gases and vapors, on the other hand, fill their containers completely. When energy is removed from the molecules of a liquid or a gas, the molecules settle into a rigid three-dimensional framework called the solid phase. It has been said that if liquids had never been observed, scientists would never have believed in the possibility of their existence.

Vapor Pressure

Even if the temperature is not high enough for boiling, some of the molecules in a sample of liquid or solid escape into the gas phase. If the substance is fairly volatile and if it is not enclosed in a container, the liquid or solid phase may disappear completely as a result of evaporation. The water in your automobile battery, for example, slowly evaporates through the air vent although it never boils. Another example is that of ice and snow which evaporate at temperatures of 0°C and below, a process known as sublimation.
Sublimation:

$$H_2O(s) + 12.1 \text{ kcal} \longrightarrow H_2O(g) \quad \text{at } 0°C \text{ and } 4.6 \text{ mm Hg}$$

The dark lustrous crystals of pure iodine produce enough vapor at room temperature for you to see its color (violet).
Sublimation:

$$I_2(s) + 14.9 \text{ kcal} \longrightarrow I_2(g) \quad \text{at } 25°C \text{ and } 0.3 \text{ mm Hg}$$

Several methods are available for the measurement of vapor pressure. One method makes use of the ordinary barometer discussed in Section 3.2, as shown in Figure 7.3. A properly constructed barometer has a vacuum space at the upper end of the glass tubing and a mercury height equal to P_{bar} in centimeters (barometric pressure). A sample of the liquid to be measured is introduced into the barometer tube by means of a syringe and flexible tubing in such a way that no air is allowed to enter. The liquid rises because it is less dense than the mercury, and it evaporates rapidly as it enters the lower pressure region at the upper end. The vapor enters the vacuum space and its pressure forces the mercury level down. A point is soon reached at which it appears that the evaporation ceases because the pressure remains constant. Actually evaporation of molecules in the liquid phase continues, but molecules in the gas phase are condensing and taking their place.

DEFINITION: *A pure substance is at **equilibrium** when it is enclosed in a given space and when its temperature and pressure remain constant. Although it appears that no changes are taking place, equilibrium is frequently maintained by molecules moving at equal rates between two phases.*

A broader definition of equilibrium will be given in Chapter 9, in which mixtures of pure substances are considered as well as chemical reactions.

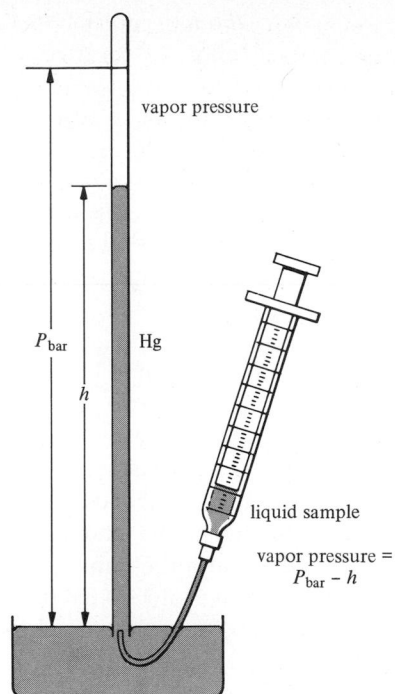

vapor pressure

P_{bar} Hg

h

liquid sample

vapor pressure =
$P_{bar} - h$

Figure 7.3. Use of a barometer to measure the
vapor pressure of a liquid.

The pressure exerted by a vapor when it is in equilibrium with its liquid phase at a given temperature is called its **vapor pressure** at that temperature. For example, if a small excess of liquid water is introduced into the barometer at 25°C, the mercury level drops 24 mm. The pressure that causes this drop is due almost entirely to the vapor because the small excess of liquid water exerts negligible force. Even a 1-mm thick layer of liquid water exerts a pressure equivalent to only 0.07 mm Hg. The addition of another 1 mm thickness would not have any noticeable effect on the mercury level because there would be no additional vaporization. Since the barometric pressure outside the tube, P_{bar}, must be equal to the total pressure inside the tube at the same level, vp + h, where vp is the vapor pressure, you can write

$$vp + h = P_{bar}$$
$$vp = P_{bar} - h$$

EXAMPLE 1.   The barometric method is used for measuring the vapor pressure of pure ethyl alcohol at 25°C. If the height of the mercury column is 756 mm before addition of the alcohol and 697 mm afterwards, calculate the vapor pressure. Only a thin layer of liquid alcohol is visible on top of the mercury column.

Solution: No correction is necessary for the weight of the layer of liquid alcohol. The height of the mercury column before addition of the alcohol,

756 mm, must be equal to P_{bar}. Vapor pressure can be calculated as follows:

$$vp = P_{bar} - h$$
$$= 756 - 697$$
$$vp = 59 \text{ mm Hg}$$

Ethyl alcohol has more than twice the vapor pressure of water at room temperature.

Changes of Vapor Pressure with Temperature

The equipment shown in Figure 7.4 can be used to measure vapor pressure of water at different temperatures. The long two-legged tube is called an open-ended **manometer.** A manometer works on the same principle as a barometer, but it can be used to measure the gas pressure in a closed vessel, such as the large spherical bulb in Figure 7.4. The pear-shaped bulb holds a reservoir of mercury which can be raised or lowered to change the amount of mercury in the manometer. It is also open to the atmosphere. Since the right leg of the manometer is open to the atmosphere, barometric pressure is always exerted on the surface of its mercury column. When the spherical bulb is evacuated and the water sample bulb is kept closed off from it by means of a stopcock (Figure 7.4a), the mercury column rises in the left leg to a level equal to barometric pressure. In this illustration barometric pressure is 756 mm Hg.

Vapor pressure is determined by temperature alone and not by the volume of the spherical bulb. Of course, if the volume of the bulb is so large that all the liquid evaporates, Boyle's law comes into play and pressure becomes dependent on volume.

Figure 7.4.   An apparatus for measuring vapor pressures of liquids at different temperatures.

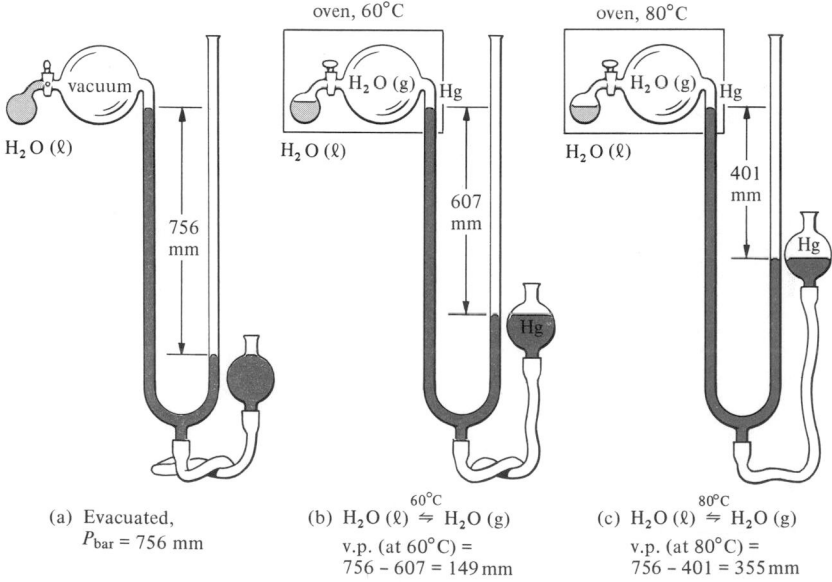

(a) Evacuated,
$P_{bar} = 756$ mm

(b) H_2O (ℓ) $\leftrightharpoons$ H_2O (g)
60°C
v.p. (at 60°C) =
756 − 607 = 149 mm

(c) H_2O (ℓ) $\leftrightharpoons$ H_2O (g)
80°C
v.p. (at 80°C) =
756 − 401 = 355 mm

In Figure 7.4b, an oven is placed around part of the equipment to control the temperature, and the stopcock is opened. If the temperature is held constant, an equilibrium is rapidly established.

$$H_2O(l) \rightleftharpoons H_2O(g) \quad \text{at } 60°C$$

This equation indicates simultaneous occurrence of evaporation and condensation at equal rates. The pressure builds up in the spherical bulb until it reaches the vapor pressure, and the reservoir is raised enough to keep the mercury in the left leg at the same level. The pressure on the mercury surface in the right leg is still 756 mm, and it is balanced by the height of mercury in the left leg, 607 mm, plus the vapor pressure of the water.

$$607 + vp = 756$$
$$vp = 756 - 607$$
$$vp = 149 \text{ mm Hg} \quad \text{at } 60°C$$

In Figure 7.4c, the procedure is repeated to measure the vapor pressure at 80°C.

Boiling Point

If vapor pressures are measured over a range of temperature and tabulated, as in Table 7.1, an interesting result occurs at 100°C. The vapor pressure reaches the value of 760 mm Hg or 1 atm. If the water was in an open container and the barometric pressure was 1 atm, the water would begin boiling as more heat was added.

The graphical representation of the vapor pressure of water in Figure 7.5 is a good

TABLE 7.1 Vapor pressure of $H_2O(l)$

Temperature (°C)	20	25	30	40	60	80	100	120
Vapor pressure (mm Hg)	18	24	32	55	149	355	**760**	1489

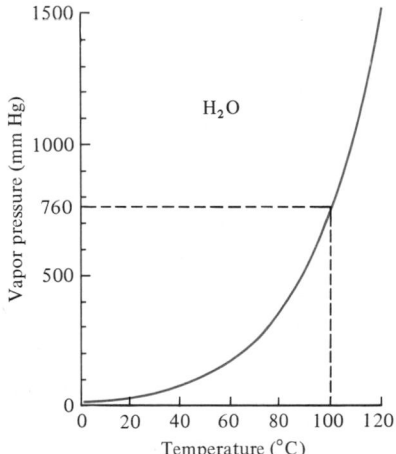

Figure 7.5. Graphical representation of the dependence on temperature of the vapor pressure of $H_2O(l)$. Note the intersection of the curve with the 760 mm line at 100°C.

example of a nonlinear plot. The increase in vapor pressure for a given increment in temperature is greater than that expected for a sample of vapor alone. The curve reaches 760 mm Hg vapor pressure when the temperature reaches 100°C, the normal boiling point of water.

DEFINITION: *The normal boiling point of a liquid is defined as the temperature at which its vapor pressure is equal to 760 mm Hg.*

Every liquid has its own characteristic vapor pressure at each temperature. Under reduced pressures liquids boil at temperatures lower than their normal boiling points. For example, on a 12,000-ft peak where the barometric pressure is 500 mm Hg, water boils at 89°C. However, changes in pressure do not affect the melting point very much. At 0.01 atm the melting point of ice has increased by an amount less than 0.01°C. Therefore, it is not so important to know the pressure for the melting point of a substance as it is for the boiling point.

EXAMPLE 2. Use the graph in Figure 7.5 to estimate the boiling point of water on the top of Mt. Everest ($P_{bar} = \frac{1}{3}$ atm).

Solution: Change the pressure to millimeters of mercury by using the equality, 1 atm = 760 mm.

$$x \text{ mm} = (\tfrac{1}{3} \text{ atm})\left(\frac{760 \text{ mm}}{1 \text{ atm}}\right)$$

$$= (\tfrac{1}{3})(760) \text{ mm}$$
$$= 250 \text{ mm (approximately)}$$

Find 250 mm on the vertical axis (halfway between 0 and 500 mm) and move horizontally to the right until you intersect the curved line. Drop vertically to the horizontal axis and estimate the temperature, 70°C.

A varied list of normal boiling points, sublimation points, and melting points is given in Table 7.2.

Vapor Pressure as a Partial Pressure

The same vapor pressure of a liquid or solid is exerted on the walls of the container even if other gases are present. This is because the distance between gas molecules is so great that they do not interfere with one another. If a glass plate is put on top of a beaker containing water, the gas phase soon becomes saturated with water vapor. The pressure exerted by the water vapor depends only on the temperature. The two examples in Figure 7.6 will illustrate the point.

In Figure 7.6 the partial pressure of $H_2O(g)$ is obtained from the table of vapor pressures at the correct temperatures (Table 7.1) and the partial pressure of air is calculated by subtracting vapor pressure from the total pressure. The air is "saturated" with water vapor in both of these examples which means that the vapor pressures have their maximum values for the given temperatures.

	Melting Point (°C)		Boiling Point (°C)	
$H_2(s)$	-259 $\rightleftharpoons$	$H_2(l)$	-253 $\rightleftharpoons$	$H_2(g)$
$N_2(s)$	-210 $\rightleftharpoons$	$N_2(l)$	-196 $\rightleftharpoons$	$N_2(g)$
$O_2(s)$	-218 $\rightleftharpoons$	$O_2(l)$	-183 $\rightleftharpoons$	$O_2(g)$
$Hg(s)$	-39 $\rightleftharpoons$	$Hg(l)$	357 $\rightleftharpoons$	$Hg(g)$
$CO_2(s)$		-79 (sub.pt.) $\rightleftharpoons$		$CO_2(g)$
$Na(s)$	98 $\rightleftharpoons$	$Na(l)$	892 $\rightleftharpoons$	$Na(g)$
$NaCl(s)$	808 $\rightleftharpoons$	$NaCl(l)$	1465 $\rightleftharpoons$	$NaCl(g)$
$Fe(s)$	1535 $\rightleftharpoons$	$Fe(l)$	3000 $\rightleftharpoons$	$Fe(g)$

EXAMPLE 3. What is the partial pressure of air in a container over the surface of vigorously boiling water?

Solution: If the water is boiling, the vapor pressure of the water must be equal to the barometric pressure.

$$\text{partial pressure of air} = P_{bar} - vp$$
$$\text{partial pressure of air} = 0$$

Thus if the water vapor is escaping fast enough to prevent air molecules from "wandering in by accident," there is no partial pressure of air in the container.

Figure 7.6. Partial pressures of air and water vapor in saturated air.

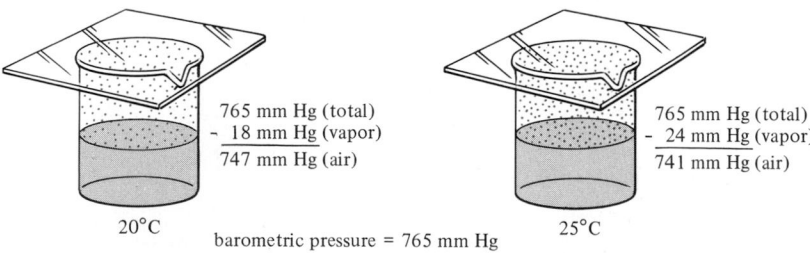

765 mm Hg (total)
− 18 mm Hg (vapor)
747 mm Hg (air)

765 mm Hg (total)
− 24 mm Hg (vapor)
741 mm Hg (air)

20°C   25°C
barometric pressure = 765 mm Hg

An interesting experiment can be performed that demonstrates not only the strong dependence of vapor pressure on temperature but also the tremendous force exerted by the atmosphere. In Figure 7.7a, a small amount of water is added to an empty 1-gal can. The partial pressure of water vapor inside the can at this temperature is 24 mm Hg (Table 7.1), but since the total pressure inside the open can must be the same as barometric pressure, say 760 mm Hg, there are no unbalanced forces exerted on the walls. The can is then placed over a burner (Figure 7.7b) and the water is boiled vigorously for several minutes. The total pressure inside the can is still 760 mm, but the effect of boiling is to sweep out all the air molecules and increase the partial pressure of $H_2O(g)$ to 760 mm. Hold these conditions constant and quickly stopper the can (Figure 7.7c) to prevent the return of air. Cool to 25°C (Figure 7.7d). A change in state occurs, $H_2O(g) \longrightarrow H_2O(l)$, until the pressure inside the can is reduced to the vapor pressure of $H_2O(l)$ at 25°C, or 24 mm Hg. Since there is no other

Figure 7.7. Collapsing a can with barometric pressure.

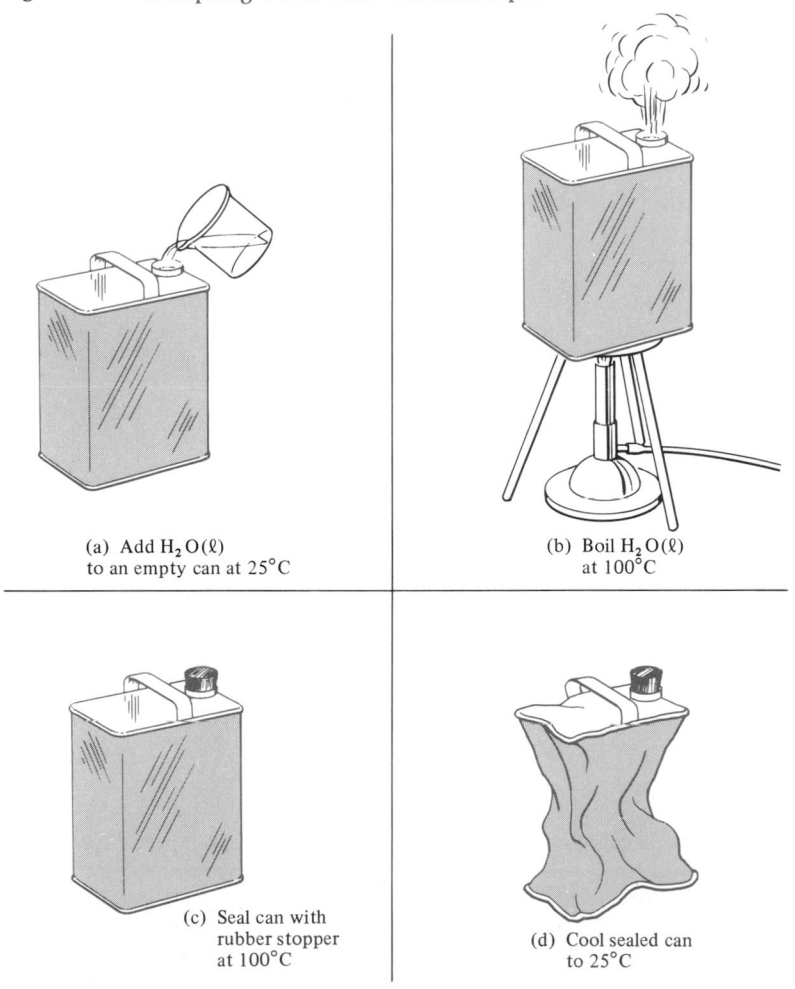

(a) Add $H_2O(\ell)$ to an empty can at 25°C

(b) Boil $H_2O(\ell)$ at 100°C

(c) Seal can with rubber stopper at 100°C

(d) Cool sealed can to 25°C

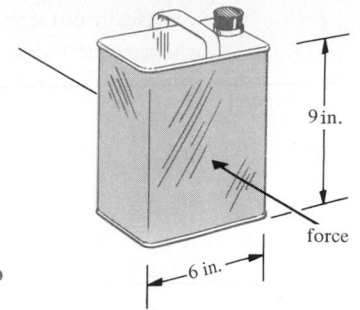

Figure 7.8. Calculation of the unbalanced force due to barometric pressure.

gas in the can, this is the total pressure, and the unbalanced pressure of the earth's atmosphere causes the can to collapse.

How great is this unbalanced pressure? The atmospheric pressure, 760 mm Hg, is equivalent to 14.7 lb in.$^{-2}$ (Sections 3.2 and 3.4). The internal pressure expressed in pounds per square inch is directly proportional to the vapor pressure, 24 mm Hg. Using the equality above to provide the necessary one-factor you may calculate the internal pressure in pounds per square inch.

$$x \text{ lb in.}^{-2} = (24 \text{ mm Hg}) \left(\frac{14.7 \text{ lb in.}^{-2}}{760 \text{ mm Hg}} \right)$$
$$= 0.46 \text{ lb in.}^{-2}$$

The unbalanced pressure is the difference between the internal pressure and the external pressure.

$$\text{unbalanced pressure} = (14.7 - 0.46) \text{ lb in.}^{-2}$$
$$= 14.2 \text{ lb in.}^{-2}$$

How great is the unbalanced force that this pressure produces? Consider the unbalanced forces exerted on the two large faces of the can, which have the dimensions shown in Figure 7.8. The force in pounds exerted on each face is directly proportional to the unbalanced pressure, 14.2 lb in.$^{-2}$ and to the area of the face, 9 in. $\times$ 6 in.

$$x \text{ lb} = \left(14.2 \frac{\text{lb}}{\text{in.}^2} \right) (9 \text{ in.})(6 \text{ in.})$$
$$= (14.2)(9)(6) \text{ lb}$$
$$= 8 \times 10^2 \text{ lb}$$

An unbalanced force of about 800 lb is exerted on each side of the can!

You live in an atmosphere that exerts enormous forces, but since these forces are usually balanced by internal forces of the same magnitude you are not aware of them.

EXAMPLE 4.   Suppose the collapsed can in Figure 7.7, which still contains a considerable amount of $H_2O(l)$, is reheated. Use the curve of vapor pressure versus temperature in Figure 7.5 to predict the temperature at which the pressure inside the can will be just twice that outside the can.

Solution: The pressure needed is $2 \times 760 = 1520$ mm Hg. Although this value lies just off the graph in Figure 7.5, the temperature required is apparently only slightly above 120°C. In Table 7.1 a value of 1489 mm Hg is given for 120°C, and in a more extended table* the value 1537 mm Hg is given for 121°C.

7.6 HEAT AND WORK: SOURCES OF ENERGY

A very useful application of heat energy, and one on which the industrialized world depends, is its conversion into work. Heat and work are equivalent to one another, and both can be expressed in calories. For example, heat can be used to boil water and develop high-pressure steam, which can be made to operate a reciprocating steam engine or steam turbine. Gasoline can be burned to produce hot gases that drive the pistons in an internal combustion engine. Energy from the metabolism of sugar in the human body can be converted into work by the muscles.

> **Historical Note:** In the years 1844 and 1845 the physician John Gorrie† discovered a method of *removing* heat by the performance of work. He used heat to run an engine that compressed air and also heated it. Then the compressed air was water cooled and, when it was allowed to expand, it became colder. This process provided air conditioning for treatment of malaria fever patients in Apalachicola, Florida. Within a year he had produced the first sample of artificial ice. Further development of Gorrie's method led to the liquefaction of air in this century. By fractional distillation of liquid air, pure oxygen can be separated and used in medical treatment, rocket propulsion, and many other ways that would have surprised the inventor of refrigeration.

Although heat is never converted 100% into work, it is useful to compare the amounts of heat available from different fuels.

EXAMPLE 1.   The principal source of heat energy for plants that generate electricity in the United States is coal. A good grade of bituminous coal found around Birmingham, which is one of the best grades in the nation, will produce 14,600 British thermal units of heat (Btu) per pound when burned completely in excess air. (*Note:* 1 Btu = 252 cal) Coal is nearly 100% carbon, and many of its properties are like those of graphite. Using the one-factor method, calculate how many Btu are produced from complete

---

* Handbook of Chemistry and Physics. Chemical Rubber Publishing Co., Cleveland, Ohio.

†Raymond B. Becker, *John Gorrie, M.D.: Father of Air Conditioning and Mechanical Refrigeration.* Carlton Press, Inc., New York, N.Y. (1972).

combustion of 1.00 lb C(graphite), and compare it with that from 1.00 lb coal.

Solution: Since the amount of heat, x Btu, is directly proportional to the amount of C(graphite), you introduce 1 lb C in the numerator. You learned in Section 7.3 that 94.1 kcal of heat is produced by the combustion of 1 mole C(graphite) in excess O_2. The amount of heat produced in Btu is directly proportional to this number of kilocalories so that the latter goes in the numerator also.

$$x \text{ Btu} = (1 \text{ lb C}) \left(\frac{94.1}{1} \frac{\text{kcal}}{\text{mole C}} \right) \cdots$$

The other equalities that you will need, and their respective one-factors, are as follows: 1 mole C = 12.0 g C, 454 g = 1 lb, 1000 cal = 1 kcal, and 1 Btu = 252 cal.

$$x \text{ Btu} = (1 \text{ lb C}) \left(\frac{454}{1} \frac{\text{g C}}{\text{lb C}} \right) \left(\frac{1}{12.0} \frac{\text{mole C}}{\text{g C}} \right) \left(\frac{94.1}{1} \frac{\text{kcal}}{\text{mole C}} \right)$$
$$\left(\frac{1000}{1} \frac{\text{cal}}{\text{kcal}} \right) \left(\frac{1}{252} \frac{\text{Btu}}{\text{cal}} \right)$$

$$= \frac{(454)(94.1)(1000)}{(12.0)(252)}$$

$$= 1.41 \times 10^4 \text{ Btu}$$

(Actually, 1.46×10^4 Btu for 1 lb coal rather than graphite.)

Environmental Note: The value for coal is a few per cent higher than for graphite. Coal contains some hydrogen, which burns to H_2O, and some sulfur, which burns to SO_2 (sulfur dioxide). The formation of the latter is unfortunate because SO_2 is very irritating to the lungs. In London, where a considerable amount of coal had been used for home heating, a deadly smog descended in 1952 which resulted in the deaths of 3500 persons from respiratory complications. Gasoline, fuel oil, and natural gas also contain sulfur, but it can be removed during the refining process. Another air pollutant derived from the combustion of coal is polycyclic organic matter (POM), which has not been completely oxidized. The National Research Council has statistical evidence that links POM with the higher incidence of deaths from lung cancer in urban areas compared with rural areas.

As a nation we must learn to live within our means of energy resources. In 1973 a plan was proposed by Dixy Lee Ray, Chairman of the Atomic Energy Commission (AEC), to achieve self-sufficiency in energy by 1980, although her panels of experts believe 1985 to be a more realistic date.* A breakdown of the proposed expenditures for research and development to achieve this goal is shown in Table 7.3. Spending at a rate of $\$2 \times 10^9$ per year from 1975 to 1980 would cost $\$10 \times 10^9$.

*The nation's energy future. *Chemical and Engineering News,* December 17, 1973.

TABLE 7.3 Research and development budget proposed to reach energy self-sufficiency in the United States by 1980

	Billions of Dollars
1. Validate the nuclear option (fission)	4.1
2. Substitute coal for oil and gas	2.2
3. Exploit renewable energy sources (solar, fusion, geothermal)	1.8
4. Conserve energy and energy resources	1.4
5. Increase domestic production of oil and gas	.5
Total	10.0

In terms of national priorities the panels recommend the following: (1) that there should be an equal effort made (a) to validate the nuclear option, (b) to substitute coal for oil and gas, and (c) to conserve energy and energy resources as short-term practical goals; and (2) that there should be a smaller but continuing effort to exploit renewable energy sources such as fusion, solar, and geothermal sources, as ideal long-term solutions.

A change in emphasis was proposed in 1975 by the ERDA*, which has now taken the place of the AEC and several other agencies. Less would be spent during the first year on development of nuclear fission and more on the disposal of radioactive wastes. There would be greater efforts directed toward solar energy, nuclear fusion, and energy conservation, which are attractive because they are nonpolluting. A number of these alternatives will be discussed in the remainder of the chapter.

Validate the Nuclear Option

About 50 nuclear fission plants are operating in this country today. Ralph Nader speaks for some concerned citizens when he points out the dangers of accident at these plants and the need for further testing, safety measures, and judicious choice of plant locations away from population centers. Nader argues that "If the public knew what the facts were and if they had to choose between nuclear reactors and candles, they would choose candles." Other equally concerned citizens such as Ralph Lapp argue that we are running out of natural gas and petroleum and that the increasing use of coal will have unacceptably harmful effects on the environment. He calculates that the chance of a nuclear accident is one in a million. "People accept the risk of death involved in traveling on a common carrier, such as an airline," he says . . . "*that* risk is about one in a million, for each flight." But even in the absence of nuclear accident there remains the serious problem of how to dispose of radioactive waste.

The "life-and-death" decisions of today are complex and many are technological in nature. A complete understanding of the problems is impossible without a knowledge of chemistry.

Substitute Coal for Oil and Gas

While our gas and oil reserves face imminent depletion, our national resources in coal are expected to last several hundred years. Considerable effort will be directed toward improvement of mining. Strip mining destroys the landscape and underground

*Energy Research and Development Administration: Energy: ERDA stresses multiple sources and conservation. *Science*, **189:369** (1975).

mining is a dangerous undertaking. In this century more than 100,000 coal miners have lost their lives in underground mines.

Many types of heat engines require liquid or gaseous fuels, and will not operate on solid coal; thus, much of the effort will be directed toward coal gasification and liquefaction. These processes have the additional advantage that sulfur, which is a health hazard, can be removed. Furthermore, cheap transportation of the liquid or gas is possible by pipeline.

Conserve Energy and Energy Resources

Aside from using less fuel oil, gas, and electricity in our homes and industries, the principal area for energy conservation is in transportation. In 1970, one quarter of all the energy consumed in the United States was consumed in transportation. Petroleum furnished 96% of this energy and petroleum, as a national resource, faces depletion.

The problem of conserving energy in transportation could be solved if we would use more efficient forms of transportation. In 1970, 55% of the transportation energy was used by automobiles. In 1971, 55% of automobile transportation energy went for short urban trips of 10 miles or less, and 56% of all commuting was done by automobiles containing only one occupant, the driver. Clearly adaptation to new living styles is essential to energy conservation.

Energy Cost of Transport

The amount of energy in kilocalories consumed by an animal or machine in moving a distance of one kilometer per kilogram weight of the animal or machine is called the **energy cost of transport.**

EXAMPLE 2. A person weighing 70 kg consumes 256 kcal of energy in walking a distance of 5.0 km. What is the cost of transport?

Solution: $$\text{Cost of transport} = 256 \text{ kcal}/(70 \text{ kg} \times 5.0 \text{ km})$$
$$= 0.73 \text{ kcal kg}^{-1} \text{ km}^{-1}$$

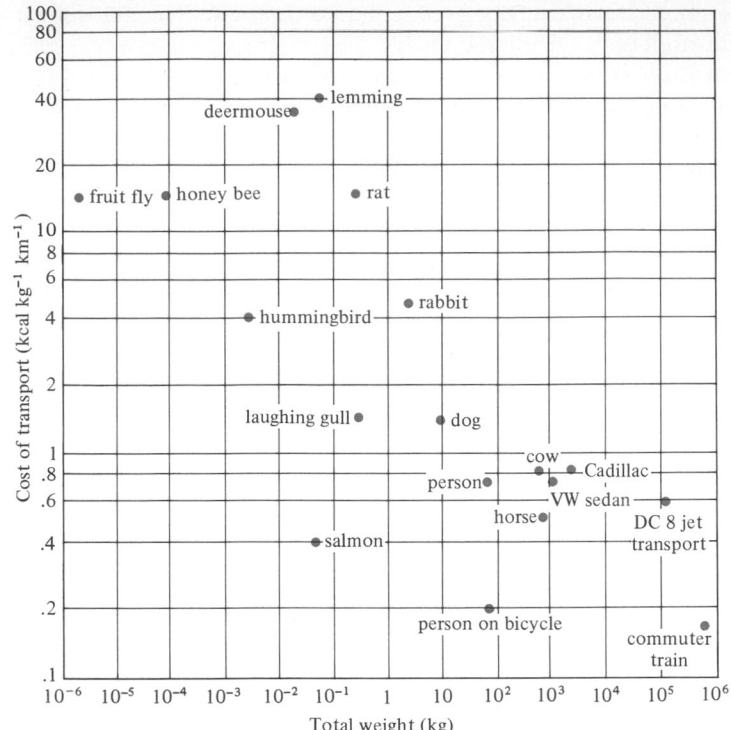

Figure 7.9. Energy cost of transport in animals and vehicles. [Adapted from S. S. Wilson: Bicycle technology. *Sci. Amer.*, 228:90 (March, 1973); and V. A. Tucker: Energetic cost of locomotion in animals. *Comp. Biochem. Physiol.*, 34:841 (1970).]

The *logarithmic* graph in Figure 7.9 gives a comparison of animals, engine-powered machines, and one animal-powered machine (the bicycle) on a double basis; the horizontal axis gives the weight in kilograms and the vertical axis gives the corresponding energy cost of transport in kilocalories per kilogram kilometer. A logarithmic graph is one in which the units are equally spaced in powers of ten; for example, . . . , 0.1, 1, 10, 100, . . . are equally spaced and . . . , 10^{-2}, 10^{-1}, 1, 10, 10^2, . . . are equally spaced. It is apparent that the heavier vehicles and animals have a smaller cost of transport but, of course, they have a greater weight to transport. It is interesting to note that a person on a bicycle (cost = 0.20 kcal kg^{-1} km^{-1})* uses only about one quarter the energy of a person walking (cost = 0.73 kcal kg^{-1} km^{-1}) to cover the same distance.

EXAMPLE 3. Compare the energy required to drive a VW sedan 1 km with the energy required to drive a Cadillac the same distance.

*Vance A. Tucker, Duke University, private communication.

Solution: From the graph in Figure 7.9, the approximate weights are 1000 kg for the VW and 2000 kg for the Cadillac. The energy costs are approximately 0.75 and 0.80 kcal kg^{-1} km^{-1} for the VW and Cadillac, respectively. Calculate the number of kilocalories, x, to move the VW 1 km. It is directly proportional to the energy cost, 0.75 kcal kg^{-1} km^{-1}, and to the distance, 1 km.

$$x \text{ kcal} = \left(0.75 \frac{\text{kcal}}{\text{kg km}}\right)(1 \text{ km}) \cdots$$

The energy, x, is also directly proportional to the weight because the heavier the vehicle the more energy is needed to move it. It is apparent also that "kg" is needed in the numerator to cancel with "kg" already in the denominator.

$$x \text{ kcal} = \left(0.75 \frac{\text{kcal}}{\text{kg km}}\right)(1 \text{ km})(1000 \text{ kg})$$

$$= 750 \text{ kcal for the VW sedan to travel 1 km}$$

You can solve for the Cadillac in a similar manner.

$$x \text{ kcal} = \left(0.80 \frac{\text{kcal}}{\text{kg km}}\right)(1 \text{ km})(2000 \text{ kg})$$

$$= 1600 \text{ kcal for the Cadillac to travel 1 km}$$

In order to compare the amount of energy on a passenger–distance basis, it is necessary to divide results like those in Example 7.13 by the number of passengers in the vehicle. In Table 7.4 a number of conveyances are compared on an energy per passenger–distance basis.

An advantage of the commuter train is that it can be converted to the use of electrical power which at least keeps the pollution centralized at the power plant. Of course, the *bicyclist* is already nonpolluting.

TABLE 7.4 Energy requirements for moving one passenger a distance of 1 km*

Conveyance	Weight (kg)	Number of Passengers	Energy Cost of Transport (kcal kg^{-1} km^{-1})	Energy per Passenger-Kilometer (kcal psngr^{-1} km^{-1})
DC 8 jet transport	1.07×10^5	195	0.57	310
Cadillac	2.21×10^3	6	0.83	310
VW sedan	1.01×10^3	4	0.73	180
commuter train (diesel)	8.6×10^5	700	0.15	180
person on a bicycle	90	1	0.20	18

*Adapted from V. A. Tucker: *Energetic cost of locomotion in animals.* Comp. Biochem. Physiol., **34**:841 (*1970*).

EXAMPLE 4. It would be interesting to compare how far a bicyclist could go on a 5-lb bag of sugar with the distance a car goes on 1 gal of gasoline. The latter has a weight of approximately 5.6 lb.

Solution: Consider a 120-lb coed using a 30-lb bicycle and use 1350 kcal mole^{-1} for the heat of combustion of sugar (sucrose) from Example 2 (page 149) to obtain the following setup for distance traveled.

$$x \text{ mi} = (5 \text{ lb sugar})\left(\frac{454 \text{ g}}{1 \text{ lb}}\right)\left(\frac{1 \text{ mole}}{342 \text{ g}}\right)\left(\frac{1350 \text{ kcal}}{1 \text{ mole}}\right)\left(\frac{\text{kg km}}{0.20 \text{ kcal}}\right)\left(\frac{5 \text{ mi}}{8 \text{ km}}\right)$$
$$\left(\frac{1 \text{ lb}}{0.454 \text{ kg}}\right)\left(\frac{1}{150 \text{ lb}}\right)$$

$$x = 410 \text{ mi}$$

This is considerably farther than she could go on 1 gal of gasoline! Note the useful equality, 5.0 mi = 8.0 km.

**Conservation
of Energy:
a Solar Water
Heater**

The amount of solar energy received in the United States each year is 9×10^{15} kilowatt-hours (kw-hr), which is equivalent to the energy obtainable from 1.2×10^{12} tons of coal. Since the total amount of coal reserves in the United States and North America is only 1.6×10^9 tons[*], it is obvious that if we could convert only a small fraction of the sun's energy into usable forms, many of our energy problems would be solved.

Solar energy is available only part of the day; thus, it is necessary to find ways of storing it for use at other times. The home water heater is based on the principal of energy storage. Normally gas or electricity is used to produce the energy, which is stored in the water to be used when needed.

A low cost "do-it-yourself" design for a solar water heater is shown in Figure 7.10. The solar heater is mounted on the roof of the house. A southerly slant about 10° to 20° greater than the angle of latitude is the most desirable tilt.[†] Water is circulated between the heater and the storage tank, which is located at a higher level on the roof. When water is drawn out of the old water heater, it is replaced by solar heated water from the storage tank. At the same time water from the city supply or your well moves into the storage tank and solar heater ready for circulation.

The principal advantages and design features of this system are

1. The solar heater consists of a sheet of copper that collects the sun's radiation and passes it into copper tubing, which is soldered in place for good thermal contact. Water circulates through the copper tubing. Sheet and tubing are painted black (heat resistant paint) for more efficient absorption of the sun's rays. The box enclosure is covered with glass to prevent heat loss by radiation and air circulation.

---

[*] Hans Rau: *Solar Energy.* Macmillan, New York 1964, p. 164.

[†] J. Richard Williams: *Solar Energy Technology and Applications.* Ann Arbor Science, Ann Arbor, Mich. 1974, p. 26.

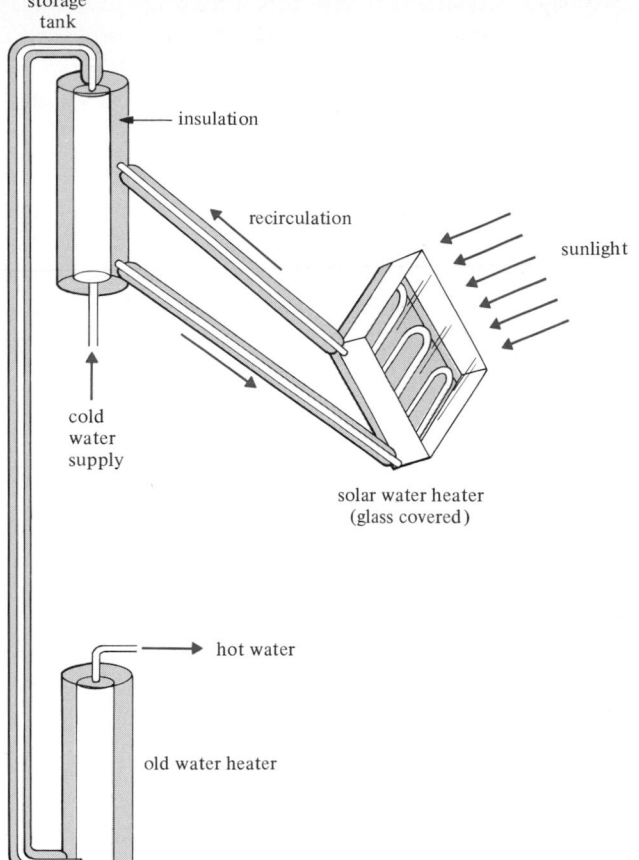

storage
tank

insulation

recirculation

sunlight

cold
water
supply

solar water heater
(glass covered)

hot water

old water heater

Figure 7.10. Solar water heater.

2. No water pump is needed because flow is accomplished by differences in density. The hot water in the solar heater flows to the upper part of the storage tank because hot water is less dense than cold water. The coolest water in the tank flows out the bottom into the lower side of the solar heater. Recirculation continues as long as the sun shines.

3. Circulation stops automatically when the sun is not shining, and the water in the solar heater becomes cooler than the water in the storage tank. Hot water stays in the highest part of the tank that it can reach.

4. When hot water is drawn out of the old water heater, it is replaced by water from the upper portion of the storage tank, where it is the hottest. Cold water from the city or your well enters the lower part of the storage tank.

5. If all your hot water is depleted because of weather conditions or unusually large usage, the thermostat which is installed in your old water

heater automatically turns on your old heating system. A well-insulated water heater normally remains turned off once the temperature reaches the value set on your thermostat.

6. Good insulation around pipes and behind the solar collector will minimize heat losses to the surroundings.

A much greater potential exists for solar energy than simply storing its heat. There will be many research projects directed toward converting solar energy into electricity both directly and through the use of wind power and warm ocean layers, which depend on the sun. A number of special laboratories such as the Florida Solar Energy Center at Cape Canaveral, Florida, and individual scientists are hopeful that solar research will help us become self-sufficient in energy.

1. A Fahrenheit thermometer in a calorimeter undergoes a temperature change of 8.5°F during a chemical reaction. How much is this increase on the Celsius scale? (*Hint:* From the graph in Figure 2.5 it is apparent that 100 degree change on the Celsius scale is equivalent to 180 degree change on the Fahrenheit scale. Use this equivalency to set up a one-factor.)

2. A 10-g sample of $H_2O(l)$ has its temperature raised 10°C by 100 cal heat. How much is this temperature increase on the Fahrenheit scale? (See hint in problem 1.)

3. How much heat must be added to change the temperature of each of the following. (a) 500 g liquid water from 0 to 100°C. (b) 500 g liquid metallic sodium from 100 to 200°C (specific heat = 0.32 cal g^{-1} °C^{-1}).

4. How much heat must be removed to change the temperature of each of the following. (a) 2 kg liquid water from 25 to 0°C. (b) 2 kg liquid chloroform from 25 to 0°C (specific heat = 0.23 cal g^{-1} °C^{-1}).

5. If exactly 3000 cal of electrical heat was released in a certain bomb calorimeter, the temperature increased 1.532°C. In order to determine the heat value of a sample of oak wood, a moist sample weighing 1.75 g was burned under a high pressure of oxygen in the calorimeter. The heat of reaction caused a 3.57°C rise in temperature. Calculate the heat of combustion of the moist oak sample in kilocalories per gram.

6. The electrical calibration of a certain bomb calorimeter indicates that its temperature increases exactly 1.000°C for every 1850 cal of heat added. It is desired to measure the heat value of a North Dakota lignite, a low grade coal in which the pattern of the wood grain is still visible. If the combustion of a 1.20-g sample causes the temperature in the same calorimeter to rise 2.16°C, calculate the heat value of lignite in kilocalories per gram.

7. The combustion of glucose, also called dextrose ($C_6H_{12}O_6$), produces 673 kcal mole^{-1}. Calculate the heat of combustion of glucose in kilocalories per gram and compare with the heat of combustion of 1.00 g sucrose (Section 7.2, Example 2).

8. The combustion of ethanol, or ethyl alcohol (C_2H_5OH), produces 328 kcal mole^{-1}. Calculate the heat of combustion of ethanol in kilocalories per gram and compare with the heat of combustion of 1.00 g sucrose (Section 7.2, Example 2).

9. From the following heats of reaction,

$$2 H_2(g) + O_2(g) \longrightarrow 2 H_2O(l) + 136.6 \text{ kcal}$$

$$H_2(g) + O_2(g) \longrightarrow H_2O_2(l) + 44.9 \text{ kcal}$$

calculate the heat of the following reaction.

$$2 H_2O_2(l) \longrightarrow 2 H_2O(l) + O_2(g)$$

Is this reaction endothermic or exothermic? Would you expect $H_2O_2(l)$, which is called hydrogen peroxide, to decompose into $2 H_2O(l)$ and $O_2(g)$ on the basis of heat of reaction alone? Explain.

10. From the following heats of reaction,

$$S(s) + O_2(g) \longrightarrow SO_2(g) + 70.9 \text{ kcal}$$

$$2 S(s) + 3 O_2(g) \longrightarrow 2 SO_3(g) + 189.2 \text{ kcal}$$

calculate the heat of the following reaction.

$$2 SO_2(g) + O_2(g) \longrightarrow 2 SO_3(g)$$

Considering only the heat of this reaction, would you predict that $SO_2(g)$ could be converted to $SO_3(g)$ in this way? Explain.

11. Consider the graph of temperature versus heat absorbed by water in Figure 7.2. (a) What point on the graph corresponds to a mixture of $\frac{1}{2}$ g $H_2O(l)$ and $\frac{1}{2}$ g $H_2O(g)$? (b) Use the horizontal axis to estimate how much heat must be added to 1 g $H_2O(s)$ at $0°C$ to reach the point of the mixture in part (a) above.

12. The words "hot" and "cold" refer to temperature, but it is possible to receive different "degree" burns by coming in contact with substances having the *same* temperature. (a) In what sense is 1 g $H_2O(g)$ at $100°C$ "hotter" than 1 g $H_2O(l)$ at $100°C$? (b) In what sense is one just as "hot" as the other?

13. The barometer in Figure 7.3 is used on a very warm day ($35°C$) when the barometric pressure is 758 mm Hg to measure the vapor pressure of H_2O. (a) If $h = 716$ mm Hg, what is the vapor pressure? (b) Compare this value with those from Table 7.1 at $30°C$ and $40°C$. Is your answer reasonable?

14. Diethyl ether, commonly used as an anesthetic, has a vapor pressure of 537 mm Hg at $25°C$. (a) Use the barometric method in Figure 7.3 to measure this vapor pressure. What would be the height of the Hg column, h, if barometric pressure is 762 mm Hg? (b) Could this method be used in an airplane in which the barometric pressure is 500 mm Hg? Explain. Assume a temperature of $25°C$ in parts (a) and (b).

15. Use the open-ended manometer illustrated in Figure 7.4 to recalculate the differences in Hg levels in experiments (a), (b), and (c) on a day when the barometric pressure is 768 mm instead of 756 mm.

16. Use the open-ended manometer illustrated in Figure 7.4 to measure vapor pressure and estimate the difference in height of Hg columns at the following temperatures. (a) $70°C$ (b) $100°C$. Use the graph in Figure 7.5 to obtain any vapor pressures needed. Assume a barometric pressure of 756 mm.

17. Use the vapor pressure data for chloroform given in the following table to find the approximate normal boiling point.

18. Use the vapor pressure data for diethyl ether given in the following table, to find the approximate normal boiling point.

chloroform	
t (°C)	vp (mm Hg)
20	160
30	246
40	366
50	526
60	740
70	1020

diethyl ether	
t (°C)	vp (mm Hg)
10	292
20	442
30	647
40	921
50	1277

19. Use the graph in Figure 7.5 to estimate the vapor pressure of water (a) at 70°C, (b) at 90°C, and (c) at 110°C.

20. Use the graph in Figure 7.5 to estimate the boiling point of water at altitudes having barometric pressures of (a) 600 mm, and (b) 500 mm.

21. A beaker partly filled with $H_2O(l)$ is covered loosely with a glass plate so that the air space is saturated with water vapor. If the barometric pressure is 752 mm Hg and the temperature is 40°C, what is the partial pressure of air in the beaker?

22. Same as problem 21, but at 80°C.

★**23.** How many grams of ice at 0°C can be melted by adding it to 500 g $H_2O(l)$ at 100°C? *Hint:* What will be the final temperature?

★**24.** The density of NaCl(s), sodium chloride, is 2.17 g cm^{-3}. (a) Calculate the volume occupied by 1 mole NaCl(s). (b) What is the volume occupied by each atom? Remember that 1 mole NaCl contains 2 moles of atoms. (c) Each atom in NaCl can be considered to be located at the center of a small cube surrounded on all sides by cubes occupied similarly by other atoms. The distance between atoms is simply the length of this cube's edge. Calculate the distance between atoms in Ångstroms. The value determined by x-ray diffraction measurements is 2.81 Å.

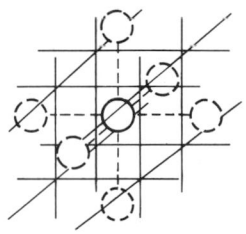

★**25.** How many miles can a 170-lb man go on a 30-lb bicycle using the energy from one tablespoon of honey? The "calorie chart" of the nutritionist† gives the heat value for this amount of honey as 100 Calories, which is, of course, equivalent to 100 kcal in the chemist's terminology.

† Irma S. Rombauer and Marion Rombauer Becker: *Joy of Cooking.* Bobbs Merrill Co., Inc., New York, 1952.

8
SOLUTIONS

Water is a good solvent, but not a universal solvent. The run-off of rain from the land has carried into the oceans vast quantities of valuable assets, as well as harmful wastes, in the form of solutes. But obviously not everything is soluble in water. Why do some substances dissolve in water readily and others only slightly?

DEFINITIONS: *A **solution** is a homogeneous mixture of two or more substances* (see Figure 1.4). *The **solvent** is one of the substances, usually the one in excess. The **solutes** are the other substances. An **aqueous solution** is one in which water is the solvent.*

It is important to know the amount of solute relative to the amount of solvent, called the **concentration** of the solution. If you are driving a car, your BAC (blood alcohol concentration) could involve you in an accident or an unpleasant encounter with the law. If you are farming in the Mexicali Valley of Mexico, the salinity of the water which the Americans send to you down the Colorado River will determine whether your crops are damaged or sustained. If you are adding fluoride solution to your drinking water, the concentration will make the difference between fluorosis and normally white decay-resistant teeth.

8.1 KINDS OF SOLUTIONS

There are several kinds of homogeneous mixtures of substances based on the three physical states of matter. The simplest case is that of a **mixture of gases.** Air, for example, is a homogeneous mixture of gases and could be called a solution. The same is true for any mixture of gases. Solutions of **gases in liquids** are well known, and the greater the pressure of the gas the greater is its concentration in the liquid. At a given pressure of 1 atm, different gases have different **solubilities** in the same liquid.

DEFINITION: **Solubility** *is the maximum concentration of a solute in a given solvent. It is attained when an excess of solute has been added and is in contact with the solution. The solution is then called* **saturated.**

The remarkably different solubilities of $NH_3(g)$ and $O_2(g)$ in aqueous solutions will be discussed in Section 8.2. Solutions of **solids in liquids** and **liquids in liquids** are too numerous to mention, but solutions of **solids in solids** are rare. British sterling silver is a solid solution, an alloy of copper in silver in the proportion of 7.5 g copper to 100.0 g sterling.

8.2 SOLUBILITY

The solubility of $O_2(g)$ in water is important to fish and other aquatic animals, which depend on it for their respiration. This solubility is very small—only 30 ml $O_2(g)$, approximately, per liter of water under 1 atm of pure $O_2(g)$. On a weight basis this is only 0.004 weight per cent (0.004 g O_2 per 100 g solution). Nevertheless, this small concentration of oxygen is crucial, and a lake which has "died" by *eutrophication* has lost its dissolved oxygen.

Another gas, ammonia (NH_3), is quite soluble in water. Such solutions are commonly used as household cleaners. Under 1 atm of pure NH_3, 1 l of water will dissolve approximately 700,000 ml ammonia gas. Why is NH_3 so much more soluble than O_2 in water?

The significant solubility of ammonia in water can be explained in terms of energy: the process of dissolving is *exothermic*.

$$NH_3(g) + aq \longrightarrow NH_3(aq) + 8.5 \text{ kcal} \qquad (8.1)$$

The symbol "aq" stands for an infinite amount of water as solvent. It is understood, as usual, that the 8.5 kcal is for 1 mole of NH_3.

Dissolving $O_2(g)$ in water produces an almost negligible amount of heat. As discussed in Section 7.3, strongly exothermic processes are usually strongly spontaneous, and the example in equation (8.1) is in agreement with the general rule.

The reason for the exothermic heat of solution in equation (8.1) is that NH_3 and H_2O are both polar molecules (See Figure 6.4). There is an attraction between solute and solvent molecules resulting from the positive side of one molecule rotating to meet the negative side of the other molecule, and a large amount of heat is given off. On the other hand, O_2, which is nonpolar, is not strongly attracted to the polar H_2O molecule, and only a small amount of heat is produced.

It is generally true that polar solutes tend to dissolve in polar liquids and nonpolar solutes do not. Paraffin (typically $C_{25}H_{52}$) is a long chain molecule consisting of units of

$$-\overset{\displaystyle H}{\underset{\displaystyle H}{\overset{|}{\underset{|}{C}}}}-$$

Since carbon and hydrogen lie so close together in metallic/nonmetallic properties (Section 6.6 Polar Bonds), the paraffin molecule is nonpolar. As a result paraffin is not soluble in water. Sucrose ($C_{12}H_{22}O_{11}$), on the other hand, consists of repeating

$$-\overset{\displaystyle O-H}{\underset{\displaystyle H}{\overset{|}{\underset{|}{C}}}}-$$

and the polar character of the O—H bonds makes the molecule polar. Sucrose, which is cane or beet sugar, is strongly soluble in water.

EXAMPLE 1. Predict the solubility of liquid CCl_4 (carbon tetrachloride), $N_2(g)$, $HCN(g)$ (hydrogen cyanide), and liquid H_3COH (methyl alcohol) in water.

  Solution: CCl_4 and N_2 are only slightly soluble because they are nonpolar. HCN and H_3COH are soluble because they are polar. (The latter is polar because of the polar O—H bond.)

The energy effect of dissolving is not enough to explain all the facts known about dissolving, or about other spontaneous processes in nature. This becomes apparent when you consider the number of examples of solvent-solute interactions that are *endothermic*. There are many examples among aqueous solutions of ionic compounds. Although some solutes dissolve in water and produce such a large amount of heat that the solution begins to boil (for example, $AlCl_3$), other heats of solution are so endothermic that moisture from the air condenses on the outside walls of the container and eventually forms frost (for example, NH_4NO_3, ammonium nitrate).

Water is a good solvent for ionic compounds. In the absence of water a very high temperature is necessary to make the ions of an ionic solid conduct electricity (Figure 6.7). However, that same solid becomes a good conductor of electricity when dissolved in water at room temperature.

In addition to ionic compounds, some polar covalent compounds are good conductors of electricity when they are dissolved in water. Compounds are classified as electrolytes or nonelectrolytes based on their ability or inability to conduct electricity when dissolved in water.

DEFINITION: *An **electrolyte** is a compound that conducts electricity in aqueous solution.*

Most ionic compounds, like the white solids NaCl, $CaCl_2$, $AlCl_3$, and AgCl, are classified as **strong electrolytes;** that is, they are good conductors of electricity relative to the amount of solid that is dissolved. In some cases, like that of the very slightly soluble silver chloride (AgCl), there is hardly any conduction of electricity because there are so few ions, but the small amount that does dissolve dissociates completely into ions and conducts electricity as well as strong electrolytes at the same level of concentration. Therefore, AgCl is considered a strong electrolyte.

Some aqueous solutions, such as sugar and alcohol in water, do not conduct electricity at all. Others, such as acetic acid and hydrogen sulfide in water, conduct electricity poorly, even at high concentrations. Although these molecules are quite soluble in water, the sugar and alcohol do not break down or dissociate into ions and, in the cases of acetic acid and hydrogen sulfide, only a small percentage of molecules dissociate. Dipping the probes of the conductivity tester (Figure 6.6) into one of these solutions will cause only a faint glow or no light at all. If the solute dissolves in water but does not conduct, it is called a **nonelectrolyte;** if it conducts poorly, it is called a **weak electrolyte.** This information is summarized in Table 8.1.

TABLE 8.1 Electrolytes versus nonelectrolytes

Type of Solute	Ability of Aqueous Solutions to Conduct Electricity	Examples
strong electrolyte	good	NaCl, $CaCl_2$, $AlCl_3$, AgCl (very slightly soluble)
weak electrolyte	poor	CH_3COOH (acetic acid), NH_3, H_2S, H_2O
nonelectrolyte	none	$C_{12}H_{22}O_{11}$ (sucrose) CH_3CH_2OH (ethyl alcohol)

Effects of Disorder and the Theory of Strong Electrolytes

Around 1890 chemists were surprised to hear from the Swedish chemist Arrhenius that strong electrolytes such as NaCl dissolve in $H_2O(l)$ by dissociating 100% into Na^+ ions and Cl^- ions. You have already seen (Section 6.8) that in the absence of water NaCl(s) must be heated to 808°C before it melts and only then do the ions become free to move and conduct electricity. This high temperature is necessary because melting is an endothermic process.

$$6.8 \text{ kcal} + NaCl(s) \longrightarrow Na^+(l) + Cl^-(l) \tag{8.2}$$

In $H_2O(l)$ the dissociation into ions is still endothermic,

$$1.0 \text{ kcal} + NaCl(s) + aq \longrightarrow Na^+(aq) + Cl^-(aq) \tag{8.3}$$

but not nearly so endothermic as in the absence of $H_2O(l)$. The symbol "aq" stands for a large excess of $H_2O(l)$ and in this case implies an infinitely dilute solution of NaCl. The endothermic heat is small because the interaction of the Na^+ and Cl^- ions with the H_2O, once they are dissociated, is exothermic. Since the H_2O molecule is polar, there is an attraction between it and the ions. The positive side of the molecule attracts the Cl^- ion and the negative side attracts the Na^+ ion. In spite of this attraction, dissociation in water would not occur spontaneously if there were not another phenomenon besides heat which affects the occurrence of natural processes. Something in nature likes a solution.

This additional "driving force" is **disorder.** If enough disorder is produced in a process, it will occur even if it is endothermic. Breaking down the crystal lattice of NaCl(s) and the subsequent free movement of the ions introduces considerable disorder in the system, enough to overcome the small endothermic heat. For the same reason an easily vaporized liquid, like ether, evaporates spontaneously and rapidly at temperatures below its boiling point even though it becomes cooler in the process. The more freedom the molecules have to move about, the greater is their disorder. This "disorder" has a scientific name and a precise mathematical definition. It is called **entropy** and it is defined in terms of probability.

EXAMPLE 2. Predict the relative solubilities of NaCl(s), $CaCl_2$(s), and $AlCl_3$(s) in water based on their respective heats of solution 1.0 kcal $mole^{-1}$ (endothermic), 18.0 kcal $mole^{-1}$ (exothermic), and 79.9 kcal $mole^{-1}$ (exothermic) and relative changes in disorder.

Solution: Since the greater the amount of heat given off in a process the greater is its tendency to occur, energy effects alone would predict the following sequence of increasing solubilities: $NaCl < CaCl_2 < AlCl_3$. The amount of disorder produced by dissolving increases in the same sequence since the number of moles of ions dissociated per mole of solid is 2, 3, and 4, respectively. Therefore from the changes in disorder, you would predict the same sequence of increasing solubilities. This sequence has been confirmed by experiment.

The theory of strong electrolytes may be summarized as follows: When strong electrolytes dissolve, they dissociate completely into ions. For the theory of weak electrolytes the following may be given: When weak electrolytes dissolve, they dissociate only partially into ions. This theory of Arrhenius gradually gained acceptance over a 10-year period after it was introduced, and is now generally accepted by chemists everywhere. Frequently, however, it is necessary to make corrections that take into consideration the attraction that positive and negative ions have for one another even after dissociating.

To detect the ions produced by very slightly soluble compounds, such as AgCl, and unusually weak electrolytes, special types of conductivity testers are needed. Pure water, for example, will conduct electricity, but not enough to cause a glow in the light globe. Water is considered a weak electrolyte. The weak electrolytes and their ionic interactions will be discussed in Chapters 10 and 11.

Other Examples of Disorder and Solubility

An ionic compound, such as NaCl, is practically insoluble in a nonpolar solvent, such as carbon tetrachloride or gasoline (a hydrocarbon). The advantage of the disorder produced by dissociating the NaCl into free ions is not enough to overcome the disadvantage of the large endothermic heat that must be provided. On the other hand nonpolar liquids are always **miscible** with (soluble in) other nonpolar liquids because there is more disorder in the mixture than in the separate pure liquids and there are no large energy effects.

> **Thought question:** Can you imagine any disadvantage of storing gasoline (typically C_8H_{18}) in polyethylene containers? Polyethylene is a plastic consisting of very long molecules, called polymers, composed of repeating $-CH_2-$ units.

8.3   WORKING WITH SOLUTIONS: DENSITY, WEIGHT PER CENT, AND MOLARITY

Chemical reactions are usually carried out in solution because of the ease with which you can mix the reactants completely and proportionately. A large family of chemical glassware has been developed over the years to measure the volumes of solutions accurately and to transfer known amounts from one place to another. A few typical pieces of apparatus are shown in Figure 8.1. In addition to this equipment, a balance such as that shown in Figure 1.2 is often needed for weighing. When chemists work with solids, they know how many moles they have by measuring the weight of the solid on a balance. The number of moles of a pure liquid can be determined from its volume, if the density is known. For solutions, both the density of the solution and the weight per cent of the solute are needed in order to calculate the number of moles from the volume.

Weight Per Cent

You should learn how to calculate the weights needed to prepare solutions of desired weight per cent.

EXAMPLE 1.   How would you prepare 400 g British sterling silver from pure copper (Cu) and silver (Ag)? This alloy has the composition 7.5% Cu and 92.5% Ag by weight.

Solution: There are two unknowns in this problem: x g Cu and y g Ag. First you may calculate the weight of copper needed, which is directly proportional to the weight of alloy to be prepared.

$$x \text{ g Cu} = (400 \text{ g alloy}) \cdots$$

The only one-factor needed comes from the equivalency of 7.5 g Cu and 100 g alloy.

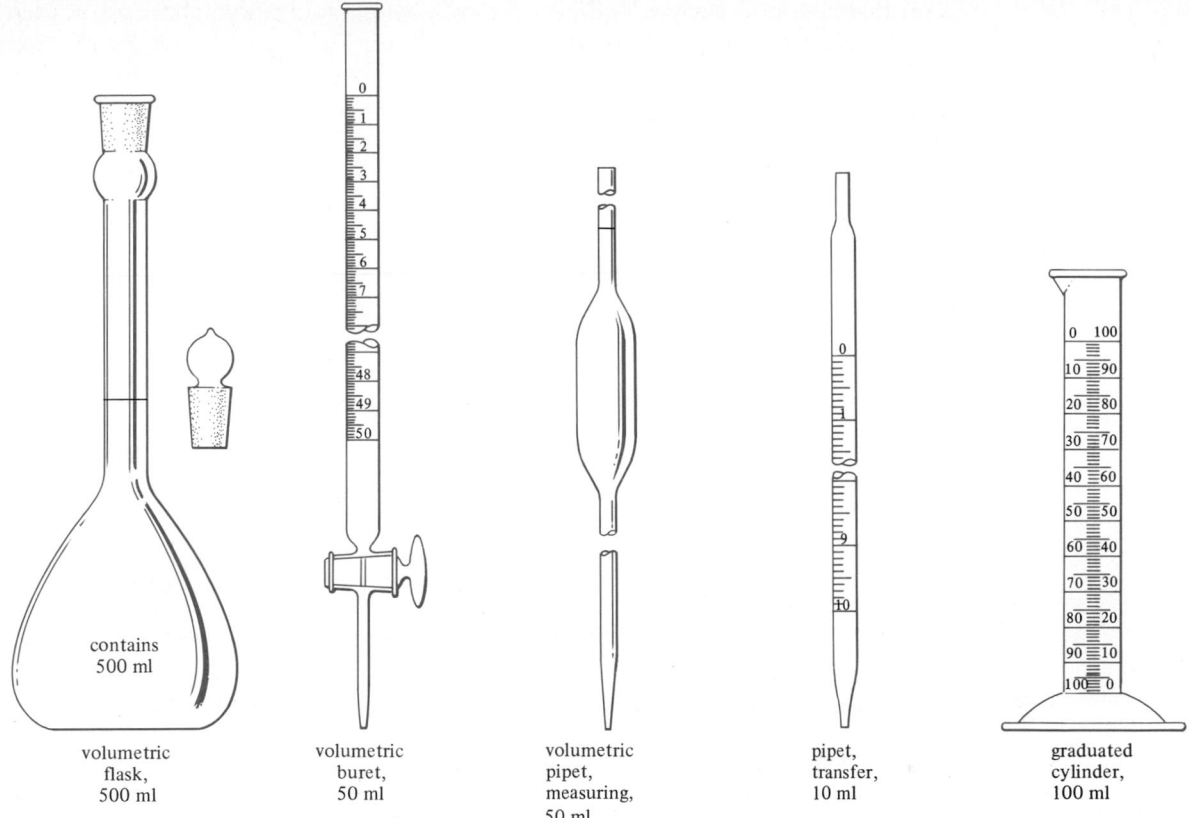

| volumetric flask, 500 ml | volumetric buret, 50 ml | volumetric pipet, measuring, 50 ml | pipet, transfer, 10 ml | graduated cylinder, 100 ml |

Figure 8.1. Volumetric glassware for preparing and dispensing solutions.

$$x \text{ g Cu} = (400 \text{ g alloy}) \left(\frac{7.5}{100} \frac{\text{g Cu}}{\text{g alloy}} \right)$$

$$= \frac{(400)(7.5)}{100} \text{ g Cu}$$

$$= 30 \text{ g Cu}$$

The setup for calculating the weight of silver is

$$y \text{ g Ag} = (400 \text{ g alloy}) \left(\frac{92.5}{100} \frac{\text{g Ag}}{\text{g alloy}} \right)$$

$$= 370 \text{ g Ag}.$$

As a check show that the total weight is 400 g.

$$x + y = 400 \text{ g}$$
$$30 \text{ g} + 370 \text{ g} = 400 \text{ g}$$
$$400 \text{ g} = 400 \text{ g}$$

In the next example you will be working with an aqueous solution.

EXAMPLE 2. If you were found to be DWI (driving while intoxicated) in England, the mandatory punishment would be a 1-year suspension of your driver's license and possibly a $250.00 fine or 4 months imprisonment or both. The proof of intoxication is based on the BAC (blood alcohol concentration) in weight per cent; if it is over 0.08%, you are legally intoxicated. The average 160-lb person has 14 lb blood. How many ounces alcohol in the blood of a 160-lb person will cause intoxication?

Solution: The unknown quantity is x oz alcohol, which is directly proportional to the weight of blood.

$$x \text{ oz alcohol} = (14 \text{ lb blood}) \cdots$$

Use the equivalency of 0.08 oz alcohol and 100 oz blood* and the equality, 16 oz = 1 lb, for the needed one-factors.

$$x \text{ oz alcohol} = (14 \; \cancel{\text{lb blood}}) \left(\frac{0.08 \text{ oz alcohol}}{100 \; \cancel{\text{oz blood}}} \right) \left(\frac{16 \; \cancel{\text{oz blood}}}{1 \; \cancel{\text{lb blood}}} \right)$$

$$= \frac{(14)(0.08)(16)}{100} \text{ oz alcohol}$$

$$= 0.18 \text{ oz alcohol}$$

Note on the Importance of BAC: Statistics on the possible causes of automobile accidents have been maintained for a number of years. A significant study in New York City showed that nearly half of all drivers in *fatal* crashes had BAC levels of 0.25% or higher. The statistical risk of a driver becoming involved in a crash begins to increase at about 0.05% BAC. Since only about 15% of alcohol consumed finds its way into the blood stream, a 160-lb person must drink two 12-oz cans of beer in an hour to reach this level. The per cent alcohol vapor in the breath is proportional to the BAC, a fact that has made it possible for police departments to use the convenient breath test. The breath test is not too reliable unless it is made on alveolar air (the last puff of exhaled breath), which is representative of alcohol vapor in the lungs at the interface with the blood stream. Ethyl alcohol is considered lethal at a BAC of 0.55%.

In the following example the solution is considerably more concentrated than in Example 2, and weight of *solution* must be used instead of *solvent*.

EXAMPLE 3. Calculate the weight of ammonia (NH_3) gas dissolved in 1.000 kg aqueous solution which is 29.5% by weight NH_3. This is considered a "concentrated" aqueous solution of ammonia because it is approaching the solubility of ammonia under 1 atm pressure at room temperature. Ammonia is frequently sold in this form because of the convenience of

---

*To be exact 0.08 oz of alcohol is equivalent to 100 oz of *solution* (blood + alcohol), but with so little alcohol the approximation is justified.

183

8.3
*Working with
Solutions: Density,
Weight Per Cent,
and Molarity*

handling. If pure NH_3 is desired, it must be purchased in a gas cylinder under pressure.

Solution: The weight of NH_3 needed is directly proportional to the weight of solution.

$$x \text{ g } NH_3 = (1.000 \text{ kg solution}) \cdots$$

The weight per cent means that 29.5 kg NH_3 is equivalent to 100 kg solution, which provides the needed one-factor.

$$x \text{ kg } NH_3 = (1.000 \text{ kg solution}) \left(\frac{29.5}{100} \frac{\text{kg } NH_3}{\text{kg solution}} \right)$$

$$= 0.295 \text{ kg } NH_3$$

Density

An aqueous solution of NH_3 or some other solute, prepared commercially for distribution to chemical laboratories, will have a label on the bottle giving the weight per cent, sometimes called the "assay." In addition to this the density or "specific gravity," is usually given. From these data, a known weight of solute can be calculated from a measured amount of volume. Suppose that this information were not given. Could you determine the density yourself for future reference? Which piece of equipment illustrated in Figure 8.1 would you use?

EXAMPLE 4. If you weigh the 500-ml volumetric flask shown in Figure 8.1 before and after filling with the 29.5% NH_3 solution in Example 3, you will obtain a solution weight of 449 g. What is the density of 29.5% NH_3 in water? Could other equipment be used for the same determination?

Solution: The definition of density is given in Section 2.4 and you may substitute in it as follows.

$$\text{density} = \frac{\text{weight}}{\text{volume}} = \frac{449}{500} \frac{\text{g}}{\text{ml}} = 0.898 \text{ g ml}^{-1}$$

Yes, other equipment could be used. See the discussion in the following paragraph.

Use of Volumetric Glassware

It should be noted when selecting volumetric glassware whether it is marked TC (to contain) as the volumetric flask, or TD (to deliver) as other pieces are frequently marked. When water is drained from a clean glass container, a thin unbroken film of liquid clings to the glass. In the case of the 100 ml graduated cylinder this could amount to as much as 0.2–0.3 ml. The correction is not important unless the highest precision is desired. Pipets and burets are usually made "to deliver." Thus the 50-ml buret, the 50-ml volumetric pipet, and the 10-ml pipet could all be used in Example 4, but the ammonia solution would have to be emptied into a previously weighed container. Graduated cylinders are usually made "to deliver" and the one marked TD in Figure 8.1 is no exception.

Exercise 1: If the 100-ml graduated cylinder in Figure 8.1 were used to determine the NH_3 solution density in Example 4 in the same manner as the volumetric flask, would the calculated density be too high or too low? Explain.

Answer: ·əɯnloʌ pəlǝqɐl ǝɥʇ uɐɥʇ ǝɹoɯ ɹoɟ sı pǝɹnsɐǝɯ ʇɥᵷıǝʍ ǝɥ⊥ ·ɥᵷıɥ oo⊥

Using Density and Weight Per Cent Together

Both density and weight per cent of a solution are needed when you want to calculate the number of moles of solute in a given volume of solution.

EXAMPLE 5. When HCl(g) (hydrogen chloride) is dissolved in H_2O, the solution is known as hydrochloric acid. In the usual preparation, a concentrated aqueous solution containing 37% HCl by weight is obtained, which has a solution density of 1.19 g ml^{-1}. Calculate (a) the weight of 1.00 l concentrated HCl, (b) the weight of pure HCl in 1.00 l concentrated HCl, and (c) the numbers of moles HCl in 1.00 l concentrated HCl.

Solution: When you use the one-factor method it is important to differentiate between *pure HCl* (a molecule) and *concentrated HCl* (a solution).

(a) The weight of concentrated HCl, x g, is proportional to its volume.

$$x \text{ g conc. HCl} = (1.00 \ l \text{ conc. HCl}) \cdots$$

The equivalency of 1.19 g and 1 ml concentrated HCl, from the solution density, and the equality of 1 l and 1000 ml, give the needed one-factors.

$$x \text{ g conc. HCl} = (1.00 \ \cancel{l \text{ conc. HCl}})\left(\frac{1000 \ \cancel{\text{ml conc. HCl}}}{1 \ \cancel{l \text{ conc. HCl}}}\right)\left(\frac{1.19 \ \text{g conc. HCl}}{1 \ \cancel{\text{ml conc. HCl}}}\right)$$
$$= (1000)(1.19) \text{ g conc. HCl}$$
$$= 1.19 \times 10^3 \text{ g conc. HCl}$$

(b) Begin with the weight of concentrated HCl obtained in part (a), which is proportional to y g pure HCl.

$$y \text{ g pure HCl} = (1.19 \times 10^3 \text{ g conc. HCl}) \cdots$$

Remember that per cent means "parts per 100 parts." The equivalency of 37 g pure HCl and 100 g concentrated HCl gives the needed one-factor.

$$y \text{ g pure HCl} = (1.19 \times 10^3 \ \cancel{\text{g conc. HCl}})\left(\frac{37 \ \text{g pure HCl}}{100 \ \cancel{\text{g conc. HCl}}}\right)$$
$$= \frac{(1.19)(10^3)(37)}{100} \text{ g pure HCl}$$
$$= 4.4 \times 10^2 \text{ g pure HCl}$$

185

8.3
*Working with
Solutions: Density,
Weight Per Cent,
and Molarity*

(c) Since all three parts are based on the same 1.00 l concentrated HCl, you may begin this part with the weight of pure HCl obtained in part (b). The number of moles of pure HCl, z, is proportional to the number of grams of pure HCl.

$$z \text{ mole pure HCl} = (4.4 \times 10^2 \text{ g pure HCl}) \cdots$$

The molecular weight of pure HCl is, $1.01 + 35.5 = 36.5$ g mole^{-1}, which provides the needed one-factor.

$$z \text{ mole pure HCl} = (4.4 \times 10^2 \text{ g pure HCl}) \left(\frac{1}{36.5} \frac{\text{mole pure HCl}}{\text{g pure HCl}} \right)$$

$$= \frac{4.4 \times 10^2}{36.5} \text{ mole pure HCl}$$

$$= 12.0 \text{ mole pure HCl}$$

This number of moles is useful. It is based on 1 l of solution. Using the delivery type volumetric glassware in Figure 8.1, any number of moles HCl could be measured and transferred for reaction with a given number of moles of another reactant.

EXAMPLE 6. How would you measure exactly 0.100 mole HCl using the solution in Example 5.

Solution: The easiest way to accomplish this is to measure a calculated volume of the concentrated HCl. By using the one-factor method and the results obtained in part (c) of Example 5, you can convert 0.100 mole HCl to w ml concentrated HCl, since the two quantities are directly proportional.

$$w \text{ ml conc. HCl} = (0.100 \text{ mole HCl}) \cdots$$

The necessary one-factors are provided by the equivalency of 12.0 mole HCl and 1.00 l concentrated HCl, and the equality, 1 l = 1000 ml.

$$w \text{ ml conc. HCl} = (0.100 \text{ mole HCl}) \left(\frac{1}{12.0} \frac{l \text{ conc. HCl}}{\text{mole HCl}} \right) \left(\frac{1000}{1} \frac{\text{ml conc. HCl}}{l \text{ conc. HCl}} \right)$$

$$= \frac{(0.100)(1000)}{12.0} \text{ ml conc. HCl}$$

$$= 8.3 \text{ ml conc. HCl}$$

The most convenient piece of glassware to measure this volume is a 10-ml pipet of the transfer type (Figure 8.1). Unlike the volumetric pipet of the measuring type, it is marked over the whole 10-ml range at every 0.1 ml.

The important ratio obtained in Example 5, part (c) is called the molarity. Among the methods used by chemists to express the concentrations of solutions it is the most popular. This solution is said to be 12.0 **molar** (M) **in HCl.**

Molarity

DEFINITION: *The **molarity** of a solution is the number of moles of solute contained in one liter of the solution. If X represents the solute, the following symbols and equalities are applicable.*

$$\text{molarity of X} = \text{moles of X per liter solution}$$
$$= \text{M X}$$
$$= [\text{X}]$$

EXAMPLE 7. "The molarity of X is 3," can be restated in each of the following ways.

Solution: (a) There are 3 moles X per liter of solution containing X.
(b) The molar concentration of X is 3 M or 3 moles ℓ^{-1}.
(c) $[\text{X}] = 3 \text{ M}$ or 3 mole ℓ^{-1}

The most suitable piece of equipment for preparing a solution of accurately known molarity is a volumetric flask, such as that shown in Figure 8.1. Since molarity is based on *solution* volume and not *solvent* volume, the solvent and solute are mixed together in the flask and the final volume of the mixture is measured. Figure 8.2 and Example 8 deal with a solid solute, but a liquid solute would be handled in a similar manner.

EXAMPLE 8. Starting with the anhydrous white solid, $CaCl_2$, describe how you would prepare 500 ml 0.300 M $CaCl_2$ using water as the solvent.

Solution: Assuming that a 500-ml volumetric flask is available, you may prepare the exact amount needed. The molecular weight of $CaCl_2$ is $40.1 + 2 \times 35.5 = 111.1$ g mole^{-1}. The number of grams, x, of $CaCl_2$ needed is directly proportional to the volume of solution and the number of moles per liter.

$$x \text{ g CaCl}_2 = (500 \text{ ml solution})\left(0.300\frac{\text{mole CaCl}_2}{\ell \text{ solution}}\right)\cdots$$

The appropriate one-factors are provided by the equalities, $1 \ell = 1000$ ml, and 1 mole $CaCl_2 = 111.1$ g $CaCl_2$.

$$x \text{ g CaCl}_2 = (500 \text{ ml solution})\left(0.300\frac{\text{mole CaCl}_2}{\ell \text{ solution}}\right)\left(\frac{1}{1000}\frac{\ell \text{ solution}}{\text{ml solution}}\right)$$
$$\left(\frac{111.1}{1}\frac{\text{g CaCl}_2}{\text{mole CaCl}_2}\right)$$
$$= \frac{(500)(0.300)(111.1)}{1000} \text{ g CaCl}_2$$
$$= 16.69 \text{ g CaCl}_2$$

Add 16.69 g $CaCl_2$ to the 500 ml flask and fill to the mark with water after dissolving as illustrated in Figure 8.2.

187

8.3
Working with
Solutions: Density,
Weight Per Cent,
and Molarity

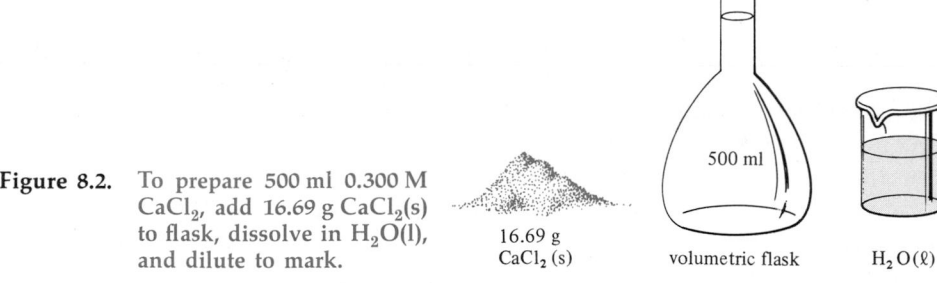

Figure 8.2. To prepare 500 ml 0.300 M CaCl$_2$, add 16.69 g CaCl$_2$(s) to flask, dissolve in H$_2$O(l), and dilute to mark.

16.69 g
CaCl$_2$ (s)

volumetric flask

H$_2$O(l)

Exercise 2. Is the method used in Figure 8.2 the same as adding 500 ml H$_2$O(l) to 16.69 g CaCl$_2$? If not, which is more dilute?

Answer: No. The second one is more dilute.

    In actual practice you may sometimes find it desirable to dissolve the solute in a little solvent in another container before adding it to the volumetric flask. In this case you should take care not to use more than enough solvent to fill the flask and to wash all the solute out of the container and into the flask.

A problem similar to that in Example 8 is that of calculating the molar concentration from the weight and formula of the solute and the final volume of the solution. **Calculating Molarity**

EXAMPLE 9. If 5.0 g of the lustrous black solid, copper(II) bromide (CuBr$_2$), is dissolved in a little water, it forms a blue solution. If it is then diluted to exactly 250 ml in a volumetric flask, it becomes green. What is the molarity of the final solution?

Solution: The concentration, x, in moles CuBr$_2$ per liter of solution is directly proportional to the number of grams of CuBr$_2$. However, x is *inversely* proportional to the number of milliliters of final solution. For example, if the solution is diluted with enough water to make the final volume twice as great, the concentration is one-half as great, and so on.

$$x \frac{\text{mole CuBr}_2}{\text{liter solution}} = (5.0 \text{ g CuBr}_2)\left(\frac{1}{250 \text{ ml solution}}\right)\cdots$$

Use the following equalities to provide the appropriate one-factors: 1 mole CuBr$_2$ = 63.5 + 2(79.9) = 223.3 g CuBr$_2$, and 1 l = 1000 ml.

$$x \frac{\text{mole CuBr}_2}{\text{liter solution}} = \left(\frac{5.0}{250} \frac{\text{g CuBr}_2}{\text{ml solution}} \right) \left(\frac{1}{223.3} \frac{\text{mole CuBr}_2}{\text{g CuBr}_2} \right) \left(\frac{1000}{1} \frac{\text{ml solution}}{\ell \text{ solution}} \right)$$

$$= \frac{(5.0)(1000)}{(250)(223.3)} \frac{\text{mole CuBr}_2}{\ell \text{ solution}}$$

$$= 0.090 \text{ M CuBr}_2$$

Very Dilute Solutions, Parts per Million (ppm)

As chemical instrumentation for analysis has become more sophisticated, it has become routine to determine extremely minute quantities of dissolved substances. These small concentrations are particularly important in health related fields because toxic quantities of some contaminants can be accumulated over an extended period of time through intake of environmental water or air. Our food sources can also accumulate toxic materials through exposure to the environment. The concentrations of such dilute solutions are frequently expressed in parts per million, which is similar to weight per cent.

DEFINITION: *The number of parts per million (ppm) is the number of grams solute per million grams solution. As in per cent, other consistent units may be used.*

EXAMPLE 10. In 1970 a Canadian student in zoology analyzed a fish for the highly toxic element Hg and found it to be 7 ppm. That amount was 35 times greater than the allowable limit in Canada. As a result of his discovery the Canadian government forced certain chloralkali manufacturing plants in the area to change their mode of operation, which was to discharge mercury wastes into waters eventually draining into Lake Erie. How many milligrams Hg would be consumed if a person ate 500 g (about 1 lb) of the contaminated fish?

Solution: The amount consumed, x mg Hg, is directly proportional to the weight of fish eaten.

$$x \text{ mg Hg} = (500 \text{ g fish}) \cdots$$

Eating 10^6 g fish is equivalent to eating 7 g Hg, and 1 mg $= 10^{-3}$ g. These facts lead to the needed one-factors.

$$x \text{ mg Hg} = (500 \text{ g fish}) \left(\frac{7}{10^6} \frac{\text{g Hg}}{\text{g fish}} \right) \left(\frac{1}{10^{-3}} \frac{\text{mg Hg}}{\text{g Hg}} \right)$$

$$= \frac{(500)(7)}{(10^6)(10^{-3})}$$

$$= 4 \text{ mg Hg}$$

Sometimes you will find it necessary to change a concentration from per cent to parts per million. In the following example it is apparent that per cent could be called "parts per hundred."

EXAMPLE 11. The legal definition of intoxication in Utah is set at a level of 0.08% BAC (blood alcohol concentration), which is the lowest of any state in the United States. What is this concentration expressed in parts per million?

Solution: The concentration is x ppm alcohol, or x g alcohol per million g blood; x is directly proportional to per cent alcohol expressed in grams alcohol per 100 g blood.

$$\frac{x \text{ g alcohol}}{10^6 \text{ g blood}} = \left(\frac{0.08 \text{ g alcohol}}{100 \text{ g blood}}\right) \cdots$$

As a one-factor, simply use the ratio 10^6 g blood/10^6 g blood in order to eliminate "g blood" in the denominator and replace it with "10^6 g blood."

$$\frac{x \text{ g alcohol}}{10^6 \text{ g blood}} = \left(\frac{0.08 \text{ g alcohol}}{100 \text{ g blood}}\right)\left(\frac{10^6 \text{ g blood}}{10^6 \text{ g blood}}\right)$$

$$= \frac{(0.08)(10^6)}{(100)} \frac{\text{g alcohol}}{10^6 \text{ g alcohol}}$$

$$= 800 \frac{\text{g alcohol}}{10^6 \text{ g alcohol}}$$

$$= 800 \text{ ppm alcohol}$$

Long term exposure to some chemical compounds, even at concentrations in parts per million, are known to cause cancer. Such compounds are called carcinogens. They are present in cigarette smoke, but they are also present in the polluted atmosphere. As early as 1972 the National Research Council warned that the difference between death rates (deaths per 100,000 population) due to lung cancer in urban and rural areas could not be accounted for by cigarette smoking alone. Now there is important new evidence that other factors may also be involved. The National Cancer Institute has made a statistical study at the county level of 6 million deaths caused by cancer between 1950 and 1969.* Some of the results are shown in Figure 8.3, for two kinds of cancer—lung cancer and skin cancer.

There is a high incidence of lung and throat cancer in the urban areas such as the northeast, but the highest incidence is in a group of parishes (counties) in Louisiana. Of the 31 United States counties (top 1%) with the highest rates, 13 are in Louisiana. Seven of the remaining top 1% are along the Gulf coast and along the Atlantic coast of north Florida, Georgia, and South Carolina.

Melanoma, a type of skin cancer that causes dark splotches, is not a common cause of death by cancer (40,000 out of 6,000,000), but the statistics clearly indicate an environmental cause—sunlight. It is only in the Sunbelt states, where the sunlight is most intense, that any counties have the highest rate of incidence.

* *Atlas of Cancer Mortality*, U.S. Department of Health, Education, and Welfare, Publication No. 75-780.

**Lung and throat, highest
10% of counties.**

**Melanoma of skin, highest
10% of state economic areas.**

Figure 8.3. Cancer mortality rates of white males and females, 1950–1969.

The federal government has set a goal to conquer cancer and has appropriated large sums of money to finance research (Conquest of Cancer Act of 1971). Some progress has been made, especially in the detection and treatment of cancer, but the cure for cancer seems to be as elusive as ever. The conquest of cancer may lie in its prevention, and one important goal should be preventing human exposure to environmental carcinogens.

PROBLEMS

1. Do *all* solid compounds that are classified as strong electrolytes make good conductors of electricity when large amounts are added to water? If not, why are they classified as strong?

2. Chemical compounds and solutions have a number of similarities as well as differences. Give two similarities and two differences.

3. Although there is no such thing as a truly universal solvent, the concept is interesting. Can you think of any disadvantage of having some of this solvent around the laboratory?

4. The sample containers for liquid solutions in the infrared spectrophotometer have windows made of rock salt (NaCl) for efficient transmission of the light beam. Can you suggest certain solvents to be avoided and others to be recommended for preparing the solutions?

5. Carbon tetrachloride, CCl_4, and gasoline, typically C_8H_{18}, are completely soluble in (miscible with) one another although the mixing process is not exothermic. Explain why it occurs.

6. Water is practically insoluble in (immiscible with) carbon tetrachloride, CCl_4, although the process of dissolving would result in a considerable increase in disorder. Explain why it does not occur.

7. Which of the following processes are associated with an increase in disorder and which with a decrease?
   (a) ice $\longrightarrow$ water
   (b) $N_2(g) + 3H_2(g) \longrightarrow 2 NH_3(g)$
    ammonia

8. Same as problem 7.

   (a) $H_2(g) \longrightarrow 2 H(g)$
   (b) $CO_2(g) \longrightarrow CO_2(s)$
    dry ice

9. Which piece of volumetric glassware illustrated in Figure 8.1 should be used to prepare a solution of known molarity? Why?

10 Which piece of volumetric glassware illustrated in Figure 8.1 should be used to transfer 49 ml of a solution most precisely? Why?

11. Gold, Au, forms solid solutions with copper, Cu, which are harder than either of the metals in their pure form. The fineness of such alloys is measured on a carat scale in which 24 carat fineness represents 100% Au. A certain 23-g gold ring is 18 carat fine, which means 18/24 or 75% by weight, Au. What is the weight in grams of pure Au in the ring?

12. One liter of sea water weighs 1020 g and consists of 3.5% by weight NaCl, approximately. What weight in grams of NaCl can be recovered by evaporating 1 l of sea water?

13. (a) How would you prepare 500 ml 3.0 M NH_4NO_3 in aqueous solution using the minimum amount possible of solid ammonium nitrate, NH_4NO_3? (b) If a volumetric pipet is used to transfer exactly 10.0 ml of this solution, how many moles NH_4NO_3 is transferred?

14. (a) Describe the preparation of exactly 500 ml 0.200 M NaOH in aqueous solution from solid NaOH. You may use a balance and whatever volumetric equipment is convenient. Use the minimum amount possible of NaOH (sodium hydroxide). (b) If exactly 10.0 ml of this solution is transferred in a volumetric pipet, how many moles NaOH is transferred?

15. A solution of NaCl is 26% NaCl by weight and has a density of 1.20 g ml^{-1}. What volume is needed to prepare exactly 1 l of 1.00 M NaCl?

16. A certain concentrated solution of NaOH is 40% pure by weight. If this solution has a density of 1.43 g ml^{-1} how many milliliters should you take to make exactly 1 l of 1.00 M NaOH?

17. Starting with a concentrated aqueous solution of H_2SO_4 (sulfuric acid) which is 85% pure by weight and has a density of 1.78 g ml^{-1}, how would you prepare 1.00 l 6.00 M H_2SO_4? Express your answer in milliliters concentrated sulfuric acid.

18. Commercial grade HNO_3 (nitric acid) is 70% pure by weight. (a) What is the molarity of this acid if its density is 1.42 g ml^{-1}? (b) How would you prepare exactly 100 ml of 3.0 M HNO_3 from the concentrated (commercial grade) solution using only volumetric glassware? Express your answer in milliliters concentrated nitric acid.

19. One form of water pollution in agricultural areas where nitrogen fertilizers are used extensively is the nitrate ion, NO_3^-. There is evidence that it increases the infant mortality rate among female babies due to a blood disorder called methemoglobinemia.* The acceptable nitrate level for drinking water is 45 ppm. How many milligrams NO_3^- is ingested if you consume 1 l (assume 1 kg) of water containing the acceptable level?

20. Water under 1 atm of pure O_2 gas contains 0.004% dissolved O_2 by weight. Under 1 atm of ordinary air, which is only $\frac{1}{5} O_2$, the concentration of dissolved O_2 is $\frac{1}{5} \times 0.004 = 0.0008\%$ by weight. What is the amount of dissolved O_2 under normal conditions in air, expressed in parts per million by weight?

*21. The highest concentration of ethyl alcohol that can be distilled directly from an aqueous solution is 95% by weight. As a bactericide, a 70% solution is more effective than one higher or lower in concentration. Show that to make 95 g 70% solution you may take 70 g 95% solution and add water until the total weight is 95 g.

*22. Mole per cent is the number of moles of solute per 100 moles solution. Consider a sample of dry air (no water vapor) to be a solution and its three principal constituent gases to be solutes. The sample has the following composition in *mole per cent:* 78.1% N_2, 21.0% O_2, and 0.9% Ar. What is the composition of dry air in *weight per cent?*

*Barry Commoner: *The Closing Circle.* Alfred A. Knopf, Inc., New York, 1971.

*23. A certain small amount of fluoride ion in drinking water is natural and produces decay-resistant teeth containing fluorapatite, $Ca_5(PO_4)_3F$, in the enamel. Too much fluoride in drinking water contaminated by phosphate mining or originating naturally can cause a mottled appearance of the teeth called fluorosis. The recommended daily intake of the element by an adult is 1 mg as the F^- ion. Assuming a daily consumption of 2 l water by an adult, at what level in parts per million F^- should the water supply be controlled to provide 1 mg daily?

*24. Sea water contains approximately 4 parts per trillion (10^{12}) Au (gold) in solution. Assuming a solution density of 1.0 g ml^{-1}, how many kilograms Au is contained in a cubic kilometer of sea water?

*25. Some undernourished children suffer from *pica*, a craving for unnatural food, such as wall plaster or flakes of paint. Since old apartment buildings may have been painted with Pb (lead) based paint, a toxic element, these children are in danger of developing lead poisoning. A sensitive analytical procedure called atomic absorption, AA, can be used to detect as little as 0.02 ppm Pb. Approximately 0.2-ml sample solution is needed for each determination. A new technique depends on obtaining a small drop of blood for analysis from a pinprick in the child's finger. If the amount of blood obtained for testing is 50 mg, and the sample solution is made by diluting it with an appropriate solvent for use in the AA analyzer, how small an amount of Pb in parts per million can be detected in the blood?

9

EQUILIBRIUM

The net effect of *all* change is that the universe tends toward equilibrium. This is a fundamental law of nature. Many examples of this process occur around you every day: gas molecules flow from regions of high pressure to low pressure, heat flows from areas of high temperature to low temperature, and solutes in solution diffuse from regions of high concentration to low concentration. The state of equilibrium is defined in terms of three conditions.

DEFINITIONS: **Equilibrium** *in a closed system is a state in which the conditions of pressure, temperature, and concentration are the same in all parts of the system and remain fixed. A* **system** *is simply a sample of matter enclosed by recognized boundaries, and a* **closed system** *is one in which matter neither leaves nor enters.*

The universe certainly qualifies as a closed system and the scientific law enunciated in the first sentence of the definition presents a rather gloomy prospect. After final equilibrium has been reached, no other change will be possible. The student may wonder whether that final equilibrium temperature will be pleasant. However, the idea of equilibrium as applied to smaller closed systems can be informative and useful. In any closed system the direction of *spontaneous* change will be toward equilibrium.

Why is it important to study equilibrium theory? Primarily it is because the theory furnishes an explanation of why *some* physical changes and chemical reactions are possible and *others* are impossible. Can coal be converted into gasoline and diamond? Should you try to develop a catalytic converter that changes the nitrogen oxides in the exhaust of an automobile back into the harmless N_2 and O_2 from which they came? A **catalyst** is a substance that is not consumed by a chemical reaction but permits it to reach equilibrium more rapidly. The important question is to what extent a reaction goes to completion before it reaches equilibrium—is it 90%? is it 10%? is it 1%? The answers to all these questions are given by equilibrium theory.

EXAMPLE 1. Which of the following systems are closed systems and which are open?
(a) In Figure 7.7d, the crushed can has an internal pressure of 24 mm Hg at 25°C.
(b) The number of automobiles on a narrow bridge remains constant because the number coming on is equal to the number going off.
(c) A man remains at constant weight from day to day at a pressure of 1 atm and a temperature of 37°C.
(d) In Figure 4.1, the unreacted quantity of C(s) remains in contact with the product, CO(g).
(e) In Figure 7.2, the water in the beaker at 100°C remains at that temperature while 540 cal heat is absorbed and while the pressure remains at 1 atm.

Solution: (a) closed, (b) open, (c) open, (d) closed, (e) open.

9.1 REVERSIBLE PROCESSES

The fact that a system has reached equilibrium does not mean that individual molecules have stopped their activity. In fact the most interesting equilibrium systems will be those in which molecules are continuously changing; however, while some are changing in one direction others are changing in the opposite direction so that the entire system does not change. With such systems it is possible to have a reversible process.

DEFINITION: *A **reversible process** is a change in a system that occurs in such a way that the system is continuously at equilibrium. At any moment the direction of the change may be reversed by a very small alteration in conditions such as pressure, temperature, or concentration.*

None of the examples of processes given in the introductory paragraph of this chapter is a reversible process. If it were, it would already *be* at equilibrium. In each case the process is unrestrained and the change occurs too rapidly for reversibility. When you studied phase changes (Sections 7.3 and 7.4), you considered several processes of this type. In the apparatus for measuring vapor pressure of liquids, Figure 7.4a, there is a sudden change when the stopcock is first opened. The water vapor causes a dramatic surge of pressure in the left leg of the mercury manometer due to the process

$$H_2O(l) \longrightarrow H_2O(g).$$

However, the large and small bulbs containing the H_2O are a closed system and, since the change occurs spontaneously, the water system must approach equilibrium. The evaporation process, which is not reversible at first, predominates until a certain pressure is reached. This pressure depends on the temperature and on the temperature alone, as shown in Table 7.1. Thus, if the temperature is kept constant, the pressure will remain constant. This condition of constant temperature and constant pressure is called **equilibrium.** It is not necessary to specify constant concentration because liquid and vapor phases are both 100% H_2O and there is no possibility of varying the concentration. At 25°C equilibrium is attained at 24 mm Hg. At this pressure the number of $H_2O(g)$ molecules striking the surface of $H_2O(l)$ and condensing

$$H_2O(g) \longrightarrow H_2O(l)$$

becomes as great as the number leaving $H_2O(l)$. This is sometimes called *dynamic equilibrium* to emphasize the events happening to individual molecules. It is symbolized in equations by using arrows in both directions.

$$H_2O(l) \rightleftharpoons H_2O(g) \quad (P = 24 \text{ mm Hg}; \; t = 25°C)$$

Under these conditions the system is at equilibrium and the process can be undertaken reversibly. A small increase or decrease in pressure or temperature will cause the process to produce either more $H_2O(l)$ or more $H_2O(g)$. In this sense the double arrow symbolizes reversibility of the process.

There is a principle that applies to all systems at equilibrium and, although it is only qualitative, it is very useful in predicting the directions of change. It is named after Henry Le Chatelier (1850–1936), a professor in Paris. One way of stating the principle is

Le Chatelier Principle

DEFINITION: *The **Le Chatelier principle** states that, if the conditions of pressure, temperature, or concentrations of a system at equilibrium are changed, a process will occur which tends to restore the original conditions.*

In the following sections you will see how the principle of Le Chatelier is applied to all the processes we have discussed that can be made to occur reversibly: phase change, formation of solution, and chemical reaction.

9.2 PHASE EQUILIBRIA

If a substance does not decompose chemically on heating, the solid can be changed to liquid or gas. Conditions of pressure and temperature can usually be attained under which any phase is in equilibrium with another. A number of examples of such phase equilibria, all at 1 atm pressure, are given in Table 7.2 (normal melting, boiling, and sublimation points).

The equilibrium described in Section 9.1, of $H_2O(g)$ and $H_2O(l)$, is a simple example for which you already know the answer to the question—What will be the effect of increasing the temperature? In order to consider temperature effects by Le Chatelier's principle it is necessary to know whether a process is endothermic or exothermic.

Effect of Changing Temperature

EXAMPLE 1. In the steps outlined the effect of increasing temperature on vapor pressure of water will be predicted by the Le Chatelier principle. Since evaporation is an *endothermic* process the heat in the first equation appears on the left hand side.

1. An equilibrium exists:

$$H_2O(l) + heat \rightleftharpoons H_2O(g)$$

2. A condition is changed: temperature is increased.
3. A process occurs: either,

   (a) $H_2O(l) + heat \longrightarrow H_2O(g)$, or
   (b) $H_2O(g) \longrightarrow H_2O(l) + heat$

4. which tends to restore the original condition: The process tends to decrease temperature.
5. Process (a) is endothermic and tends to decrease temperature. Process (b) is exothermic and tends to increase temperature.
6. Therefore, process (a) occurs; more $H_2O(g)$ is formed, and the vapor pressure is higher at the higher temperature.

This is, of course, in agreement with the experimental facts as shown in Table 7.1 of vapor pressures and should not come as a surprise. However, you may not already know the answer to Example 2.

Effect of Changing Pressure

EXAMPLE 2. What happens to an equilibrium mixture of ice and water when the pressure exerted on the system is increased? The only fact you need to have is that ice floats on water. The analysis proceeds as follows.

1. An equilibrium exists:

$$H_2O(l) \rightleftharpoons H_2O(s)$$

2. A condition is changed: pressure is increased.
3. A process occurs: either,
   (a) $H_2O(l) \longrightarrow H_2O(s)$, or
   (b) $H_2O(s) \longrightarrow H_2O(l)$

4. which tends to restore the original condition: The process tends to decrease pressure.
5. Step (a) is an increase in volume, because $H_2O(s)$ is less dense than $H_2O(l)$ (ice floats on water.). If the system increases in volume (a) it tends to increase the pressure. If it decreases in volume (b) it tends to decrease the pressure.
6. Therefore, process (b) occurs, and $H_2O(s)$ melts. The $H_2O(l)$ may be frozen again, at the higher pressure, but only at a lower temperature. For example, at 100 atm pressure the freezing point of $H_2O(l)$ is $-0.8°C$.

The changes in phase equilibria in Examples 1 and 2 can be illustrated very effectively on a **phase diagram.** Such a diagram for H_2O is shown in Figure 9.1. The lines on the diagram separate it into areas of $H_2O(s)$, $H_2O(l)$, and $H_2O(g)$, and represent the series of temperature-pressure points at which the two phases of H_2O on opposite sides of the line are in equilibrium. Part of this curve is a reproduction of the curve in Figure 7.5, the vapor pressure of $H_2O(l)$, but this diagram is more complete since it contains information on solid $\rightleftharpoons$ gas and solid $\rightleftharpoons$ liquid

Figure 9.1. Phase diagram of H_2O. The pressure and temperature axes are not drawn to scale.

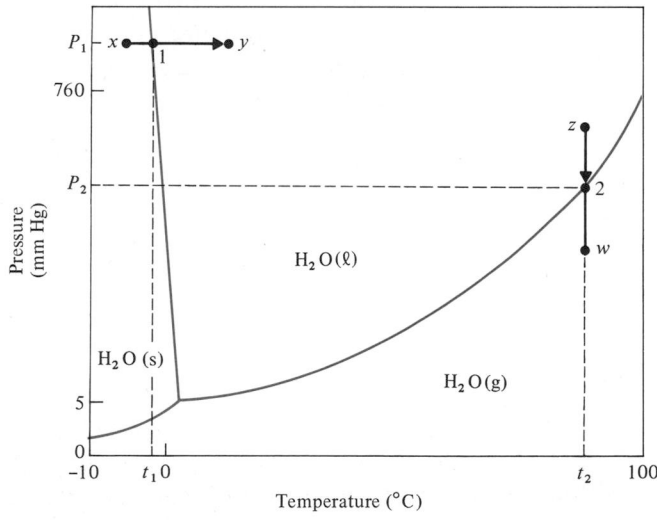

equilibria as well. The point at which the three lines intersect gives the temperature and pressure at which all three phases (l, s, and g) of H_2O are at equilibrium with one another. You can see at a glance that the vapor pressure of $H_2O(l)$, given by the $H_2O(l) \rightleftharpoons H_2O(g)$ line, increases with temperature as predicted by Example 1.

You will understand the phase diagram more clearly if you consider what happens when you cross one of the lines at constant pressure or at constant temperature. It is assumed that all changes are performed reversibly. The two points at x and y connect a constant pressure (P_1) line which crosses from the solid (ice) phase into the liquid (water) phase. At point x you observe a sample of ice at some pressure, P_1, above 1 atm. This pressure is held constant while adding heat to increase the temperature. When your horizontal path intersects the (s, l) line at point 1, the ice begins to melt. You will note that this melting temperature, t_1, is less than 0°C as indicated by the point on the temperature axis where the vertical dotted line intersects. This fact was predicted in Example 2. It is for this reason that the (s, l) line is drawn sloping to the left as it rises. Now as the ice-water mixture continues to absorb heat, the temperature remains fixed at t_1. Finally all the ice is melted and you are permitted to enter the $H_2O(l)$ region. The temperature continues to increase until you reach point y, a system of liquid water.

In crossing the (l, g) line you will begin at point z, which represents liquid water at some pressure less than 1 atm and at a temperature, t_2, less than 100°C. You can move vertically toward point w by holding the temperature constant and reducing the applied pressure (slight expansion). When you reach point z, the water begins to vaporize (boil) and to expand enormously. You will not be able to reduce the applied pressure, P_2, at this temperature until all the water has evaporated (boiled away). At that time you will be able to enter the $H_2O(g)$ region of the phase diagram at a pressure lower than P_2 where only vapor is present. Then you continue decreasing the pressure (further expansion) until you reach point w.

EXAMPLE 3. Explain the operation of the coffee percolator in Figure 9.2 in terms of the phase diagram for water in Figure 9.1.

Solution: The center stem of a coffee percolator is filled with $H_2O(l)$, and the bottom of this stem is designed to gather most of the heat from the heat source, such as the electric range in Figure 9.2. The pressure in the stem

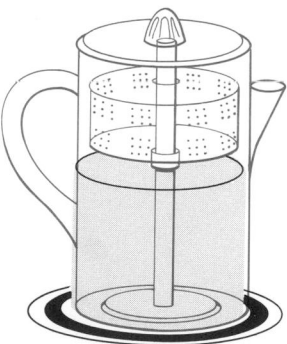

Figure 9.2. Explain the operation of a coffee percolator using the phase diagram of H_2O.

increases with the depth of the water, so that the highest pressure is at the bottom. From the phase diagram (Figure 9.1) it is obvious that the boiling point, which must be located on the (l, g) line, is increased if the applied pressure is increased. The water with the highest boiling point is located at the bottom of the stem where the pressure is highest. As the water is heated, it will rise due to expansion, but the water at the bottom is always the hottest. Eventually the water in the upper part of the stem reaches 100°C, which is its boiling point since it is not under any extra pressure. If the percolator is well designed, the water in the lower part of the stem reaches its boiling point (higher than 100°C) at the same time and a little hot water geyser gurgles out of the top of the stem as the whole column of water erupts.

Example 4 shows the effect of changing pressure on phase equilibria, and a solid-solid phase transition occurs.

EXAMPLE 4.   For the synthesis of diamond from graphite the phase change is simply from one crystalline state to another. In crystalline graphite the carbon atoms are hexagonally arranged (Figure 9.3) in rather loosely stacked sheets, whereas in diamond the atoms are held together in a rather more tightly packed tetrahedral arrangement.

We know that diamond is more dense than graphite. What condition of pressure would you select for the manufacture of diamonds?

Solution: You may think of C(graphite) $\rightleftharpoons$ C(diamond) as an equilibrium just like any other and ask what happens when conditions of pressure change. By Example 1, you know that an increase in pressure favors the more dense phase. Therefore, the C(graphite) $\longrightarrow$ C(diamond) phase change should proceed at high pressures.

It is necessary also to increase the temperature because, no matter how much pressure is applied at ordinary temperatures, the change is too slow to be practical. It is a well known fact that chemical changes, in this case breaking of C—C bonds and forming new ones, occur faster at higher temperatures than at lower.

Figure 9.3   Crystal structure of (a) graphite and (b) diamond. Note that graphite appears less dense than diamond.

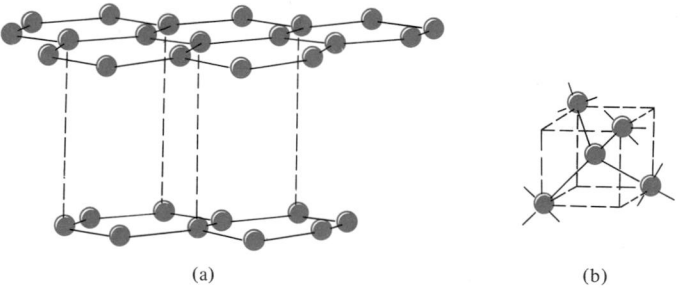

(a)   (b)

In 1954 the right conditions of pressure (100,000 atm) and temperature (2500°C) were attained to produce the first man-made diamonds. They are used mainly as grit in cutting and grinding wheels.

9.3 SOLUBILITY EQUILIBRIA

Another important type of equilibrium is that which exists in a saturated solution. In Example 1 ordinary table salt is used as the solute and water as the solvent.

EXAMPLE 1. When $NaCl(s)$ and $H_2O(l)$ are mixed, the salt dissolves spontaneously.

$$NaCl(s) + aq + heat \longrightarrow Na^+(aq) + Cl^-(aq)$$

Here "aq" represents an unspecified amount of $H_2O(l)$. Even though the process is endothermic, it will continue spontaneously until equilibrium is reached. At this point the solution will be saturated. What will happen to the solubility of $NaCl(s)$ if the temperature is lowered?

Solution:

1. An equilibrium exists:

$$NaCl(s) + aq + heat \rightleftharpoons Na^+(aq) + Cl^-(aq)$$

2. A condition is changed: temperature is decreased.
3. A process occurs: either,
   (a) $NaCl(s) + aq + heat \longrightarrow Na^+(aq) + Cl^-(aq)$, or
   (b) $Na^+(aq) + Cl^-(aq) \longrightarrow NaCl(s) + aq + heat$
4. which tends to restore the original condition: The process tends to increase temperature.
5. Process (a) is endothermic and tends to decrease temperature. Process (b) is exothermic and tends to increase temperature.
6. Therefore, process (b) occurs, some of the Na^+ and Cl^- in solution crystallizes as $NaCl(s)$, and the solubility is lower at the lower temperature.

The variation in the solubility of NaCl in water is shown in Figure 9.4. Notice that its solubility increases slightly with temperature, but that another salt, KCl, has a considerably greater increase. The reason for this is that the heat of solution of KCl is more endothermic than that of NaCl.

Example 2 demonstrates how the change in concentration of a product moves a system to a new equilibrium.

EXAMPLE 2. Suppose that $KCl(s)$ is added to a saturated solution of $NaCl(s)$. What will happen to the solubility of the NaCl?

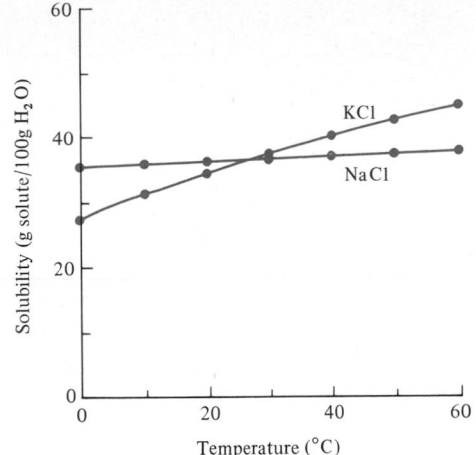

Figure 9.4. Solubility of KCl and NaCl in aqueous solution at different temperatures.

Solution:

1. An equilibrium exists:

$$NaCl(s) + aq \rightleftharpoons Na^+(aq) + Cl^-(aq)$$

2. A condition is changed. The concentration of $Cl^-(aq)$ is increased as KCl dissolves

$$KCl(s) + aq \longrightarrow K^+(aq) + Cl^-(aq)$$

3. A process occurs: either,
   (a) $NaCl(s) + aq \longrightarrow Na^+(aq) + Cl^-(aq)$, or
   (b) $Na^+(aq) + Cl^-(aq) \longrightarrow NaCl(s) + aq$
4. which tends to restore the original conditions: The process tends to decrease the concentration of $Cl^-(aq)$.
5. Process (a) tends to increase the concentration of $Cl^-(aq)$. Process (b) tends to decrease it.
6. Therefore, process (b) occurs. The concentration of $Na^+(aq)$ decreases as more $NaCl(s)$ is formed, and the solubility of $NaCl(s)$ is decreased.

The solubility of $NaCl(s)$ cannot be increased by adding $NaCl(s)$ to a saturated solution because, as a pure solid, its "concentration" is constant. There is fixed spacing between ions in the crystal.

The following experiment gives some very interesting quantitative results as well as agreeing with the qualitative results predicted by the Le Chatelier principle.

The ionic compound silver acetate (AgOAc) is not very soluble in water. A saturated solution at 25°C contains about 1% dissolved material in the form of silver

K_{sp}:
an Equilibrium
Constant

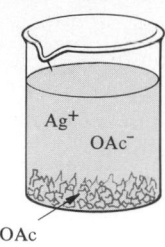

AgOAc

Figure 9.5. A saturated solution of AgOAc (silver acetate).

ion (Ag^+) and acetate ion (OAc^-). The structure of the compound, showing its ions, and a commonly used abbreviation for the acetate ion are shown below.

$$Ag^+ \quad {}^-O-\overset{\overset{O}{\|}}{C}-CH_3 \equiv \quad AgOAc$$

silver   acetate   silver acetate
ion   ion

In a saturated solution of AgOAc (Figure 9.5) there is an equilibrium between the solid and the ions, which is a reversible process.

$$AgOAc(s) + aq \rightleftharpoons Ag^+(aq) + OAc^-(aq) \tag{9.1}$$

The results of changing ion concentrations by adding new substances are shown in Figure 9.6. In Example 2, NaCl(s) was precipitated from a saturated solution by adding KCl(s). In a similar way the precipitate in Figure 9.6a is formed in order to decrease the concentration of the added Ag^+ and in Figure 9.6b in order to decrease the concentration of added OAc^- ion. The purpose of the test in Figure 9.6c is to prove that neither NH_4^+ nor NO_3^- ion was responsible for precipitation in the previous cases.

Figure 9.6. Effects of adding concentrated solutions of various strong electrolytes to saturated AgOAc (silver acetate).

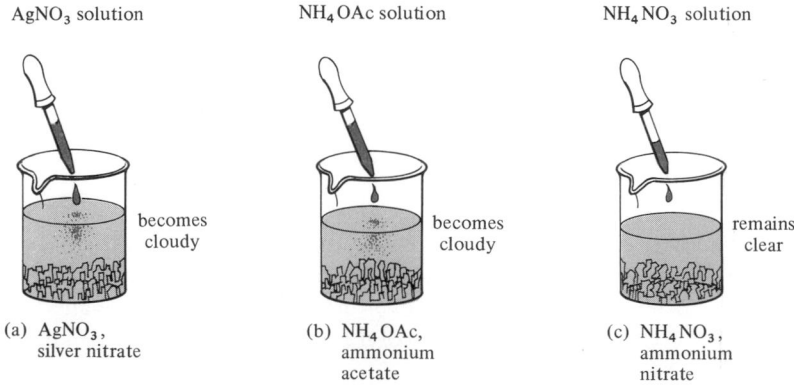

AgNO$_3$ solution   NH$_4$OAc solution   NH$_4$NO$_3$ solution

becomes cloudy   becomes cloudy   remains clear

(a) AgNO$_3$,   (b) NH$_4$OAc,   (c) NH$_4$NO$_3$,
    silver nitrate   ammonium   ammonium
               acetate   nitrate

If you want to obtain *quantitative* information about these processes, it is best to express concentrations in units of moles per liter. For example, the solubility of pure AgOAc expressed in moles per liter is 0.060 M at 25°C. A series of quantitative measurements on the experimental solutions in Figure 9.6 will give some interesting results and illustrate an important principle of equilibrium. Table 9.1 shows the results of measuring the solubility of AgOAc at different levels of Ag^+ and OAc^- obtained by adding concentrated $AgNO_3$ and NH_4OAc.

TABLE 9.1 Calculations based on the solubility of AgOAc in the presence of $AgNO_3$ and NH_4OAc at 25°C

Mixture	Solubility (M)	Ion Concentration		Ion Product $[Ag^+] \times [OAc^-]$
		$[Ag^+]$	$[OAc^-]$	
Figure 9.5: pure saturated AgOAc	0.060	0.060	0.060	0.0036
Figure 9.6a: add $AgNO_3$ until $[Ag^+] = 0.10$	0.036	0.10	0.036	0.0036
Figure 9.6b: add NH_4OAc until $[OAc^-] = 0.10$	0.036	0.036	0.10	0.0036
Figure 9.6a: add $AgNO_3$ until $[Ag^+] = 0.20$	0.018	0.20	0.018	0.0036

In each of the three mixtures to which more Ag^+ or OAc^- has been added from an external source, the solubility is lowered and AgOAc must precipitate, which is in agreement with the Le Chatelier principle. If you compare the two mixtures which have been treated with different amounts of $AgNO_3$, you will see that adding a greater amount of Ag^+ causes a greater decrease in solubility. In the two columns under "Ion Concentration" the total molarity of each dissolved ion from all sources is recorded, based on experimental measurements. In the last column under "Ion Product" an interesting fact is revealed—the product of the two ionic concentrations has a constant value, 0.0036. This is a very useful constant because, given its value and the concentration of either of the ions, the other ion concentration can be calculated, or one ionic concentration can be controlled by varying another.

EXAMPLE 3. In a saturated solution of AgOAc which contains enough NaOAc (sodium acetate) to make the OAc^- concentration 0.30 M, (a) calculate the concentration of Ag^+ ion. (b) How could the concentration of Ag^+ ion be decreased to $\frac{1}{10}$ the value in part (a)?

Solution: Use the ion product for saturated solutions from Table 9.1 and substitute the known value $[OAc^-]$, which is 0.30 M.

(a)
$$[Ag^+][OAc^-] = 0.0036$$
$$[Ag^+](0.30) = 0.0036$$
$$[Ag^+] = 0.0036/0.30$$
$$= 0.012 \text{ M}$$

(b) In this part you wish to decrease $[Ag^+]$ by the factor $\frac{1}{10}$.

$$[Ag^+] = \tfrac{1}{10} \times 0.012$$
$$= 0.0012$$

Use the constant ion product as before, but this time $[OAc^-]$ is the unknown.

$$[Ag^+][OAc^-] = 0.0036$$
$$(0.0012)[OAc^-] = 0.0036$$
$$[OAc^-] = 0.0036/0.0012$$
$$= 3.0 \text{ M}$$

By increasing the acetate concentration to 3.0 M, the Ag^+ ion concentration can be decreased to 0.0012 M.

The constant 0.0036 has a special name in equilibrium theory. It is called K_{sp}, the solubility product constant for AgOAc at 25°C. Since solubility changes with temperature, K_{sp} changes also; but as far as variations in ion concentrations are concerned, K_{sp} *is* constant.

The K_{sp} for different ionic compounds has different numerical values, of course. You can identify compounds with slight solubilities by their small solubility product constants. A partial list of solubility product constants is given in Table 9.2.

TABLE 9.2 Values of K_{sp} for some 1–1*
electrolytes

Compound	K_{sp} at 25°C
AgF (silver fluoride)	very large
AgOAc (silver acetate)	3.6×10^{-3}
CaSO$_4$ (calcium sulfate)	2.4×10^{-5}
BaSO$_4$ (barium sulfate)	1.5×10^{-9}
CaCO$_3$ (calcium carbonate)	4.7×10^{-9}
AgCl (silver chloride)	1.7×10^{-10}
AgBr (silver bromide)	5.0×10^{-13}
AgI (silver iodide)	8.5×10^{-17}
CuS (copper(II) sulfide)	8×10^{-37}
HgS (mercury(II) sulfide)	1.6×10^{-54}

*1–1 electrolytes are those which dissociate into one
ion of each type.*

In this series of compounds, each of which dissociates into two ions, the solubilities can be calculated by taking the square root of K_{sp}. They vary all the way from AgF, which has K_{sp} too large to be practically useful, to HgS, which has a solubility so small as to be practically meaningless. There is little chance of finding even one pair of Hg^{2+} and S^{2-} ions in a liter of the saturated solution!

EXAMPLE 4.   Use K_{sp} for BaSO$_4$ from Table 9.2 to calculate its solubility.

Solution: In a saturated aqueous solution of pure BaSO$_4$, the dissolved ions are equal in concentration.

$$[Ba^{2+}] = [SO_4^{2-}] = \text{number moles BaSO}_4 \text{ dissolved per liter}$$
$$= \text{solubility}$$

The product of these concentrations equals K_{sp}.

$$[Ba^{2+}][SO_4^{2-}] = 1.5 \times 10^{-9} \quad \text{(Table 9.2)}$$
$$(\text{solubility})^2 = 1.5 \times 10^{-9}$$
$$\text{solubility} = \sqrt{1.5 \times 10^{-9}}$$
$$= \sqrt{15 \times 10^{-10}}$$
$$= \sqrt{15} \times \sqrt{10^{-10}}$$
$$= 3.9 \times 10^{-5}\,M$$

This concentration of ions is very small and a saturated solution of $BaSO_4$ is a poor conductor of electricity (Figure 11.5).

The solubility product constant includes the concentrations of the ions but does not include the amount of solid present. It is a well known fact that after a solution has been saturated with a given solid, the solubility cannot be increased by adding more solid. This means that the solid behaves as if it had a constant concentration, regardless of how much of it is present. The constant for the solid is included in the K_{sp} so that its concentration can be set equal to 1. This convention is used not only for pure solids but also for pure liquids involved in equilibria. In dilute solutions, the solvent is almost pure and this convention, which is a great convenience, may be applied.

Effect on Solubility of Adding More Solid

EXAMPLE 5. Use equation (9.1), which represents the equilibrium between AgOAc(s) and its ions in aqueous solution, to explain why the solid and "aq" are left out of the equation for K_{sp}.

Solution: The following equilibrium exists.

$$AgOAc(s) + aq \rightleftharpoons Ag^+(aq) + OAc^-(aq) \quad (9.1)$$

By convention,
$$[AgOAc(s)] = 1$$

because it is a pure solid (that is, not a solid solutic 1) and as such has constant concentration; furthermore,

$$[aq] = 1$$

because the aqueous phase is almost pure water, and as such has constant concentration. Although the amount of dissolved AgOAc is 0.06 M, this is a negligible concentration compared with that of the water molecules themselves (see problem *28 of Chapter 4 and Section 10.3). Therefore, the only concentrations involved in the equilibrium expression for K_{sp} are as follows.

$$[Ag^+(aq)][OAc^-(aq)] = 0.0036$$

The "true values" of [AgOAc(s)] and [aq] have been incorporated into K_{sp} as appropriate factors.

Effects of Pressure on Solubility The solubility of a solid in a liquid does not change much if the pressure is changed because there is such a small change in volume when the solution is formed. However, this is not the case when you consider the solubility of a *gas* in a liquid. At STP, 1 mole of NH_3 occupies 22.4 l as a gas; but, when NH_3 dissolves in 1 l of water, it increases the water volume by only 0.02 l.

$$NH_3(g) + aq \rightleftharpoons NH_3(aq)$$

As in Examples 2 and 4, of Section 9.2 the denser phase is favored by increased pressure so that the solubility of $NH_3(g)$ increases as its pressure increases. The relationship is a direct proportion so that, if the pressure is increased by a factor of two, the solubility is increased by a factor of two, and so on. (See problem 20 of Chapter 8.)

9.4 CHEMICAL EQUILIBRIA

It is of interest to the chemist to increase the amount of desired product starting with the available reactants. Changing the conditions of temperature, pressure, and concentration can often produce a more favorable equilibrium.

A simple yet interesting example of a chemical equilibrium is that between two forms of a gas notorious for air pollution, N_2O_4 (nitrogen tetroxide) and NO_2 (nitrogen dioxide). The electronic structure of the NO_2 molecule is an exception to the octet rule (Section 6.5) since it has an odd number of electrons. The odd electron is responsible for the deep red-brown color of this substance. The lower energy colorless form, N_2O_4, has all its octets filled. At any temperature and pressure there is a certain amount of both gases present in an equilibrium mixture.

$$\underset{\text{colorless}}{N_2O_4(g)} + \text{heat} \rightleftharpoons \underset{\text{red}}{2\,NO_2(g)}$$

Temperature Effect on Chemical Equilibrium The glass retort in Figure 9.7 contains a system of NO_2–N_2O_4. It has been closed by melting the glass tip in a flame. The mixture has a red color, the intensity of which depends on the number of NO_2 molecules per unit volume. Since the volume of the container is constant, an increase in color intensity must be due to the formation of more NO_2 molecules.

In the application of Le Chatelier's principle illustrated in Example 1, the effect of temperature on the observed color intensity in Figure 9.7 will be explained.

EXAMPLE 1.   How will the color intensity of a constant volume bulb containing an equilibrium mixture of NO_2–N_2O_4 change if the temperature is increased?

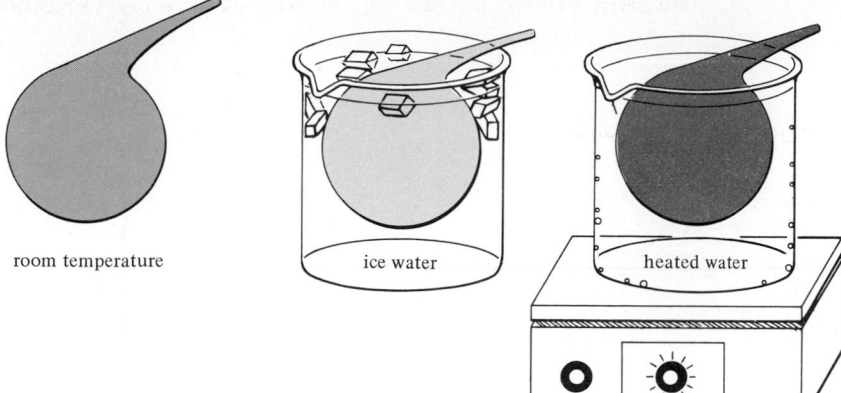

(a) 25°C, red–brown due to presence of NO_2 (g)

(b) 0°C, becomes lighter; some NO_2 (g) disappears

(c) 60°C, becomes dark; some NO_2 (g) produced

(Care! The glass will not stand more than 2 atm pressure internally.)

Figure 9.7. Effect of temperature on the $N_2O_4(g) +$ heat $\rightleftharpoons 2\,NO_2(g)$ equilibrium.

Solution:

1. An equilibrium exists:

$$N_2O_4(g) + \text{heat} \rightleftharpoons 2\,NO_2(g)$$

2. A condition is changed: temperature is increased.
3. A process occurs: either,
   (a) $N_2O_4(g) + \text{heat} \longrightarrow 2\,NO_2(g)$, or
   (b) $2\,NO_2(g) \longrightarrow N_2O_4(g) + \text{heat}$
4. which tends to restore the original conditions: The process tends to decrease temperature.
5. Process (a) is endothermic and tends to decrease temperature. Process (b) is exothermic and tends to increase temperature.
6. Therefore, process (a) occurs; some of the $N_2O_4(g)$ dissociates and the higher concentration of $NO_2(g)$ causes a *darker color.*

If the system is cooled as in Figure 9.7b, instead of being heated, the reasoning of Example 1 leads to the *lighter color,* which is experimentally observed.

There is a chemical equilibrium in aqueous solution that produces a red color indicative of the concentration of the reactive species. The pale yellow compound iron(III) chloride is an ionic solid having the formula $FeCl_3$. It dissociates into Fe^{3+} in dilute aqueous solution and has the same pale yellow color. Potassium thiocyanate is a colorless ionic compound with the formula KCNS. It dissociates into K^+ and CNS^-

Concentration Effect on Chemical Equilibrium

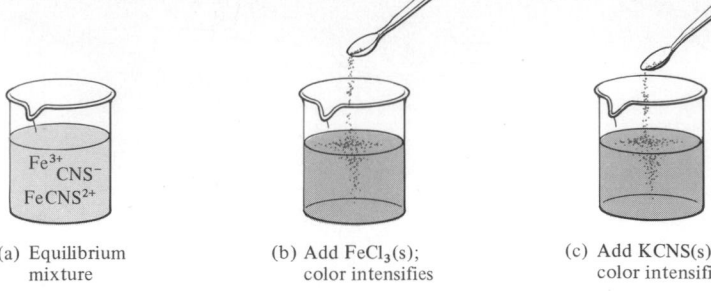

(a) Equilibrium
     mixture

(b) Add $FeCl_3(s)$;
     color intensifies

(c) Add $KCNS(s)$;
     color intensifies

Figure 9.8. Effect of increasing $[Fe^{3+}]$ or $[CNS^-]$ on the blood-red
color of the $FeCNS^{2+}$ ion.

(thiocyanate ion) in aqueous solution. When mixed, these solutions produce an
intense blood-red color due to the formation of $FeCNS^{2+}$.

The $FeCNS^{2+}$ ion is referred to as a **complex ion.** Although a complex ion can be
simply an ion composed of two or more different elements, it usually means a
combination of two or more ions or an ion and one or more neutral molecules.

The color of $FeCNS^{2+}$ ion is so intense that it can be detected visually at concentrations down to 10^{-5} M. It is one of the best qualitative tests for the presence of Fe^{3+}
in the analysis of an unknown material.

In Figure 9.8a a solution of the complex ion has been prepared by mixing dilute
solutions of the two compounds. It is obvious from the reddish color that $FeCNS^{2+}$
ions exist. We can demonstrate that unreacted Fe^{3+} and CNS^- ions both remain in the
following way. Addition of $FeCl_3(s)$ produces a darker red color (Figure 9.8b) which
indicates that not all the CNS^- ions had reacted. Addition of $KCNS(s)$ likewise
produces a darker color (Figure 9.8c), which proves that there must have been some
Fe^{3+} ions still present. In other words, Figure 9.8a represents a chemical equilibrium
that is easily reversible and can be studied quantitatively for concentration effects.

$$FeCNS^{2+}(aq) \rightleftharpoons Fe^{3+}(aq) + CNS^-(aq)$$

Table 9.3 shows the experimentally determined concentration of each of three ions
in four different solutions. It is obvious from these measurements that when concentrations of dissociation products, $[Fe^{3+}]$ and $[CNS^-]$, are multiplied together, the
product is *not* constant. In the column "Ion Product of Products" this quantity varies
from 10^{-6} to 10^{-4}. On the other hand, if each one of these products is divided by the
corresponding concentration of the reactant* $FeCNS^{2+}$, a constant quantity, 1×10^{-3},
is obtained, within experimental error. The results of these calculations are shown in
the last column of Table 9.3. The quotient

$$\frac{[Fe^{3+}][CNS^-]}{[FeCNS^{2+}]}$$

*This is the usual convention—substances on the left of a chemical equation are called **reactants** and those
on the right **products.**

TABLE 9.3 Calculations based on the equilibrium of $FeCNS^{2+}(aq)$ with its dissociation products $Fe^{3+}(aq)$ and $CNS^-(aq)$ at 25°C.

Solution	Ion Concentration			Ion Product of Products $[Fe^{3+}] \times [CNS^-]$	Ion Product of Products Over Reactant $\dfrac{[Fe^{3+}] \times [CNS^-]}{[FeCNS^{2+}]}$
	$[Fe^{3+}]$	$[CNS^-]$	$[FeCNS^{2+}]$		
Figure 9.8a, original equilibrium mixture	10^{-3}	10^{-3}	10^{-3}	10^{-6}	1×10^{-3}
Figure 9.8b, $[Fe^{3+}]$ increased to 10^{-2} M	10^{-2}	2×10^{-4}	1.8×10^{-3}	2×10^{-6}	1.1×10^{-3}
Figure 9.8c, $[CNS^-]$ increased to 2×10^{-2} M	10^{-4}	2×10^{-2}	1.9×10^{-3}	2×10^{-6}	1.1×10^{-3}
Both $[Fe^{3+}]$ and $[CNS^-]$ increased to 10^{-2} M	10^{-2}	10^{-2}	10^{-1}	10^{-4}	1×10^{-3}

is called the equilibrium expression for reaction (9.2) when equilibrium concentrations are used.

$$FeCNS^{2+}(aq) \rightleftharpoons Fe^{3+}(aq) + CNS^-(aq) \qquad (9.2)$$

The numerical value of the equilibrium expression is called the equilibrium constant, K.

$$\frac{[Fe^{3+}][CNS^-]}{[FeCNS^{2+}]} = K = 1 \times 10^{-3} \qquad (25°C)$$

The convention of writing products over reactants in the equilibrium expression is universally accepted by chemists. A reaction can, of course, be written in reverse, provided that you take the *reciprocal* of the equilibrium constant.

$$Fe^{3+}(aq) + CNS^-(aq) \rightleftharpoons FeCNS^{2+}(aq) \qquad (9.3)$$

$$\frac{[FeCNS^{2+}]}{[Fe^{3+}][CNS^-]} = \frac{1}{K} = K' = 1000 \qquad (25°C)$$

The new equilibrium constant is K'. If there could be any doubt about the meaning of an equilibrium constant you should write down the reaction along with the constant. It is always necessary to specify the *temperature* of reaction because the equilibrium is usually quite sensitive to temperature changes.

EXAMPLE 2. Vinegar is an aqueous solution of 5% acetic acid, by weight. The acetic acid molecule, HOAc, ionizes in water to form H^+ and OAc^- (acetate ion). The per cent ionization of HOAc is not large, which makes it a weak electrolyte. Equilibrium molar concentrations of each molecule and ion are

$$HOAc(aq) \rightleftharpoons H^+(aq) + OAC^-(aq)$$
$$0.8\ M \qquad 4 \times 10^{-3}M \quad 4 \times 10^{-3}M \qquad (25°C).$$

(a) Write the equilibrium expression and calculate the equilibrium constant.

(b) Calculate the equilibrium constant for the reverse reaction.

Solution:

(a) $$K = \frac{[H^+][OAc^-]}{[HOAc]}$$

$$= \frac{(4 \times 10^{-3})(4 \times 10^{-3})}{0.8} = 2 \times 10^{-5} \qquad (25°C)$$

(b) $$H^+(aq) + OAc^-(aq) \rightleftharpoons HOAc(aq)$$

$$K' = \frac{[HOAc]}{[H^+][OAc^-]}$$

$$= \frac{0.8}{(4 \times 10^{-3})(4 \times 10^{-3})} = 50{,}000 \qquad (25°C)$$

You will notice that small equilibrium constants mean that the corresponding reactions produce very little product and large equilibrium constants mean the opposite.

Pressure Effect on Chemical Equilibrium: More on Equilibrium Constants

Pressure has little effect on reactions in which none of the reactants or products is a gas because there is too small a volume change during reaction to relieve any added pressure. Of course the effects of increased pressure can be predicted by Le Chatelier's principle if the increase is great enough (for example, 100,000 atm in the transformation of graphite to diamond, Section 9.2, Example 4.

Earlier in this section you saw the effect of temperature on the $N_2O_4(g) \rightleftharpoons 2\ NO_2(g)$ equilibrium. Now consider the effect of pressure, which should be considerable on this gas-phase reaction.

EXAMPLE 3. The total number of gas molecules in the N_2O_4–NO_2 system changes as the reaction proceeds, so the equilibrium should respond to pressure. In this example you must imagine a container whose volume may be changed, such as the piston and cylinder arrangement shown in Figure 9.9. Which is increased when the total pressure is decreased, the number of moles of N_2O_4 or the number of moles of NO_2?

Solution: Le Chatelier's principle is applied as follows:

1. An equilibrium exists in Figure 9.9a such that

$$N_2O_4(g) \rightleftharpoons 2\ NO_2(g) \qquad (9.4)$$

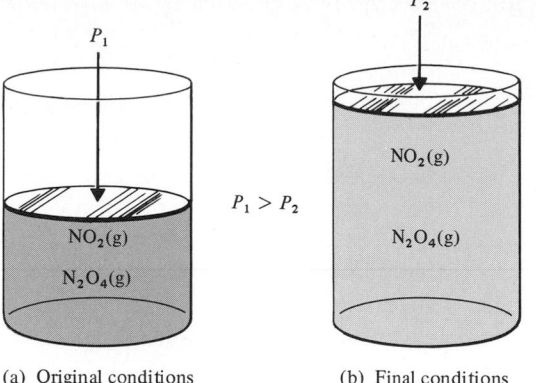

(a) Original conditions   (b) Final conditions

Figure 9.9. Decreasing the pressure on an equilibrium mixture of $NO_2(g)$ and
$N_2O_4(g)$.

2. A condition is changed in Figure 9.9b: pressure is decreased.
3. A process occurs: either,
  (a) $N_2O_4(g) \longrightarrow 2\,NO_2(g)$, or
  (b) $2\,NO_2(g) \longrightarrow N_2O_4(g)$
4. which tends to restore the original conditions: The process tends to increase pressure.
5. Process (a) produces two molecules for every one that is consumed and tends to increase pressure. Process (b) produces one molecule for every two that are consumed and tends to decrease pressure.
6. Therefore, process (a) occurs. The number of moles of N_2O_4 decreases and the number of moles of NO_2 increases.

Equilibrium (9.4) can be expressed in terms of an equilibrium constant like those equilibria for aqueous solutions. There is a certain number of moles of each gas in 1 ℓ of mixture, which represents the molarity of that substance. The concentrations are variable and their experimental measurement under a variety of pressures leads to the following interesting results.

$$\frac{[NO_2]}{[N_2O_4]} \neq \text{a constant} \qquad \text{but} \qquad \frac{[NO_2]^2}{[N_2O_4]} = \text{a constant}$$

This constant has the value 6×10^{-3} at 25°C, meaning that, if N_2O_4 were at a concentration of 1 M (about 20 atm), only a small percentage would be in the form of NO_2 (about 8%).

It is apparent then that in the balanced chemical equation,

$$N_2O_4(g) \rightleftharpoons 2\,NO_2(g)$$

the coefficient of NO_2 becomes an exponent in the equilibrium expression,

$$\frac{[NO_2]^2}{[N_2O_4]} = K.$$

If you treated the problem in Example 3 quantitatively with the help of this equilibrium expression, you would find that the *concentration* of $NO_2(g)$ becomes smaller because of the increase in volume. The color intensity, which depends on concentration, would also decrease. However, Le Chatelier's principle does give the right answer because the *number of moles* of $NO_2(g)$ is increased, by an amount calculated using the equilibrium constant.

EXAMPLE 4. In the synthesis of $NH_3(g)$ (ammonia) by the Haber process, $N_2(g)$ and $H_2(g)$ are mixed together under pressure (about 350 atm) and at elevated temperatures (about 500°C) in the presence of a catalyst (mixture of Fe, K, and Al oxides). Write the equilibrium expression for this reaction.

Solution: The first step is to write a balanced chemical equation.

$$N_2(g) + 3\,H_2(g) \rightleftharpoons 2\,NH_3(g)$$

In the equilibrium expression the product is written in the numerator, the reactants in the denominator, and each coefficient becomes an exponent.

$$\frac{[NH_3]^2}{[N_2][H_2]^3} = K$$

**Summary:
A General
Equilibrium
Expression**

A hypothetical balanced chemical equation is shown in equation (9.5). Some substances are dissolved in dilute solution and one *solvent* molecule, C, is consumed as a reactant.

$$3\,A(\text{sol'n}) + 2\,B(s) + C \rightleftharpoons D(\text{sol'n}) + 2\,E(g) \tag{9.5}$$

The symbols (s) and (g) have their usual meaning. The equilibrium expression is set up in the following manner.

$$[B] = 1, \text{ a pure solid}$$
$$[C] = 1, \text{ the solvent}$$
$$\frac{[D][E]^2}{[A]^3} = K, \text{ the equilibrium constant}$$

This is a general equilibrium expression.

PROBLEMS

1. Identify each of the following systems and indicate whether it is "closed" or "open" according to equilibrium theory.
   (a) an anthill with constant population

   (b) a bottle of fermenting grape juice with a rubber balloon attached to trap the $CO_2(g)$ produced
   (c) one of the beakers containing water vapor in Figure 7.6.

2. Same as problem 1.

   (a) the bomb calorimeter in Figure 7.1 during the combustion of benzoic acid
   (b) a beehive with constant population

   (c) the internal combustion engine of an automobile.

3. What is wrong with the following statement? A tank of $N_2O_4(l)$ under 5 atm pressure is opened to the atmosphere and a gas, consisting primarily of NO_2, is released; this process is reversible because the $N_2O_4(l)$ can be produced again by applying pressure to the escaped gas. How could the process be carried out to make the statement correct?

4. What is wrong with the following statement? When a drop of water falls on a hot skillet it vaporizes into $H_2O(g)$; this process is reversible because, by collecting the $H_2O(g)$ and applying pressure to it, the drop of $H_2O(l)$ can be reformed. How could the process be carried out to make the statement correct?

5. Is the inside of a closed refrigerator a closed system? Is it at equilibrium? Explain.

6. Is the inside of a closed air conditioned home in which the cool air is simply recycled a closed system? Is it at equilibrium? Explain.

7. The Russian army found that the tin buttons on their uniforms suffered from "tin pest" and disintegrated in the cold Russian winters.

$$Sn \rightleftharpoons Sn$$
$$\text{(white metal)} \qquad \text{(grey powder)}$$

The transition does not occur at temperatures above 18°C, but as the temperature is decreased, it becomes quite rapid. Is the process endothermic or exothermic? Explain using Le Chatelier's principle.

8. Freeze-dried coffee is manufactured by freezing the solution and then causing the water to separate from it at a constant temperature below its melting point. Use Le Chatelier's principle to explain how this process could be accomplished. *Hint:* Identify the (s, g) curve in Figure 9.1, which represents the equilibrium between $H_2O(s)$ and $H_2O(g)$.

9. In the steam turbine engine high pressure steam impinges on a paddle wheel, which it turns, and then condenses on the other side. There are advantages to using substances with higher boiling points, like metallic Na or Hg. Discuss whether the Na and Hg systems are closed or open. Is the process they undergo reversible?

10. One of the problems of depending on solar energy to heat our homes is how to save it for a rainy day. One method is to collect the heat in water pipes outdoors, which pass through a heat storage container, and then back outdoors. Heat is stored in the container by the following process of solution.

$$16 \text{ kcal} + Na_2SO_4 \cdot 10\ H_2O(s) + aq \rightleftharpoons$$
$$Na_2SO_4(aq) \qquad \text{at } 32.3°C$$

When needed to warm the house, heat is exchanged with the air indoors [Note: $32.3°C = 90.1°F$], which reverses the process in the container. Discuss whether the heat storage container system is closed or open. Is the process reversible?

11. Make a sketch of the phase diagram of H_2O (Figure 9.1). Indicate the following processes by using short lines on the sketch with arrows to show the direction of change.
   (a) Melting of ice at a constant temperature below 0°C.
   (b) Vaporization of water at a constant pressure below 100°C.

12. Same as problem 11.

   (a) Sublimation of ice at a constant pressure.
   (b) Freezing water at a constant pressure above 1 atm.

13. Would you expect the addition of the white ionic compound $NaNO_3(s)$ to decrease the solubility of KCl in a saturated aqueous solution? Explain.

14. Would you expect the addition of the white ionic compound $KNO_3(s)$ to decrease the solubility of NaCl in a saturated aqueous solution? Explain.

15. Calculate the solubility in mole per liter of AgCl, which is given by the numerical value of $[Ag^+]$
   (a) in pure water. Use the data in Table 9.2.
   (b) What is its solubility in 0.01 M KCl?

16. Calculate the solubility in mole per liter of AgBr, which is given by the numerical value of $[Ag^+]$
   (a) in pure water. Use the data in Table 9.2.
   (b) What is its solubility in 0.001 M NaBr?

17. Choose conditions of pressure and temperature that would maximize the solubility of $CO_2(g)$ in water according to the equilibrium,

$$CO_2(g) + aq \rightleftharpoons CO_2(aq) + heat$$

How does this compare with the conditions causing flat Coca Cola (or flat beer)?

18. When divers return from deep water they must make their ascent slowly in order to avoid the bends. This malady is a result of dissolved nitrogen forming bubbles in the blood when it is released too rapidly. (a) Explain, using Le Chatelier's principle. (b) Under what conditions could air pilots experience the same sickness?

19. One of the following chemical reactions is not favored by either high pressure or low. Which one is it? Explain.

(a) $C_2H_2(g) + 2 H_2(g) \rightleftharpoons C_2H_6(g)$

(b) $H_2(g) + I_2(g) \rightleftharpoons 2 HI(g)$

20. Same as problem 19.

(a) $H_2(g) + CO_2(g) \rightleftharpoons H_2O(g) + CO(g)$

(b) $C_2H_4(g) + H_2(g) \rightleftharpoons C_2H_6(g)$

21. The following reaction offers the possibility of converting coal to a gaseous fuel, which is more convenient to use and burns with less pollution.

$$C(s) + H_2O(g) \rightleftharpoons H_2(g) + CO(g)$$

(a) Give the equilibrium expression in molarities. (b) Would you recommend looking for a catalyst to shorten the time for equilibration at 25°C? $K = 4 \times 10^{-18}$ at 25°C. Explain. (c) At high temperatures this reaction is practical and the product is called "water gas." Is the reaction endothermic or exothermic? Explain.

22. The following type of reaction offers the possibility of converting coal to hydrocarbon fuel, which is in short supply compared with world demand.

$$C(s) + 2 H_2(g) \rightleftharpoons CH_4(g)$$

(a) Give the equilibrium expression in molarities. (b) Would you recommend looking for a catalyst to shorten the time for equilibration at 25°C? $K = 2 \times 10^{10}$ at 25°C. Explain. (c) Since the reaction is exothermic would equilibrium at higher temperatures favor more product or less? Explain.

23. At 25°C the equilibrium constant for the following reaction is $K = 0.006$ in molarity.

$$N_2O_4(g) \rightleftharpoons 2 NO_2(g)$$

What is K for the reverse reaction?

$$2 NO_2(g) \rightleftharpoons N_2O_4(g)$$

24. At 25°C the equilibrium constant for the following reaction is $K = 5 \times 10^8$ in molarity.

$$N_2(g) + 3 H_2(g) \rightleftharpoons 2 NH_3(g)$$

What is K for the reverse reaction?

$$2 NH_3(g) \rightleftharpoons N_2(g) + 3 H_2(g)$$

25. Write equilibrium expressions for the following reactions. Follow the usual conventions of setting some concentrations equal to unity.

(a) $Al(OH)_3(s) + aq \rightleftharpoons Al^{3+}(aq) + 3 OH^-(aq)$

(b) $NH_3(aq) + H_2O \rightleftharpoons NH_4^+(aq) + OH^-(aq)$

(c) $Br_2(l)^* + 2 Cl^-(aq) \rightleftharpoons 2 Br^-(aq) + Cl_2(g)$

(d) $2 MnO_4^-(aq) + 3 H_2O_2(aq) \rightleftharpoons 2 MnO_2(s) + 3 O_2(g) + 2 OH^-(aq) + 2 H_2O$

26. Same as problem 25.

(a) $MgF_2(s) + aq \rightleftharpoons Mg^{2+}(aq) + 2 F^-(aq)$

(b) $HOAc(aq) + H_2O \rightleftharpoons H_3O^+(aq) + OAC^-(aq)$

(c) $I_2(s) + 2 Br^-(aq) \rightleftharpoons 2 I^-(aq) + Br_2(l)^*$

(d) $2 Cr(OH)_3(s) + IO_3^-(aq) + 4 OH^-(aq) \rightleftharpoons 2 CrO_4^{2-}(aq) + I^-(aq) + 5 H_2O$

*Note: pure liquids behave like pure solids in equilibrium expressions.

*27. Using the solubility curves for KCl and NaCl in Figure 9.4 make the following predictions:
  (a) Which salt will dissolve to a greater extent at 20°C? At 60°C?
  (b) What is the least amount of water required to dissolve 20 g KCl at 40°C?

*28. Salt is mixed with ice and water in making "home-made" ice cream. The purpose of the salt is to lower the temperature below 0°C. Explain this effect by means of Le Chatelier's principle. (*Hint:* The salt dissolves in the water and "dilutes" it, but the ice remains free of salt.)

10

ACIDS AND BASES

How do you know what reacts with what? By studying classes of chemical reactions and applying the principles covered up to this point, you will be able to predict with confidence a very large number of reactions. The chemical reactions between acids and bases belong to a large and useful class.

Many important industrial processes involve acid-base reactions. The phosphate industry in Florida brings molten sulfur (mp = 119°C) from Texas and Louisiana in insulated barges through the intracoastal waterway to the west coast of Florida where it is converted into sulfuric acid, H_2SO_4. Very slightly soluble phosphate rock, $Ca_3(PO_4)_2$, strip-mined in central Florida, is brought to the coast and treated with this acid. The result is an acid-base reaction that produces calcium dihydrogen phosphate, $Ca(H_2PO_4)_2$, a solid compound which is water soluble (2 g/100 g H_2O). The importance of this reaction to a hungry world is that the product is usable as a fertilizer. Phosphate, along with potassium and nitrogen, must be present in the soil to produce good crops.

Other important acid-base reactions are perfectly natural. The bloodstream contains a delicate balance of phosphoric acid (H_3PO_4) and carbonic acid (H_2CO_3). These acids and their products protect and facilitate the enzymes in the blood that catalyze many biochemical reactions essential to the human body. The digestion of a steak dinner requires the presence in the stomach of about 50 ml 0.08 M hydrochloric acid (HCl). Fortunately this acid does *not* react with the walls of a healthy stomach!

10.1 PROPERTIES OF ACIDS AND BASES

The word "acid" means "sour" and this property of taste is characteristic of all acids. On the other hand the most obvious characteristic of a base is provided by the sense of touch—it feels slippery or soapy. Soap itself is a base.

There was a time when chemists believed that the essential element in all acids was oxygen. In fact the German word for oxygen is "Sauerstoff," and the English word comes from the Greek words meaning "acid forming." Today chemists believe that it is *hydrogen* which is the most characteristic element in acids. This is demonstrated by the fact that an active metal displaces $H_2(g)$ from any acid. A typical reaction of an active metal (zinc, Zn) with an acid (H_2SO_4) is shown in Figure 10.1. The test tube, which has been filled previously with water to exclude air, becomes filled with $H_2(g)$ and a small amount of $H_2O(g)$ (Figure 10.1a). The mixture can be tested for combustibility with a burning splint (Figure 10.1b). Some metals, such as copper (Cu), silver (Ag), and gold (Au) do not displace $H_2(g)$, and some acids produce other gases in addition to $H_2(g)$. This reaction is not limited to acids because the combination of many metals with bases produce $H_2(g)$.

An interesting test for the presence of acid or base is the litmus test. Litmus is a dye that turns red in acid and blue in base. Such dyes are very common in nature and are called acid-base **indicators.** You have probably noticed the lighter color produced in tea when lemon juice, containing citric acid [$HOOCCH_2C(OH)(COOH)$ CH_2COOH] is added. A reaction you probably have not noticed happens when a little household ammonia (aqueous solution of NH_3, a base) is added to red grape juice—a change to green which is changed back to red by adding excess vinegar (an

Figure 10.1. Collection of $H_2(g)$ from the reaction of Zn(s) with H_2SO_4(aq).

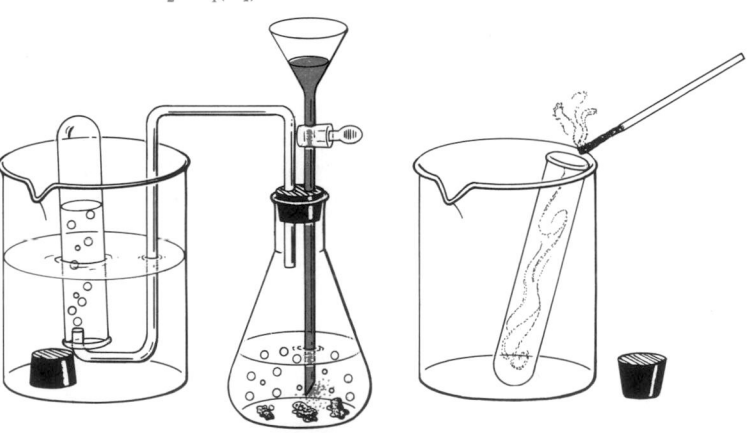

(a) Gas collection under water   (b) Test for combustibility

aqueous solution of CH_3COOH, acetic acid). These reactions provide evidence that indicators themselves are acids and bases (Section 11.4).

In addition to these properties, acids and bases conduct electricity in aqueous solution (Table 8.1). They are therefore classified as electrolytes. A strong acid is simply an acid which is a strong electrolyte—it ionizes completely when it dissolves in water. A weak acid ionizes only partially in aqueous solution. Similar statements can be made for strong and weak bases. A list of the most common acids and bases is shown in Table 10.1.

TABLE 10.1 The common acids and bases

Formula	Name	Formula	Name
		Acids	
*HCl	hydrochloric acid	†H_3PO_4	phosphoric acid
*HBr	hydrobromic acid	H_3PO_3	phosphorus acid
*HI	hydroiodic acid		
HF	hydrofluoric acid		
		*$HClO_4$	perchloric acid
H_2S	hydrogen sulfide	*$HClO_3$	chloric acid
	(hydrosulfuric acid)	$HClO_2$	chlorous acid
		$HClO$	hypochlorous acid
†*H_2SO_4	sulfuric acid	*HIO_4	periodic acid
H_2SO_3	sulfurous acid		
		CH_3COOH (HOAc)	acetic acid
*HNO_3	nitric acid	H_2CO_3	carbonic acid
HNO_2	nitrous acid		
		Bases	
*$NaOH$	sodium hydroxide	NH_3	ammonia
*KOH	potassium hydroxide	‡Na_2CO_3	sodium carbonate
*$Ca(OH)_2$	calcium hydroxide	‡$NaHCO_3$	sodium bicarbonate
*$Ba(OH)_2$	barium hydroxide		(sodium hydrogen carbonate)

Strong.
†*The first H^+ in H_3PO_4 and the second H^+ in H_2SO_4 are moderately strong.*
‡*Strong as salts, but weak as bases.*

It is important to learn which acids and bases are strong and which are weak in order to predict chemical reactivity. Although metal hydroxides like $Fe(OH)_3$ and $Cu(OH)_2$ are basic, they are *not* listed because they are very slightly soluble in water and are not useful as bases in aqueous reactions. A summary of the characteristic properties of acids and bases, which have been discussed in this section, and a new property, which will be discussed in the next section, is provided in Table 10.2.

EXAMPLE 1. A certain dye called phenolphthalein is perfectly colorless in pure water, but in the presence of ammonia (NH_3) it is a brilliant pink. How would you show that phenolphthalein is an acid-base indicator?

TABLE 10.2 Characteristic properties of acids and bases

Acids	Bases
1. Taste sour.	1. Feel slippery.
2. React with most metals to produce $H_2(g)$.	2. Some react with metals to produce $H_2(g)$.
3. Turn blue litmus and green grape juice red.	3. Turn red litmus blue and red grape juice green.
4. Are electrolytes.	4. Are electrolytes.
5. Donate H^+ when they react with bases.	5. Accept H^+ when they react with acids.

Solution: On the addition of excess acid such as vinegar or lemon juice, it should become colorless again.

As you will notice, some of the properties listed in Table 10.2 do not differentiate between an acid and a base. However, the fate of the proton when acids and bases react provides an excellent differentiation. It will explain how the same substance can behave as an acid with some bases and as a base with some acids. It further describes every acid-base reaction in such a way that the acid becomes a base and the base becomes an acid. Our goal is to predict whether the reaction takes place or not.

10.2 TRANSFER OF PROTONS: THE ACID-BASE REACTION

The last entry in Table 10.2 states that acids *donate* protons and bases *accept* protons when they react. In equation (10.1) a proton is donated by acetic acid (CH_3COOH or HOAc) and is accepted by the hydroxide ion (OH^-) of sodium hydroxide. Since acetic acid is weak (Table 10.1), it is written in the nonionized form; since sodium hydroxide, like most metal hydroxides, is strong (Table 10.1), it is written in the ionized form.

$$\underset{\substack{\text{proton} \\ \text{donor}}}{HOAc(aq)} + Na^+(aq) + \underset{\substack{\text{proton} \\ \text{acceptor}}}{OH^-(aq)} \longrightarrow Na^+(aq) + OAc^-(aq) + H_2O(l) \qquad (10.1)$$

The H_2O molecule, which is a product of the combination of H^+ and OH^-, must appear on the right in addition to the "aq" symbols, which indicate water as a solvent.

Since $Na^+(aq)$ appears on both sides of the equation, it does not enter into the reaction and is included only to show equal amounts of positive and negative charge (electroneutrality) on each side of the equation. Frequently you will find it convenient to omit ions of this kind in order to emphasize the "real" reaction.

$$HOAc(aq) + OH^-(aq) \longrightarrow OAc^-(aq) + H_2O(l) \qquad (10.2)$$

An ionic equation, like equation (10.2), is not balanced unless all atoms are balanced, *and* the charge is balanced. In this case the charge is −1 on left and right.

EXAMPLE 1.   Is the following ionic equation balanced?

$$H_2PO_4^- + 2\,OH^- \longrightarrow PO_4^{3-} + 2\,H_2O(l)$$

Solution: The numbers below correspond to the molecules and ions above.

H:	2 +	2	=		2(2)	
P:	1		=	1		
O:	4 +	2	=	4 +	2	
charge:	−1 + 2(−1)		=	−3		

since the atoms and the charges are equal on both sides of the reaction, the equation *is* balanced.

Strong Acids and the Hydronium Ion

Unlike the weak acid HOAc, the strong acid HCl must ionize completely if it is in aqueous solution. Since most of the acid-base reactions in which you will be interested are carried out in the presence of water, it is important to consider the special ions that are formed under these conditions.

When the covalent polar molecules of HCl(g) are allowed to flow into a sample of $H_2O(l)$ to become HCl(aq), the ionization of HCl is complete.

$$HCl(aq) \xrightarrow{100\%} H^+(aq) + Cl^-(aq)$$
a strong acid

The H^+ ion does not really exist in aqueous solution. Since it is only a proton, it has an extremely small diameter (Figure 5.2) and is able to approach the negative side of the water dipole very closely, giving rise to a large force of attraction. In addition to this argument you may consider the octet rule. This rule predicts that an H atom in a stable molecule or ion will have associated with it an electron *pair*. The most likely structure in aqueous solution satisfying these requirements is the **hydronium ion.** It is a pyramidal arrangement of three equivalent H atoms and an O atom as shown.

hydronium ion

The proton simply attaches itself to one of the unshared pairs of electrons in the water molecule. Symbols used in this text to represent the hydrogen ion in aqueous solution will vary depending on convenience and the emphasis required. At times all of the following symbols will be used to represent the same ion, if it is clear that a bare proton is not intended.

$$H^+ \equiv H^+(aq) \equiv H_3O^+$$

hydrogen ion   hydrogen ion in aqueous solution   hydronium ion

EXAMPLE 2. Some acid-base studies have been made using liquid ammonia (NH_3, bp = $-33°C$) as the solvent instead of water. How would you expect HCl(g) to ionize when dissolved in NH_3(l)?

Solution: This reaction, which is very vigorous, will be discussed further in Section 10.3. Since NH_3 is a polar molecule like H_2O (Figure 6.4), you would expect H^+ to attach itself to an unshared pair of electrons on the negative side. The structure is tetrahedral and all the hydrogens are equivalent.

ammonium ion

If the excess ammonia is allowed to evaporate, a white crystalline solid remains which is ammonium chloride: NH_4Cl.

The "real" process that occurs when the strong acid HCl(g) is dissolved in H_2O(l) is shown in equation (10.3).

$$HCl(g) + H_2O(l) \longrightarrow H_3O^+ + Cl^- \qquad (10.3)$$

The reaction between HCl and NaOH in aqueous solution is somewhat different from that between the weak acid HOAc and NaOH in equation (10.1) although the products are similar. The ionic equation is

$$H_3O^+(aq) + Cl^-(aq) + Na^+(aq) + OH^-(aq) \longrightarrow$$
$$Cl^-(aq) + Na^+(aq) + 2\,H_2O(l). \qquad (10.4)$$

Eliminating those ions which appear on both the left and the right gives the "real" reaction shown in equation (10.5).

$$H_3O^+(aq) + OH^-(aq) \longrightarrow 2\,H_2O(l) \qquad (10.5)$$

This equation is always a part of any reaction between a strong acid and a strong base.

One of the advantages of learning your acids and your bases (fully as important as p's and q's to the chemist) is the tremendous number of *salts* that can be predicted from their reactions. For example, ordinary table salt or rock salt (NaCl) is the product of equation (10.4). Although you were told that no products were formed in this reaction except water, it is possible to obtain another product. If the water can be separated from the Na^+ and Cl^- ions by evaporation or if the ions can be precipitated by adding either KCl (Section 9.3, Example 2) or a solvent less polar than water, a new product *is* formed—NaCl(s).

DEFINITION: A **salt** is a compound (usually ionic) that can be prepared by combining the negative ion of an acid and the positive ion of a base.

Likewise in equation (10.1), distillation of the water or precipitation of the ions produces the white ionic solid NaOAc (sodium acetate).

In general, hydroxide bases and acids produce salts plus water.

$$\text{acid} + \text{(hydroxide) base} \longrightarrow \text{a salt} + \text{water}$$

EXAMPLE 3. Write a balanced equation for the acid-base reaction between hydrobromic acid and calcium hydroxide in aqueous solution. Which salt can be recovered as a product?

Solution: Since both acid and base are strong (Table 10.1), the dissolved compounds are entirely ionic.

$$H^+ + Br^- + Ca^{2+} + 2\,OH^- \longrightarrow \text{products}$$

The real reaction is between H^+ and OH^- to form H_2O. Since there are $2\,OH^-$ ions, it is necessary to provide $2\,H^+$ ions, which are accompanied by $2\,Br^-$ ions.

$$2\,H^+ + 2\,Br^- + Ca^{2+} + 2\,OH^- \longrightarrow Ca^{2+} + 2\,Br^- + 2\,H_2O$$

The real reaction is

$$2\,H^+ + 2\,OH^- \longrightarrow 2\,H_2O.$$

Precipitation of the ionic product gives the white solid salt $CaBr_2$ (calcium bromide).

All of the binary (two-element) compounds produced from ions in Table 6.3 are considered salts except the oxides, which are a special case to be discussed in Section 11.5. In naming the binary acid corresponding to each of the negative ions, the suffix "-ic" is used (Table 10.3, page 229). Although metal hydrides, such as NaH, are considered salts, there is no acid corresponding to the hydride ion, H^-.

EXAMPLE 4. Give the formulas and names of the binary acids of F^- and S^{2-}.

Solution: HF (hydrofluoric acid); H_2S (hydrosulfuric acid, correct, but usually called hydrogen sulfide to avoid confusion with the more important sulfuric acid, H_2SO_4).

Note that when ammonia, NH_3, is the base, the product of reaction with an acid is a salt, but no water is produced. See Example 2, p. 224.

Polyprotic Acids and Stepwise Ionization

Some acids contain two, three, or even four replaceable protons. They are called **polyprotic** acids. Unlike HOAc and HCl, which have only one replaceable proton, H_2SO_4 (sulfuric acid) and H_3PO_4 (phosphoric acid) have two and three replaceable protons, respectively. Scale models of H_2SO_4 and H_3PO_4 molecules are shown in Figure 10.2. Note that the S and P atoms are centrally located (not visible in the models) and that the H atoms are attached to the O atoms. This is typical of the oxyacids, which is the name given to those acids containing oxygen. If the hydrogen is not bound to the oxygen but to the central atom, it is not released in water and it is not considered acidic.

The second proton is always considerably more difficult to remove than the first. This is because the second proton must be drawn away from an ion charged 2⁻ instead of 1⁻.

Sulfuric acid is strong with respect to its first proton but only moderately strong with respect to its second. For example, in a 0.1 M aqueous solution of H_2SO_4, the changes shown in equations (10.6) occur in reaching equilibrium.

$$H_2SO_4 \xrightarrow{\text{100%}} H^+ + HSO_4^-$$

hydrogen
sulfate ion

0.1 M H_2SO_4
originally present

$$HSO_4^- \xrightarrow{\text{10%}} H^+ + SO_4^{2-}$$

sulfate
ion

(10.6)

Figure 10.2. Scale models of the polyprotic acids H_2SO_4 (left) and H_3PO_4 (right). The S and P atoms are centrally located and hidden from view.

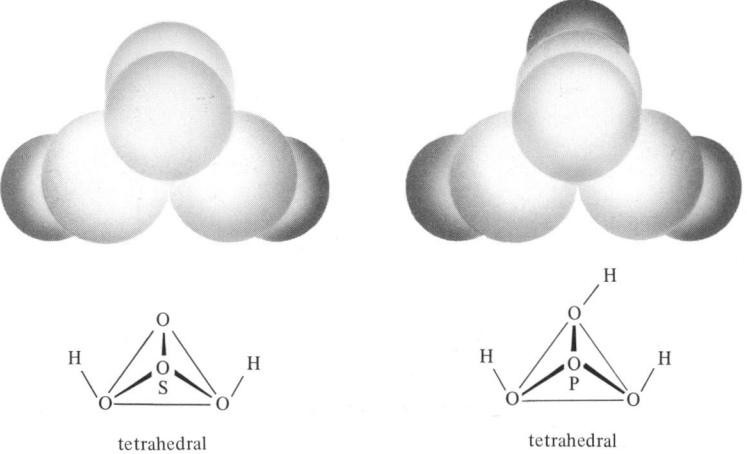

tetrahedral   tetrahedral

Although practically all of the original H_2SO_4 ionizes into HSO_4^- and H^+, only 1 in 10 of the HSO_4^- ions dissociate into SO_4^{2-} and additional H^+.

The difference in degree of proton ionization is even more striking in the case of a 0.1 M aqueous solution of H_3PO_4.

$$H_3PO_4 \xrightarrow{23\%} H^+ + \underset{\substack{\text{dihydrogen} \\ \text{phosphate ion}}}{H_2PO_4^-}$$

$$H_2PO_4^- \xrightarrow{10^{-4}\%} H^+ + \underset{\substack{\text{monohydrogen} \\ \text{phosphate ion}}}{HPO_4^{2-}} \qquad \begin{array}{c} \text{0.1 M } H_3PO_4 \\ \text{originally present} \end{array} \qquad (10.7)$$

$$HPO_4^{2-} \xrightarrow{10^{-15}\%} H^+ + \underset{\text{phosphate ion}}{PO_4^{3-}}$$

Since almost one in four H_3PO_4 molecules ionize into H^+ and $H_2PO_4^{2-}$, it is considered moderately strong with respect to the first H^+ ion. However, the $H_2PO_4^-$ and HPO_4^{2-} acids are considered weak.

Reactions of Polyprotic Acid with Base

The real reaction between a polyprotic acid and a hydroxide base is to combine one or more protons from the acid with hydroxide ions from the base to form water.

EXAMPLE 5.   Write a balanced equation in which NaOH reacts with H_2SO_4 to produce Na_2SO_4 (sodium sulfate).

Solution: During the course of reaction enough base is added to H_2SO_4 to strip it of both its protons.

$$H_2SO_4 \longrightarrow 2\,H^+ + SO_4^{2-}$$

Likewise NaOH loses its OH^- ion, and both sides are multiplied by two to provide equal numbers of H^+ and OH^- ions.

$$2\,(NaOH \longrightarrow Na^+ + OH^-)$$

Add both sides of these equations, combining $2\,H^+$ and $2\,OH^-$ to form $2\,H_2O$.

$$H_2SO_4 + 2\,NaOH \longrightarrow 2\,H_2O + Na_2SO_4$$

EXAMPLE 6.   Complete and balance the following acid-base reaction to produce the indicated salt.

$$\text{—}Ca(OH)_2 + \text{—}H_3PO_4 \longrightarrow \text{—}Ca(H_2PO_4)_2 + \text{—}\text{———}$$

Solution: A metal hydroxide reacts with an acid to give salt plus water. To determine the amount of water write the equations in ionic form.

$$Ca(OH)_2 \longrightarrow Ca^{2+} + 2\ OH^-$$

Only enough base is added to produce $H_2PO_4^-$. Multiply the equation below by 2 to balance H^+ and OH^-.

$$2\ (H_3PO_4 \longrightarrow H^+ + H_2PO_4^-)$$

Add both sides of the equations combining $2\ H^+$ and $2\ OH^-$ to form $2\ H_2O$.

$$Ca(OH)_2 + 2\ H_3PO_4 \longrightarrow Ca(H_2PO_4)_2 + 2\ H_2O$$

Exercise 1. Complete and balance the following acid-base reactions to produce the indicated salt.

(a) __NaOH + __H_3PO_4 $\longrightarrow$ __Na_2HPO_4 + __ _____

(b) __$Ba(OH)_2$ + __H_2SO_4 $\longrightarrow$ __$Ba(HSO_4)_2$ + __ _____

Answer: (a) 2, 1, 1, 2 H_2O. (b) 1, 2, 1, 2 H_2O.

The first rule of nomenclature is to provide a unique name for every compound; the second rule is to keep the name as simple as possible. Some of the specific rules of nomenclature have already been given, and they are indicated by example in Table 10.3.

There are families of oxyacids which differ only in the number of oxygen atoms they contain. The most common member of the family, called the "parent" acid, has the suffix "-ic". Increasing amounts of oxygen starting with the least amount are indicated in the following sequence of prefix-suffix combinations.

least oxygen: hypo- _____ -ous acid

_____ -ous acid

_____ -ic acid

most oxygen: per- _____ -ic acid

Examples are given in Table 10.3.

The rules for naming the salts of acids are as follows:

1. Name the metal first; NH_4^+ has the "-um" ending of a metal, ammonium.

2. For binary acids (HF, HCl, H_2S, and so on) an "-ic" acid becomes an "-ide" salt. *For example:* hydrochlor*ic* acid $\longrightarrow$ sodium chlor*ide*.

3. For oxyacids (H_2SO_4, CH_3COOH, HNO_2, and so on).

(a) An "-ic" acid becomes an "-ate" salt. *For example:* acet*ic* acid $\longrightarrow$ potassium acet*ate*

Type of Compound	Examples	
	Name	Formula
I. Monoprotic acid		
A. binary acid		
hydro(nonmetal)ic acid	hydrochloric acid	HCl
B. salt of binary acid		
(metal) (nonmetal)ide	sodium chloride	NaCl
C. oxyacid		
most, per(nonmetal)ic acid	perchloric acid	$HClO_4$
(nonmetal)ic acid	chloric acid	$HClO_3$
(nonmetal)ous acid	chlorous acid	$HClO_2$
least, hypo(nonmetal)ous acid	hypochlorous acid	HClO
D. salt of oxyacid		
most: (metal) per(nonmetal)ate	sodium perchlorate	$NaClO_4$
(metal) (nonmetal)ate	sodium chlorate	$NaClO_3$
(metal) (nonmetal)ite	sodium chlorite	$NaClO_2$
least: (metal) hypo(nonmetal)ite	sodium hypochlorite	NaClO
II. Diprotic acid		
A. binary acid		
hydro(nonmetal)ic acid	hydrosulfuric acid	H_2S
B. salt of binary acid		
1. acid salt		
(metal) hydrogen(nonmetal)ide	sodium hydrogen sulfide	NaHS
2. normal salt		
(metal) (nonmetal)ide	sodium sulfide	Na_2S
C. oxyacid		
more: (nonmetal)ic acid	sulfuric acid	H_2SO_4
less: (nonmetal)ous acid	sulfurous acid	H_2SO_3
D. salt of oxyacid		
1. acid salt		
more: (metal) hydrogen (nonmetal)ate	sodium hydrogen sulfate	$NaHSO_4$
less: (metal) hydrogen (nonmetal)ite	sodium hydrogen sulfite	$NaHSO_3$
2. Normal salt		
more: (metal) (nonmetal)ate	sodium sulfate	Na_2SO_4
less: (metal) (nonmetal)ite	sodium sulfite	Na_2SO_3
III. Triprotic oxyacid		
A. acid		
more: (nonmetal)ic acid	phosphoric acid	H_3PO_4
less: (nonmetal)ous acid	phosphorous acid	H_3PO_3
B. salt		
1. acid salt, di-		
more: (metal) dihydrogen (nonmetal)ate	sodium dihydrogen phosphate	NaH_2PO_4
less: (metal) dihydrogen (nonmetal)ite	sodium dihydrogen phosphite	NaH_2PO_3
2. acid salt, mono-		
more: (metal) monohydrogen (nonmetal)ate	sodium monohydrogen phosphate	Na_2HPO_4
less: (metal) monohydrogen (nonmetal)ite	sodium monohydrogen phosphite	Na_2HPO_3
3. normal salt		
more: (metal) (nonmetal)ate	sodium phosphate	Na_3PO_4
less: (metal) (nonmetal)ite	sodium phosphite (non-existent)	Na_3PO_3

(b) An "-ous" acid becomes an "-ite" salt. *For example:* nitrous acid $\longrightarrow$ barium nit*rite*

Other examples are given in Table 10.3.

The number of H^+ ions that have not been replaced in a salt is indicated, if there is any doubt, by monohydrogen-, dihydrogen-, trihydrogen-, . . . for one, two, three, . . . remaining H^+ ions. If only one or no H^+ ions are possible in a salt, as in a salt of H_2SO_4 or H_2CO_3, it is not necessary to use "monohydrogen-" although it would not be wrong: for example, $KHSO_4$ is called potassium hydrogen sulfate. Other examples are given in Table 10.3.

It will help you remember some of the rules of nomenclature if you will fill in the blanks in a copy of the table in Exercise 2.

Exercise 2. In the following table various positive and negative ions are indicated, each of which comes from a different acid or base. Write the correct formula of the corresponding salt and provide its chemical name.

	Cl^-	SO_4^{2-}	NO_2^-	ClO_4^-	HPO_4^{2-}
Na^+	NaCl sodium chloride	(1)	(2)	$NaClO_4$ sodium perchlorate	(3)
Ca^{2+}	(4)	(5)	$Ca(NO_2)_2$ calcium nitrite	(6)	(7)
K^+	(8)	K_2SO_4 potassium sulfate	(9)	(10)	K_2HPO_4 potassium monohydrogen phosphate

Answer:

(1) Na_2SO_4, sodium sulfate

(2) $NaNO_2$, sodium nitrite

(3) Na_2HPO_4, sodium monohydrogen phosphate

(4) $CaCl_2$, calcium chloride

(5) $CaSO_4$, calcium sulfate

(6) $Ca(ClO_4)_2$, calcium perchlorate

(7) $CaHPO_4$, calcium monohydrogen phosphate

(8) KCl, potassium chloride

(9) KNO_2, potassium nitrite

(10) $KClO_4$, potassium perchlorate

10.3 WATER AS AN ACID AND A BASE

Water is classified as a weak electrolyte (Table 8.1). Upon ionization it releases both H^+ and OH^- ions. Since the reverse of this process is simply the reaction of acid and base, it is obvious that the ionization does not proceed very far and that it reaches an equilibrium.

$$H_2O(l) \rightleftharpoons H^+(aq) + OH^-(aq) \qquad (10.8)$$

According to Le Chatelier's principle, it is impossible to alter the concentration of one of these ions after equilibrium has been reached without altering the concentration of the other.

EXAMPLE 1.   What will happen to the OH^- concentration of $H_2O(l)$ if $HCl(g)$ is added?

Solution: The direct effect of adding $HCl(g)$ to water is to produce $H^+(aq)$ as shown in equation 10.3. Since this ion is already present, the increase in concentration has an effect on the equilibrium. The Le Chatelier principle is applied as follows.

1. An equilibrium exists:

$$H_2O(l) \rightleftharpoons H^+(aq) + OH^-(aq)$$

2. A condition is changed: The concentration of $H^+(aq)$ is increased.
3. A process occurs:

$$H^+(aq) + OH^-(aq) \longrightarrow H_2O(l),$$

4. This process tends to restore the original smaller concentration of $H^+(aq)$ by forming more $H_2O(l)$.
5. Therefore the concentration of OH^- is decreased.

There is a very simple *quantitative* relationship between the concentrations of $H^+(aq)$ and $OH^-(aq)$ ions in all aqueous solutions. You can derive it from the equilibrium expression for equation (10.8). For brevity let H^+ represent $H^+(aq)$, and let OH^- represent $OH^-(aq)$.

$$K = \frac{[H^+][OH^-]}{[H_2O(l)]} \qquad (10.9)$$

For pure water the concentration of H_2O is 55.5 M (see problem *28 of Chapter 4). Whenever water solutions of acids and bases are dilute, the value of $[H_2O(l)]$ will remain practically 55.5 M over a range of values for $[H^+]$ and $[OH^-]$. For example, even for solutes as concentrated as 1.0 M, the molarity of water in a solution of NaOH is 55.6 M and in a solution of HCl it is 54.1 M. If we multiply both sides of the previous equation by 55.5, we obtain a new constant,

$$55.5 \times K = [H^+][OH^-].$$

This constant has been found by experiment *at 25° C* to have the value shown in equation (10.10).

$$1.00 \times 10^{-14} = [H^+][OH^-] \qquad (10.10)$$

First you may confirm that this expression predicts the same effect on $[OH^-]$ that the Le Chatelier principle predicts when HCl is added to H_2O. If the product $[H^+][OH^-]$ does not change but the $[H^+]$ increases, the $[OH^-]$ must decrease, the same result obtained previously. The equilibrium constant, however, will also tell you *by how much* $[OH^-]$ decreases.

EXAMPLE 2. Use the constant in equation (10.10) (a) to calculate $[OH^-]$ if $[H^+]$ is increased to 10^{-1} M, and (b) to calculate $[H^+]$ if $[OH^-]$ is increased to 10^{-2} M.

Solution: In the first step write down equation (10.10) and in the second step, substitute your known concentration.

(a)
$$[H^+][OH^-] = 1.00 \times 10^{-14}$$
$$(10^{-1})[OH^-] = 1.00 \times 10^{-14}$$
$$[OH^-] = 1.00 \times 10^{-14}/10^{-1}$$
$$[OH^-] = 10^{-13} \text{ M}$$

(b)
$$[H^+][OH^-] = 1.00 \times 10^{-14}$$
$$[H^+](10^{-2}) = 1.00 \times 10^{-14}$$
$$[H^+] = 1.00 \times 10^{-14}/10^{-2}$$
$$[H^+] = 10^{-12} \text{ M}$$

The solutions of Example 2 represent the final concentrations of both ions in aqueous solution, but there is no indication whether there has been an increase or a decrease. To answer this question you need to know the ion concentrations in water before anything is added.

The concentrations of H^+ and OH^- in pure water are calculated very easily because one H^+ ion must be produced for every OH^- ion when H_2O molecules dissociate. This provides the following equality.

$$[H^+] = [OH^-] \quad \text{in pure water}$$

By substituting this equality into equation (10.10), you obtain the following relations.

$$[H^+][H^+] = [OH^-][OH^-] = 1.00 \times 10^{-14}$$
$$[H^+]^2 = [OH^-]^2 = 1.00 \times 10^{-14}$$

Take the square root of each quantity to obtain the concentrations in pure water.

$$[H^+] = [OH^-] = \sqrt{1.00 \times 10^{-14}}$$
$$[H^+] = [OH^-] = 1.00 \times 10^{-7} \text{ M} \quad \text{in pure water (25°C)} \qquad (10.11)$$

If you compare this result with that in Example 2 you will observe that in part (a) the concentration of OH^- in pure water is decreased from 10^{-7} M to 10^{-13} M by the addition of H^+. In part (b) the concentration of H^+ is decreased from 10^{-7} to 10^{-12} M by the addition of OH^-.

It should be reemphasized that you are using different symbols to represent the same aqueous hydrogen ion concentration from time to time.

$$[H^+] \equiv [H^+(aq)] \equiv [H_3O^+]$$

The reverse of equation (10.5) for the strong acid–strong base reaction should be equivalent to equation (10.8) for the ionization of water. The reverse of equation (10.5) is

$$2\,H_2O(l) \rightleftharpoons H_3O^+(aq) + OH^-(aq).$$

Writing the equation in this way emphasizes the fact that there are more than one H_2O molecules involved in the ionization of water. In fact if there were not another molecule to accept the released proton, the loss would not occur. In this sense, water is *both acid and base*. The water molecule that donates the proton is an acid; the water molecule that accepts the proton is a base. This in no way affects the quantitative results obtained in equation (10.8), because H^+ and H_3O^+ are identical. Furthermore, replacing $[H_2O(l)]$ by $[H_2O(l)]^2$ in equation (10.9) has no effect on the final results in equation (10.10) because, by convention, $[H_2O(l)] = 1$ and, of course, $1 = (1)^2$.

Exercise 1. What is the equilibrium constant for the reaction in equation (10.5) at 25°C?

Answer: ₱IOI = [₋HO][₊O⁸H]/I

10.4 THE pH SCALE

The value of $[H^+]$ is very important in chemical reactions, especially biochemical reactions. For example, in the brewing of clear lager beer, the starch from the grain must be converted to sugar at a $[H^+]$ between 3×10^{-6} M and 6×10^{-6} M. The enzyme diastase, which is present in barley sprouts (malt), is added to break down the starch. Living yeast cells are also added, which provide another enzyme to convert the sugar into alcohol. The reaction is known as fermentation and the equation that gives reactants and products is

$$C_6H_{12}O_6 \xrightarrow{\text{an enzyme}} 2\,CO_2(g) + 2\,C_2H_5OH.$$
$$\text{a sugar} \qquad\qquad\qquad \text{ethyl alcohol}$$

Enzymes and cells are easily made inoperative if too much or little H^+ ion is developed. Sørensen, a Danish chemist working in the Carlsberg Laboratory, devised a convenient scale, called the **pH scale** for expressing these important though minute concentrations of H^+ ion. According to this scale, pH is the negative power of 10 that represents the H^+ ion concentration expressed in moles per liter.

DEFINITION: $$[H^+] = 10^{-pH} \tag{10.12}$$

EXAMPLE 1.   What is the pH of pure water?

Solution: From equation (10.11) you have the $[H^+]$.

$$[H^+] = 10^{-7}\,M \quad (25°C)$$
$$[H^+] = 10^{-pH} \tag{10.12}$$
$$pH = 7$$

Pure water has pH 7 at 25°C.

EXAMPLE 2.   Calculate pH in Example 2, parts (a) and (b), p. 232.

Solution: A knowledge of $[H^+]$ is all that you need to calculate pH. If $[OH^-]$ is given you may calculate $[H^+]$ before calculating pH.

(a) $[H^+] = 10^{-1}\ = 10^{-pH}$ $pH = 1$
(b) $[H^+] = 10^{-12} = 10^{-pH}$ $pH = 12$

In Table 10.4 a number of aqueous solutions are compared in terms of their H^+ ion concentrations and pH values. Of course it is possible to go beyond the limits of Table 10.4 and have solutions with pH < 0 (negative values) and pH > 14. However, the pH scale is not frequently used to express the acidity of such concentrated solutions.

If the H^+ ion concentration is not a whole number power of ten, the calculation of pH is the same in principle but requires the use of logarithms.

TABLE 10.4   Aqueous solutions and their pH values at 25°C

Solution	$[H^+]$	pH	Description
1 M HCl	1	0	↑
0.1 M HCl	10^{-1}	1	
0.01 M HCl	10^{-2}	2	
⋮	⋮	⋮	acidic
pure H_2O	10^{-7}	7	⟵ neutral*
⋮	⋮	⋮	basic
0.01 M NaOH	10^{-12}	12	
0.1 M NaOH	10^{-13}	13	
1 M NaOH	10^{-14}	14	↓

*An aqueous solution with an equal number of H^+ ions and OH^- ions is called **neutral**.

EXAMPLE 3. If convenient you should provide a qualitative or approximate answer to a question as in part (a) in order to check your quantitative calculation as in part (b) below.
(a) Is the pH of 2.0×10^{-7} M H$^+$ greater or less than 7?
(b) Calculate the pH.

Solution: This solution has twice the H$^+$ concentration of a neutral solution, so it must be acidic.
(a) According to Table 10.4 all acidic solutions have pH < 7.
(b) Since the H$^+$ concentration is already expressed in scientific notation as a number between 1 and 10 times a power of 10 (Section 2.5), you may proceed directly with the calculation.

$$[H^+] = 2.0 \times 10^{-7} \qquad (10.13)$$

It is necessary to write the number 2.0 as a power of 10 to complete the calculation. The log scale is perfectly suited for this operation because log 2.0 is the power of 10 which is equal to the number 2.0 (see Table 10.5). This relationship is

$$\log 2.0 = 0.30 \qquad \text{means that} \qquad 10^{0.30} = 2.0.$$

Return now to the original problem. You can make this substitution in equation (10.13) and write [H$^+$] as a negative power of 10.

$$[H^+] = 10^{0.30} \times 10^{-7}$$
$$= 10^{(-7.00+0.30)*}$$
$$[H^+] = 10^{-6.70}$$

TABLE 10.5 Logarithms of numbers from 1 through 10

Number	Logarithm	Number	Logarithm
1.0	0.00	6.0	0.78
1.5	0.18	6.5	0.81
2.0	0.30	7.0	0.85
2.5	0.40	7.5	0.88
3.0	0.48	8.0	0.90
3.5	0.54	8.5	0.93
4.0	0.60	9.0	0.95
4.5	0.65	9.5	0.98
5.0	0.70	10.0	1.00
5.5	0.74		

*See Multiplication and Division of Powers of Ten under Section 2.1 for more on this rule.

Comparing this last result with the definition of pH you can make the final calculation.

$$[H^+] = 10^{-pH} \qquad (10.12)$$
$$pH = 6.70$$

Note that pH < 7 as required in part (a) above.

Exercise 1.   What is the pH of a solution in which the H$^+$ ion concentration is 0.050 M?

Answer: $[H^+] = 5.0 \times 10^{-2}$, pH $= 1.30$.

In Example 4 only the OH$^-$ ion concentration is given, and the first step in calculating the pH is to find the H$^+$ ion concentration.

EXAMPLE 4.   What is the pH of a solution in which $[OH^-] = 3.0 \times 10^{-7}$?

Solution: You may begin by substituting the known value of $[OH^-]$ into equation (10.10).

$$[H^+][OH^-] = 1.00 \times 10^{-14} \qquad (10.10)$$
$$[H^+](3.0 \times 10^{-7}) = 1.00 \times 10^{-14}$$
$$[H^+] = 1.00 \times 10^{-14}/3.0 \times 10^{-7}$$
$$= 10.0 \times 10^{-15}/3.0 \times 10^{-7}$$
$$[H^+] = 3.3 \times 10^{-8}$$

Use the logarithms given in Table 10.5 to find the logarithm of 3.3*.

No.	Log
3.0	0.48
(3.3)	(0.52)
3.5	0.54

$$\log 3.3 = 0.52 \qquad \text{means that} \qquad 3.3 = 10^{0.52}$$

You may now write the H$^+$ ion concentration as a power of 10 and compare it with the definition of pH.

$$[H^+] = 3.3 \times 10^{-8}$$
$$= 10^{0.52} \times 10^{-8}$$

---

*Note that the number 3.3 is a little more than halfway between 3.0 and 3.5; thus, the value of its logarithm must be a little more than halfway between 0.48 and 0.54. This is called **interpolation**.

$$[H^+] = 10^{-8.00+0.52}$$
$$[H^+] = 10^{-7.48}$$
$$[H^+] = 10^{-pH} \qquad (10.12)$$
$$pH = 7.48$$

Note that this solution has pH > 7, which means that it is basic (Table 10.4). Since the OH^- ion concentration, 3.0×10^{-7} M, is three times the value of $[OH^-]$ in pure water, the solution *should* be basic.

Frequently you will want to know the $[H^+]$ or $[OH^-]$ value in a solution of given pH. You may use the same rules as in the previous examples to make this calculation.

EXAMPLE 5.   The pH of Coca Cola is 4.70. This acidity (pH < 7) is due to the presence of H_2CO_3 (carbonic acid), a weak acid which forms when $CO_2(g)$ dissolves in water. What is the H^+ ion concentration in Coca Cola?

Solution: Since the solution is acidic, you will expect an H^+ ion concentration greater than 10^{-7} M. Use the definition of pH to express $[H^+]$ as a power of ten.

$$[H^+] = 10^{-pH} \qquad (10.12)$$
$$[H^+] = 10^{-4.70}$$

You will need to use the table of logarithms, but examination of Table 10.5 reveals that all the logarithms have positive values between 0.00 and 1.00. Rewrite the exponent (-4.70) above as the sum of a positive decimal fraction between 0.00 and 1.00 and the next more negative whole number after -4.

$$[H^+] = 10^{0.30-5.00}$$
$$[H^+] = 10^{0.30} \times 10^{-5} \qquad (10.14)$$

In Table 10.5 you will find that 0.30 is the logarithm of 2.0.

$$\log 2.0 = 0.30 \qquad \text{means that} \qquad 2.0 = 10^{0.30}$$

Make this substitution in equation (10.14).

$$[H^+] = 2.0 \times 10^{-5} \text{ M}$$

This the H^+ ion concentration. Since it is greater than 10^{-7} M, the solution is acidic.

PROBLEMS

1. The ionic reaction shown occurs in aqueous solution. Eliminate all ions except those that actually take part in or result from the chemical reaction. Write the final equation in its simplest form.

$$2\,Na^+ + 2\,HSO_3^- + Ca^{2+} + 2\,OH^- \longrightarrow$$
$$2\,Na^+ + 2\,SO_3^{2-} + Ca^{2+} + 2\,H_2O$$

2. Same as problem 1.

$$2\,Na^+ + 2\,OH^- + K^+ + H_2PO_4^- \longrightarrow$$
$$2\,Na^+ + 2\,H_2O + K^+ + PO_4^{3-}$$

3. Write a balanced equation for the production of $H_2(g)$ from Al and H_2SO_4 in which both H^+ ions in H_2SO_4 are displaced. What is the name of the salt which can be produced by precipitation?

4. Write a balanced equation for the production of $H_2(g)$ from Mg and H_3PO_4 in which all three H^+ ions in H_3PO_4 are displaced. What is the name of the salt that can be produced by precipitation?

5. Is the following ionic equation balanced? If so, prove it by equating all atoms and charges on left and right.

$$H_2PO_4^- + NH_3 \longrightarrow NH_4^+ + HPO_4^{2-}$$

6. Same as problem 5.

$$CO_3^{2-} + 2\,H_3PO_4 \longrightarrow$$
$$2\,H_2PO_4^- + H_2O + CO_2$$

7. Use Table 10.1 to identify weak and strong acids and bases and write the following reactions as balanced ionic equations. Write them in such a manner that the "real" reaction is emphasized and nonparticipating ions are omitted. Assume that any salts formed are strong, soluble electrolytes.
   (a) $H_2SO_3 + 2\,NaOH \longrightarrow$
   $$Na_2SO_3 + 2\,H_2O$$

   (b) $2\,HIO_4 + Ca(OH)_2 \longrightarrow$
   $$Ca(IO_4)_2 + 2\,H_2O$$

8. Same as problem 7.

   (a) $HNO_3 + NH_3 \longrightarrow NH_4NO_3$

   (b) $2\,HClO + Ba(OH)_2 \longrightarrow$
   $$Ba(ClO)_2 + 2\,H_2O$$

9. How many moles of water can be produced from (a) 1 mole $Ca(OH)_2$ and 1 mole H_2SO_4, (b) 1 mole HClO and 1 mole $Ca(OH)_2$, and (c) 2 moles H_3PO_4 and 1 mole $Ba(OH)_2$. Explain.

10. How many moles of water can be produced from (a) 1 mole HCl and 1 mole NaOH, (b) 1 mole H_2SO_4 and 3 moles NaOH, and (c) 1 mole H_3PO_4 and 2 moles $Ca(OH)_2$? Explain.

11. Give balanced equations for the reactions (a) between $Ca(OH)_2$ and the first H^+ from H_3PO_4 and (b) between $Ca(OH)_2$ and the two replaceable H^+ ions in H_3PO_3

12. Give balanced equations for the reactions (a) between $Al(OH)_3$ and the first H^+ ion from H_2SO_4 and (b) between $Al(OH)_3$ and both H^+ ions of H_2SO_4, to form water

to form water and salts. The third H atom in H_3PO_3 is not acidic because it is bound to P instead of an O atom. Let all OH^- ions from $Ca(OH)_2$ be consumed in each reaction. (c) Give names for the salts.

and salts. Let all OH^- ions from $Al(OH)_3$ be consumed in each reaction. (c) Give names for the salts.

13. Fill in the blanks in a copy of the table shown, giving the names and formulas of the salts formed by combining the appropriate ions.

	HPO_3^{2-}	Br^-	SO_4^{2-}	ClO_4^-
Al^{3+}				
Cs^+				
Mg^{2+}				

14. Same as problem 13.

	I^-	HSO_3^-	ClO^-	PO_4^{3-}
NH_4^+				
Al^{3+}				
Ra^{2+}				

15. Calculate pH for each of the following solutions.
    (a) 0.1 M H^+
    (b) 0.0001 M H^+
    (c) 0.001 M OH^-
    (d) 0.002 M OH^-

16. Same as problem 15.
    (a) 0.01 M H^+
    (b) 0.001 M H^+
    (c) 0.01 M OH^-
    (d) 0.02 M OH^-

17. Vinegar is sold in the United States as a 5% aqueous solution of acetic acid. The corresponding molarity of HOAc is 0.8 M, but since it is a weak acid, the H^+ concentration is only 4×10^{-3} M. (a) Calculate the pH of vinegar and (b) compare it with the pH of an 0.8 M HCl solution.

18. The average human stomach contains 50 ml of 0.02 M HCl while resting. About an hour after a dinner including meat, this concentration rises to 0.08 M HCl due to gastric secretion. Calculate the pH expected for stomach fluids both before and after the meal indicated.

19. When green beans are not canned under sterile conditions, they sometimes cause botulism, a type of food poisoning. The microbe, *Clostridium botulinum*, which produces the toxin cannot grow at pH values below 4.5. What is the corresponding concentration of strong acid?

20. The pH of the aqueous part of human blood is normally between 7.30 and 7.45. The limits of blood pH compatible with life are 6.8 and 7.8. What is the lower limit of $[H^+]$ in blood compatible with life?

*21. The solubilization of phosphate rock, $Ca_3(PO_4)_2$, using H_2SO_4 leads to a mixture of $Ca(H_2PO_4)_2$ and $CaSO_4$. If the more expensive acid H_3PO_4 is used instead of H_2SO_4, the sole product is $Ca(H_2PO_4)_2$. Calculate the per cent phosphorus in each product. Which is the better fertilizer on a weight basis?

*22. In a dilute aqueous solution 1 g HNO_3 reacts with an excess of NaOH to produce 212 cal heat; 1 g of HCl under similar conditions produces 366 cal. How do these data support the theory that the real reaction between all strong acids and bases is the same ($H^+ + OH^- \longrightarrow H_2O$)? *Hint:* Compare these acids on a mole basis instead of a weight basis.

11

ACID CONSTANTS, TITRATION, AND INDICATORS

In Table 10.1 some acids and bases were noted as "strong" and the remainder were considered "weak." In the present chapter you will see that there are two acids involved in every acid-base reaction, one on the reactant side and the other on the product side. In order to predict whether or not the reaction takes place, you must know which acid is stronger and this requires that you know their acid constants. Finally when acids and bases react with one another there should be a way of knowing when the reaction is complete. This is often accomplished by the use of color indicators and the procedure is called titration. Much of what you have learned about pH will be helpful in understanding these new topics.

11.1 ACID-BASE REACTION AS A COMPETITION FOR HYDROGEN ION: THE BRØNSTED-LOWRY THEORY

Whenever an acid and a base react in water, they can be considered to form a new acid-base pair. If OAc^-, in the form of $NaOAc$, is dissolved in water, the solution becomes slightly basic as a result of the reaction

$$OAc^-(aq) + H_2O(l) \longrightarrow HOAc(aq) + OH^-(aq).$$
$$\quad\;\; \text{base} \qquad\;\; \text{acid}$$

In this equation OAc^- is a base because it accepts a proton; H_2O is an acid because it donates the proton. But you will notice that this reaction is just the reverse of equation (10.2) on page 222. You should write this equation with a double arrow to emphasize the fact that it can go in either direction.

$$HOAc(aq) + OH^-(aq) \rightleftharpoons OAc^-(aq) + H_2O(l) \qquad (10.2)$$
$$\quad \text{acid} \qquad\qquad \text{base} \qquad\qquad \text{base} \qquad\quad \text{acid}$$

This combination of an acid-base pair to form a new acid-base pair is an important part of the Brønsted-Lowry theory, which was proposed in 1923 by the Danish chemist Brønsted and the English chemist Lowry. According to this theory *every* acid-base reaction results in the formation of a new acid-base pair. The OAc^- ion is called the **conjugate base** of HOAc, and H_2O is the **conjugate acid** of OH^-. In this reaction both HOAc and H_2O could lose H^+ ions. When HOAc loses it, OAc^- is formed. When H_2O loses it, OH^- is formed. The reason that the reaction symbolized by equation (10.2) goes strongly to the right is that HOAc is a stronger acid than H_2O.

It is interesting that in the Brønsted-Lowry theory there are no salts, only acids and bases. An acid-base reaction is simply the transfer of a proton from one ion or molecule to another.

According to the Brønsted-Lowry theory, even dissolving an acid in water is an acid-base reaction. In the equation (11.1) pure sulfuric acid, a dense oily liquid, is added to water. The reaction is *exothermic*—so exothermic, in fact, that chemists warn not to add the water to the acid for fear that when the small amount of water is suddenly converted to steam, it will splatter.

$$H_2SO_4(l) + H_2O(l) \rightleftharpoons H_3O^+(aq) + HSO_4^-(aq) + \text{heat} \qquad (11.1)$$
$$\text{acid} \qquad\quad \text{base} \qquad\qquad \text{acid} \qquad\qquad \text{base}$$

Equation (11.1) is a better representation of what actually happens in the first step of ionization than equation (10.6), page 226. A proton is simply transferred from H_2SO_4 to H_2O. Although this reaction goes practically 100% to the right, it is conceivable that it could go in reverse. In this case HSO_4^- gains a proton and is considered a base. Of course, HSO_4^- can also behave like an acid and lose a proton, which is a more likely reaction.

> **Note on Solar Energy:** One method that has been proposed for storing the sun's energy* involves a system of H_2SO_4 and H_2O contained in two glass bulbs connected by a stopcock (Figure 11.1). One bulb contains an aqueous solution of H_2SO_4. On sunny days a mirror is used to collect the sun's rays and concentrate them on this bulb. The water evaporates, but concentrated 98% $H_2SO_4(aq)$ is left behind and not vaporized (bp = 330°C). The system, which contains no air, readily permits the water vapor to pass into the other bulb. In this bulb, which is not being heated externally, the water vapor condenses to a liquid. When sunshine is no longer available to heat the H_2SO_4, the stopcock is closed, leaving the two liquids separated. When the "stored heat" is needed, the stopcock is simply opened. This permits the

*Farrington Daniels: *Direct Uses of the Sun's Energy.* Yale University Press, New Haven, Conn., 1964, p. 145.

243

11.1
Acid-Base Reaction as
a Competition for
Hydrogen Ion: The
Brønsted-Lowry Theory

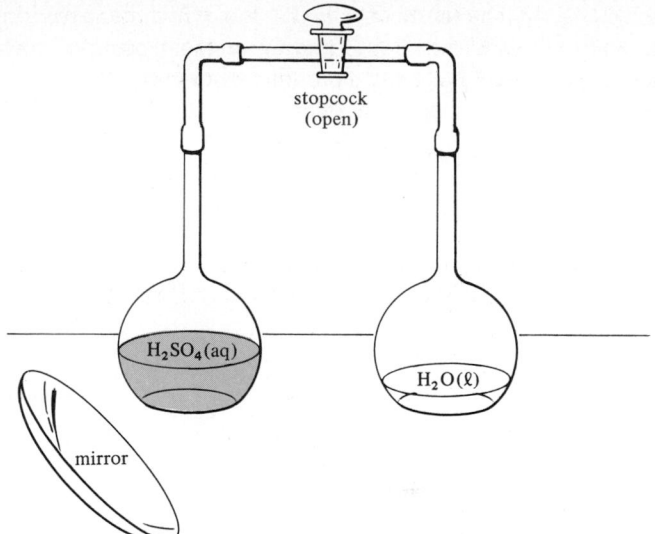

Figure 11.1. Storing solar energy by the distillation of water from aqueous sulfuric acid. H_2SO_4 (dil.) + heat $\longrightarrow$ H_2SO_4(conc.) + H_2O(g).

water vapor from the right-hand bulb (vp = 24 mm Hg at 25°C) to re-combine with the H_2SO_4 (equation 11.1) in the left-hand bulb where the heat is given off again. Concentrated sulfuric acid produces 580 cal for every gram of water vapor with which it combines.

Many bases that do not contain the OH^- ion themselves produce this ion when dissolved in water. Table 10.1 lists several common bases of this type. Their reactions in water are described in this example.

Reaction of Water with Other Nonhydroxide Bases

EXAMPLE 1.   Show by balanced equations how each of the following weak bases from Table 10.1 react with water to produce OH^- ion: (a) NH_3, (b) Na_2CO_3, and (c) $NaHCO_3$. Indicate all conjugate acids and bases according to Bronsted-Lowry theory.

Solution: In each of the following reactions, H_2O acts as an acid because it loses a proton. If the base is a salt (strong electrolyte), the first step is the complete ionization of the salt into its metal ion and associated negative ion.

(a) NH_3 (ammonia)

$$NH_3(aq) + H_2O \rightleftharpoons NH_4^+(aq) + OH^-(aq)$$

base   acid   acid   base

(ammonium ion)

The name of the NH_4^+ ion, which behaves like a metal ion in many salts, has the "-um" ending of most positive ions.

(b) Na_2CO_3 (sodium carbonate)

Step 1: A strong electrolyte dissociates in water.

$$Na_2CO_3(aq) \xrightarrow{100\%} 2\,Na^+(aq) + CO_3^{2-}(aq)$$
$$\text{carbonate ion}$$

Step 2: The negative ion reacts with water because HCO_3^- is a weak electrolyte.

$$\underset{\text{base}}{CO_3^{2-}(aq)} + \underset{\text{acid}}{H_2O} \rightleftharpoons \underset{\substack{\text{acid} \\ \text{(bicarbonate ion)}}}{HCO_3^-(aq)} + \underset{\text{base}}{OH^-(aq)}$$

(c) $NaHCO_3$ (sodium bicarbonate)

Step 1: A strong electrolyte dissociates in water.

$$NaHCO_3(aq) \xrightarrow{100\%} Na^+(aq) + HCO_3^-(aq)$$

Step 2: The negative ion reacts with water because H_2CO_3 is a weak electrolyte.

$$\underset{\text{base}}{HCO_3^-(aq)} + \underset{\text{acid}}{H_2O} \rightleftharpoons \underset{\substack{\text{acid} \\ \text{(carbonic acid)}}}{H_2CO_3(aq)} + \underset{\text{base}}{OH^-(aq)}$$

None of the reactions with water in this example goes very strongly in the direction that produces OH^- ions, but in each case the solution is slightly basic. To provide a more quantitative approach to acid-base reactions it is necessary to employ equilibrium constants.

11.2 RELATIVE STRENGTHS OF ACIDS: THE ACID CONSTANT, K_a

The equilibrium constant, K_a, furnishes you with a quantitative comparison of the strengths of different acids and a means of predicting the direction of acid-base reactions. This constant is similar to the equilibrium constants discussed in Sections 9.3 and 9.4.

For acetic acid, the acid strength is shown by the equilibrium constant for release of H^+ ions. The structure of the acetate ion is given on page 204.

$$CH_3COOH \rightleftharpoons H^+ + CH_3COO^- \qquad K_a = 1.8 \times 10^{-5} \text{ (at 25°C)} \quad (11.2)$$

Similarly, the strength of hydrofluoric acid is shown by the relationships

$$HF \rightleftharpoons H^+ + F^- \qquad K_a = 6.7 \times 10^{-4} \text{ (at } 25°C). \qquad (11.3)$$

Since the acid constant for HF is greater than for CH_3COOH, HF is the stronger acid. This can be seen by comparing the ion concentrations of two solutions of acid which are prepared at the same molecular concentration. The equilibrium expressions are

$$\frac{[H^+][CH_3COO^-]}{[CH_3COOH]} = 1.8 \times 10^{-5}$$

$$\frac{[H^+][F^-]}{[HF]} = 6.7 \times 10^{-4}$$

Suppose that $[CH_3COOH] = [HF] = 1$ M. Then the ratio of the ion product in HF to that in CH_3COOH is calculated as follows.

$$\frac{[H^+][F^-]}{[H^+][CH_3COO^-]} = \frac{6.7 \times 10^{-4}}{1.8 \times 10^{-5}} = 37$$

Obviously HF ionizes to a greater extent than CH_3COOH and is the stronger acid. This fact will be used in Example 1 to predict the outcome of a possible acid-base reaction.

EXAMPLE 1. Suppose you are considering the possible reaction between HF, an acid, and CH_3COO^-, a base. Will it take place or not?

Solution: First write the reaction down and identify the acids and the bases.

$$\underset{\text{acid}}{HF} + \underset{\text{base}}{CH_3COO^-} \rightleftharpoons \underset{\text{acid}}{CH_3COOH} + \underset{\text{base}}{F^-}$$

The acid that is stronger will release protons, and the main reaction must go in the direction in which the acid becomes a negative ion. Since the acid HF is stronger than the acid CH_3COOH the reaction must go from left to right. The answer is yes the reaction will take place.

The K_a values used in Example 1 and values for a number of other important acids are listed in Table 11.1. You will notice that some acids have K_a values too large to measure. All of these acids are considered strong electrolytes, as indicated earlier in Table 10.1. The stepwise dissociation of polyprotic acids is discussed again in the next subsection.

From Table 11.1 you can see why the oxyacids containing more than one oxygen-bound hydrogen atom react with strong bases by losing one H^+ ion at a time (page 226). Consider the series of ionization steps in H_3PO_4, phosphoric acid

Stepwise Ionization of Acids

$$H_3PO_4 \rightleftharpoons H^+ + H_2PO_4^- \qquad K_a = 7.1 \times 10^{-3}$$

TABLE 11.1   The acid constants in water at 25°C

Equilibrium	K_a
$HClO_4 \rightleftharpoons H^+ + ClO_4^-$	very large
$HNO_3 \rightleftharpoons H^+ + NO_3^-$	very large
$HCl \rightleftharpoons H^+ + Cl^-$	very large
$H_2SO_4 \rightleftharpoons H^+ + HSO_4^-$	large
$H_2SO_3 \rightleftharpoons H^+ + HSO_3^-$	1.7×10^{-2}
$HSO_4^- \rightleftharpoons H^+ + SO_4^{2-}$	1.3×10^{-2}
$H_3PO_4 \rightleftharpoons H^+ + H_2PO_4^-$	7.1×10^{-3}
$HF \rightleftharpoons H^+ + F^-$	6.7×10^{-4}
$HNO_2 \rightleftharpoons H^+ + NO_2^-$	5.1×10^{-4}
$CH_3COOH \rightleftharpoons H^+ + CH_3COO^-$	1.8×10^{-5}
$H_2CO_3 \rightleftharpoons H^+ + HCO_3^-$	4.4×10^{-7}
$H_2S \rightleftharpoons H^+ + HS^-$	1.0×10^{-7}
$H_2PO_4^- \rightleftharpoons H^+ + HPO_4^{2-}$	6.3×10^{-8}
$HSO_3^- \rightleftharpoons H^+ + SO_3^{2-}$	6.2×10^{-8}
$NH_4^+ \rightleftharpoons H^+ + NH_3$	5.7×10^{-10}
$HCO_3^- \rightleftharpoons H^+ + CO_3^{2-}$	4.7×10^{-11}
$HPO_4^{2-} \rightleftharpoons H^+ + PO_4^{3-}$	4.4×10^{-13}
$HS^- \rightleftharpoons H^+ + S^{2-}$	1.3×10^{-13}
$H_2O \rightleftharpoons H^+ + OH^-$	$\dfrac{[H^+][OH^-]}{55.5} = 1.8 \times 10^{-16}$

which is 100,000 times bigger than

$$H_2PO_4^- \rightleftharpoons H^+ + HPO_4^{2-} \qquad K_a = 6.3 \times 10^{-8}$$

which is 100,000 times bigger than

$$HPO_4^{2-} \rightleftharpoons H^+ + PO_4^{3-} \qquad K_a = 4.4 \times 10^{-13}$$

The large ratio of one constant to the next means that one acid is very much stronger than the next and reacts completely as base is added before the next acid begins to react.

Other Examples of Acid-Base Reactions

A vigorous acid-base reaction is illustrated in Figure 11.2. The acid solution is vinegar, which ordinarily contains 5% by weight acetic acid (CH_3COOH). The base is baking soda, which is pure sodium bicarbonate ($NaHCO_3$). In aqueous solution the latter forms HCO_3^- ions.

The acid-base reaction and its reverse is in equation (11.4).

$$\underset{\text{acid}}{CH_3COOH} + \underset{\text{base}}{HCO_3^-} \rightleftharpoons \underset{\text{base}}{CH_3COO^-} + \underset{\text{acid}}{H_2CO_3} \qquad (11.4)$$

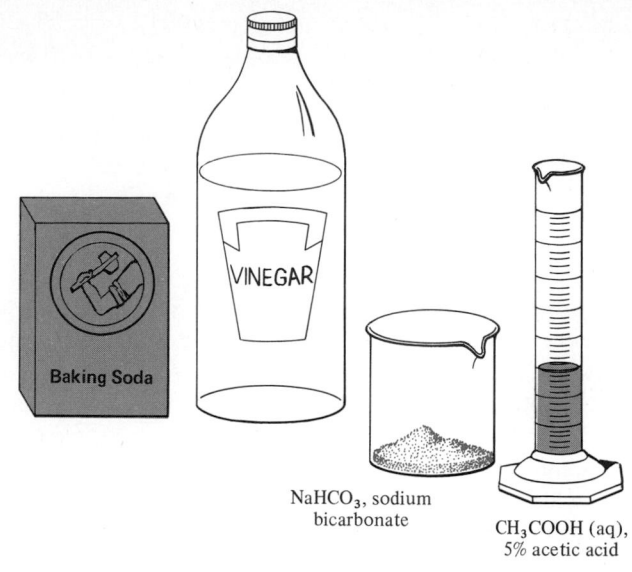

NaHCO$_3$, sodium
bicarbonate

CH$_3$COOH (aq),
5% acetic acid

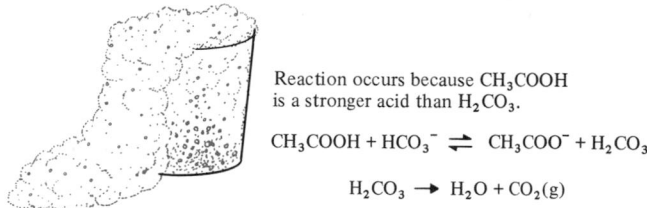

Reaction occurs because CH$_3$COOH
is a stronger acid than H$_2$CO$_3$.

$$CH_3COOH + HCO_3^- \rightleftharpoons CH_3COO^- + H_2CO_3$$

$$H_2CO_3 \longrightarrow H_2O + CO_2(g)$$

Figure 11.2. An acid-base reaction.

If CH$_3$COOH is a stronger acid than H$_2$CO$_3$, the reaction will go from left to right. You may select the following values from Table 11.1: $K_a = 1.8 \times 10^{-5}$ for CH$_3$COOH and $K_a = 4.4 \times 10^{-7}$ for H$_2$CO$_3$ (for the first H$^+$). Since $1.8 \times 10^{-5} > 4.4 \times 10^{-7}$, the reaction does go to the right. This fact is obvious when the reactants are mixed (Figure 11.2). The H$_2$CO$_3$ formed is not very soluble in water and breaks down rapidly into H$_2$O(l) and CO$_2$(g), as shown in equation (11.5).

$$H_2CO_3 \longrightarrow H_2O(l) + CO_2(g) \qquad (11.5)$$

If this kind of rapid evolution of gas takes place in an unknown solution when strong acid is added, it is considered proof of the presence of either HCO$_3^-$ ion or CO$_3^{2-}$ ion in that solution.

EXAMPLE 2. Would you expect a strong reaction of the acid-base type between aqueous solutions of NH$_4^+$ ion and HCO$_3^-$ ion?

Solution: The HCO$_3^-$ ion might conceivably release H$^+$ or absorb H$^+$, since both CO$_3^{2-}$ and H$_2$CO$_3$ are listed in Table 11.1. However, NH$_4^+$ can only release H$^+$ since NH$_5^{2+}$ is not listed and can be presumed not to exist.

$$HCO_3^- + NH_4^+ \rightleftharpoons H_2CO_3 + NH_3$$

base   acid   acid   base

The question is, which is the stronger acid, NH_4^+ or H_2CO_3? Since K_a is 4.4×10^{-7} for H_2CO_3 and 5.7×10^{-10} for NH_4^+, H_2CO_3 is stronger. The reaction from left to right will not take place. The answer to the original question is *no*.

You might be surprised to try this reaction and find that you can detect the presence of NH_3 by its odor. The relative values of K_a mean that the reaction prefers reactants to products but both reactants *and* products are in equilibrium at *some* concentration level. The human nose happens to be a very good detector of NH_3, part of which escapes from the aqueous solution as a gas.

A standard test for the presence of NH_4^+ ion in an unknown solution is to add a strong base, such as NaOH, and sniff it for NH_3. Alternatively you may hold a moistened piece of red litmus paper *above* the solution. The latter turns blue in the presence of bases (Table 10.2). The reaction of NH_4^+ ion with OH^- ion is

$$NH_4^+ + OH^- \rightleftharpoons NH_3(g) + H_2O. \tag{11.5}$$

The reaction with the moistened litmus paper is the reverse of equation (11.5), with NH_3 present in the gas phase.

Exercise 1. Identify the acid-base pairs in reaction (11.5). Use the K_a values in Table 11.1 to show that the reaction should go strongly to the right.

Answer:

acid   base   base   acid

$$NH_4^+ + OH^- \rightleftharpoons NH_3 + H_2O, \quad K(NH_4^+) \gg K(H_2O).$$

Exercise 2. Write the possible acid-base reactions between each of the following pairs and identify the Brønsted-Lowry acid and base on each side. Predict whether the reaction is favored from left to right by the K_a values in Table 11.1. (a) $CH_3COOH + HS^-$; (b) $H_3PO_4 + HPO_4^{2-}$

Answer:

acid   base   base   acid

(a) $CH_3COOH + HS^- \rightleftharpoons CH_3COO^- + H_2S.$

Yes, $K(CH_3COOH) > K(H_2S).$

acid   base   acid and base

(b) $H_3PO_4 + HPO_4^{2-} \rightleftharpoons 2 H_2PO_4^-$

Yes, $K(H_3PO_4) > K(H_2PO_4^-).$

11.3 STRONG AND WEAK BASES IN THE TABLE OF ACID CONSTANTS

Since the product of every acid in an acid-base reaction is its own *conjugate base,* the table of acid constants serves equally well as a table of base constants. You will notice that the reverse of every equation in Table 11.1 represents the acceptance of a proton by an ion or molecule, which is therefore a base. The reciprocal of K_a gives the equilibrium constant for the corresponding reaction. Therefore the smaller the value of K_a, the stronger is the conjugate base.

EXAMPLE 1.   Rank the following bases in the order of their increasing strength starting with the weakest. Explain. (a) NH_3; (b) HPO_4^{2-}; (c) HCO_3^-; (d) CO_3^{2-}; (e) OH^{1-}.

Solution: Those species appearing as *products* in the reactions listed in Table 11.1 and having the largest K_a are listed first. They are the weakest bases.

$$HCO_3^- \quad < \quad HPO_4^{2-} \quad < \quad NH_3 \quad < \quad CO_3^{2-} \quad < \quad OH^-$$
$$4.4 \times 10^{-7} > 6.3 \times 10^{-8} > 5.7 \times 10^{-10} > 4.7 \times 10^{-11} > 1.8 \times 10^{-16}$$

11.4   TITRATION

If a standard base solution is available, that is, one of known molarity, it is possible to determine the number of moles of available H^+ ion in an unknown acid. The procedure used is known as **titration.** As the last bit of H^+ ion from the acid is "neutralized" by the base, there is a sharp change in pH. A very small addition of base near the neutralization point causes a large increase in the pH. If you have some way of detecting the pH change with modest precision, you will be able to measure the volume of base equivalent to the acid with considerable accuracy.

EXAMPLE 1.   Calculate points *a, b,* and *c* for the titration shown in Figure 11.3 during which 0.20 M NaOH is added to 20 ml of HCl solution. After titrating, it is found that the original concentration was 0.20 M HCl.

Solution: Before any NaOH has been added, corresponding to 0 ml on the horizontal axis, the concentration of H^+ ion is 0.20 M because all of the HCl is in its ionic form.
(a) Calculation of the corresponding pH is

$$[H^+] = 2.0 \times 10^{-1} \qquad \cdot$$

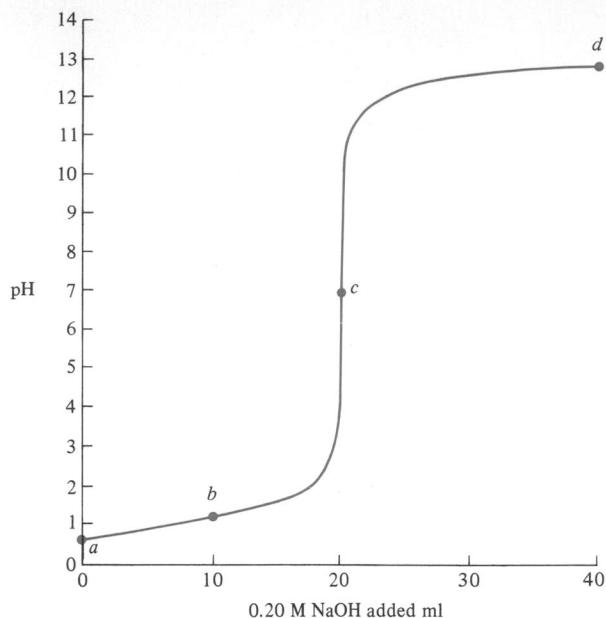

Figure 11.3. pH titration curve for the addition of 0.20 M NaOH to 20 ml 0.20 M HCl (see text for calculation of points *a*, *b*, *c*, and *d*).

$$\log 2.0 = 0.30 \qquad \text{means that} \qquad 2.0 = 10^{0.30} \text{ (from Table 10.5)}$$

$$[H^+] = 10^{0.30} \times 10^{-1}$$
$$= 10^{-0.70}$$
$$[H^+] = 10^{-pH} \qquad (10.12)$$
$$pH = 0.70$$

This point is shown at *a* on the curve in Figure 11.3.

(b) You can calculate all the other points on the titration curve by considering the extent of reaction and the final volume. Point *b* represents the pH after 10.0 ml NaOH has been added. What are the numerical values of the H⁺ concentration and the pH at this point? The HCl has reacted partially, according to the equation

$$\text{HCl} + \text{NaOH} \longrightarrow \text{NaCl} + \text{H}_2\text{O}.$$

$$\text{moles HCl reacted} = (10.0 \; \text{ml NaOH})\left(\frac{0.20 \; \text{mole NaOH}}{1 \; \text{l NaOH}}\right)\left(\frac{1 \; \text{l NaOH}}{1000 \; \text{ml NaOH}}\right)$$

$$\left(\frac{1 \text{ mole HCl reacted}}{1 \text{ mole NaOH reacted}}\right)$$

$$= 2.0 \times 10^{-3} \text{ mole HCl reacted}$$

To calculate pH you must know the number of moles of HCl remaining.

moles of HCl remaining = [orig. moles HCl] − [moles of HCl reacted]

$$= \left[(20.0 \text{ ml HCl}) \left(\frac{0.20}{1} \frac{\text{mole HCl}}{\text{l HCl}} \right) \right.$$

$$\left. \left(\frac{1}{1000} \frac{\text{l HCl}}{\text{ml HCl}} \right) \right] - [\text{moles of HCl reacted}]$$

$$= 4.0 \times 10^{-3} \text{ original mole HCl} - 2.0 \times 10^{-3}$$
$$\text{mole HCl reacted}$$

$$= 2.0 \times 10^{-3} \text{ mole HCl remaining}$$

What is the molarity of the remaining HCl? The volume of HCl is no longer 20.0 ml because 10.0 ml NaOH has been added to it.

$$x \text{ M HCl} = \frac{\text{moles HCl remaining}}{\text{liters solution}}$$

$$= \left(\frac{2.0 \times 10^{-3} \text{ mole HCl remaining}}{30.0 \text{ solution}} \right) \left(\frac{1000}{1} \frac{}{\ell} \right)$$

$$= 6.7 \times 10^{-2} \text{ M HCl}$$

What is the pH corresponding to this H^+ ion concentration? Remember that HCl is a strong acid.

$$[H^+] = 6.7 \times 10^{-2}$$

$$\log 6.7 = 0.83 \qquad \text{means that} \qquad 6.7 = 10^{0.83} \text{ (from Table 10.5)}$$

$$[H^+] = 10^{0.83} \times 10^{-2}$$
$$= 10^{-1.17}$$
$$[H^+] = 10^{-pH} \qquad (10.12)$$
$$pH = 1.17$$

This is point b on the curve in Figure 11.3. In the process of consuming half of the HCl, the NaOH has changed the pH from 0.70 to 1.17.

(c) How much has the pH changed by the time complete neutralization has been attained? At the 20.0 ml point on the horizontal axis of Figure 11.3 (point c), the OH^- ions added have neutralized all of the H^+ ions according to the reaction.

$$H^+ + OH^- \longrightarrow H_2O$$

The complete reaction of all H^+ and OH^- ions with one another leaves simply a solution of NaCl in water. Since NaCl is neither acidic or basic this solution must have the pH of pure water.

$$pH = 7.00$$

This is point c. It represents a drastic change in pH during the second half of neutralization, from pH 1.17 to pH 7.00. Thus the pH changes more rapidly as the neutralization point is approached. By calculations similar to these you may derive the entire titration curve shown in Figure 11.3.

Exercise 1.   Calculate the point on the titration curve corresponding to 40.0 ml NaOH (point d). The pH is the same as that obtained by adding 20.0 ml of 0.20 M NaOH to 40.0 ml H_2O. Why?

Answer: $pH = 12.83$

Acid-Base Indicators

One obvious indicator of the neutralization point is the pH itself. A pH meter with a single probe like that in Figure 11.4 will indicate pH to the nearest 0.01 unit and is frequently used in acid-base titrations. Other indicators of the neutralization point are dyes like litmus and phenolphthalein, which are actually weak acids and bases themselves. They are called **acid-base indicators,** and they change color in different ranges of pH (Table 11.2).

TABLE 11.2   Acid-base indicators and their range of color changes*

pH	Color in Acid	Indicator	Color in Base	pH
3.1	red	methyl orange	orange	4.4
4.4	red	methyl red	yellow	6.2
6.2	yellow	bromothymol blue	blue	7.6
6	red	litmus	blue	8
8.0	colorless	phenolphthalein	red	10.0
10.0	yellow	alizarin yellow	lilac	12.0

*N. A. Lange: Handbook of Chemistry, 10th ed. McGraw-Hill Book Co., New York, 1961.

A complete color change is usually effected by a pH change of as little as 1 or 2 units. The white solid, phenolphthalein, for example, is a weak acid having the molecular formula $C_{20}H_{14}O_4$. In neutral or acid solutions it is colorless, but when it reacts with a base, such as the strong base NaOH, it becomes a brilliant red. For brevity let the acid form of the indicator be represented by HIn and the base form by In⁻.

$$\text{HIn(aq)} \quad + \text{OH}^-\text{(aq)} \longrightarrow \text{In}^-\text{(aq)} + \text{H}_2\text{O}$$

colorless   red
(phenolphthalein)

This color change takes place in the pH range 8.0 to 10.0.

EXAMPLE 2. Suppose that phenolphthalein is used as an indicator in the titration illustrated in Figure 11.4 instead of the pH meter. Describe the color change and the point at which it occurs, using the titration curve in Figure 11.3. Would you expect the error due to the acid-base reaction between phenolphthalein and NaOH to be significant? Explain.

Solution: At the beginning of the titration the indicator, which is placed in the beaker, is colorless because the solution is acidic. After 20 ml 0.20 M NaOH has been added, the solution turns pink and finally bright red. In the absence of indicator the pH change from 7.0 to 8.0 requires an almost immeasurably small volume of NaOH because the curve is very steep in this region. In going beyond pH 8.0, the point at which HIn (colorless) begins to react to form In⁻ (red), the amount of NaOH required to give a perceptible color change depends on the intensity of the indicator's color. Of course, acid-base indicators are chosen for this

Figure 11.4. A pH titration. The voltage of the glass electrode is determined by the concentration of the H⁺ ion in which the glass membrane is immersed.

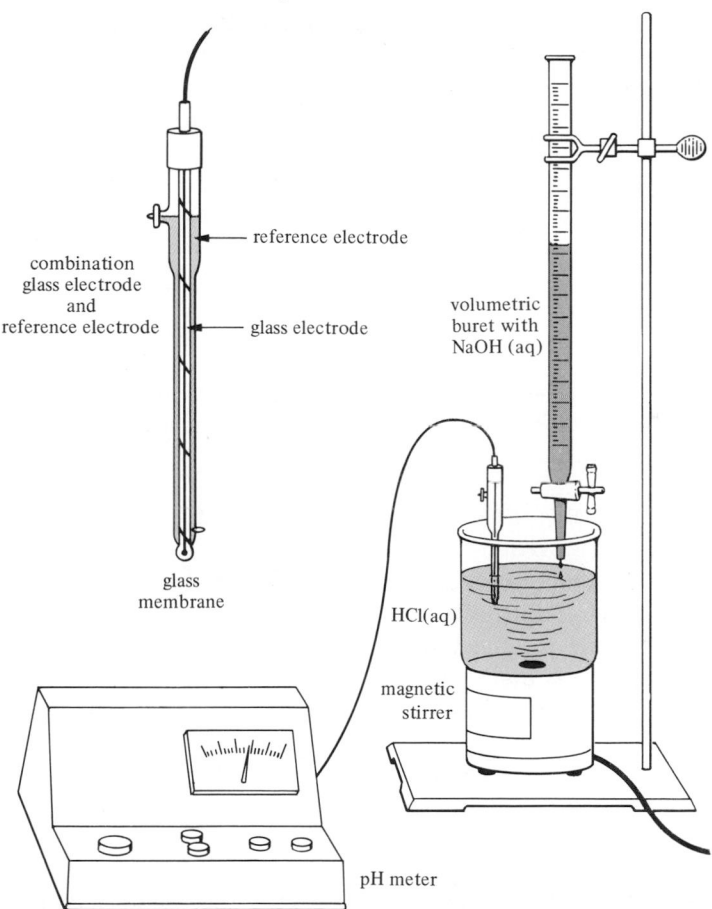

combination
glass electrode
and
reference electrode

reference electrode

glass electrode

volumetric
buret with
NaOH (aq)

glass
membrane

HCl(aq)

magnetic
stirrer

pH meter

property and the error in this instance is less than 0.1%. The point on the volume axis of the titration curve at which the color change occurs is called the **end point.**

Many acid-base titrations do not have neutralization points at pH 7 like HCl-NaOH. Furthermore, the transition in pH may not be as sharp as in the case of strong acid–strong base titrations. For this and other special purposes, it is desirable to have a variety of indicators that change colors in different pH ranges. A list of some typical acid-base indicators is given in Table 11.2 along with the range of pH applicability.

In Example 3 a typical calculation based on data from an acid-base titration is worked out.

EXAMPLE 3.   A standard base solution consists of 0.1052 M NaOH. An unknown amount of HCl, in the presence of phenolphthalein, requires the addition of 35.06 ml base to change from colorless to pink. How many grams HCl is in the unknown?

Solution: Write a balanced chemical equation.

$$NaOH + HCl \longrightarrow NaCl + H_2O$$

The number of grams HCl, x, is directly proportional to the molarity and to the volume of base required to neutralize it.

$$x \text{ g HCl} = (35.06 \text{ ml NaOH})\left(\frac{1}{1000}\frac{l\ NaOH}{ml\ NaOH}\right)\left(\frac{0.1052}{1}\frac{mole\ NaOH}{l\ NaOH}\right)$$
$$\left(\frac{1}{1}\frac{mole\ HCl}{mole\ NaOH}\right)\left(\frac{36.5}{1}\frac{\text{g HCl}}{mole\ HCl}\right)$$

$$= 0.1346 \text{ g HCl}$$

The high precision implied by the number of significant figures in the answer is not uncommon for this analytical method.

**Another
End Point
for Titration:
Electric
Current**
Any observable change that occurs abruptly at the neutralization point of an acid and a base can be used in place of the colored indicator or pH meter described previously. An interesting example of an analytical procedure is the use of an electrical conductivity tester as illustrated in Figure 11.5. A dilute aqueous solution of $Ba(OH)_2$ in the buret is titrated against one of H_2SO_4 in the beaker. Both are good conductors alone because they are strong electrolytes.

$$Ba(OH)_2(aq) \xrightarrow{100\%} Ba^{2+}(aq) + 2\ OH^-(aq)$$

$$H_2SO_4(aq) \xrightarrow{100\%} H^+(aq) + HSO_4^-(aq)$$

At the neutralization point, however, there are practically no ions present in the products to conduct electricity. The white precipitate $BaSO_4(s)$ is almost completely insoluble (Section 9.3, Example 4) and H_2O is a very weak electrolyte.

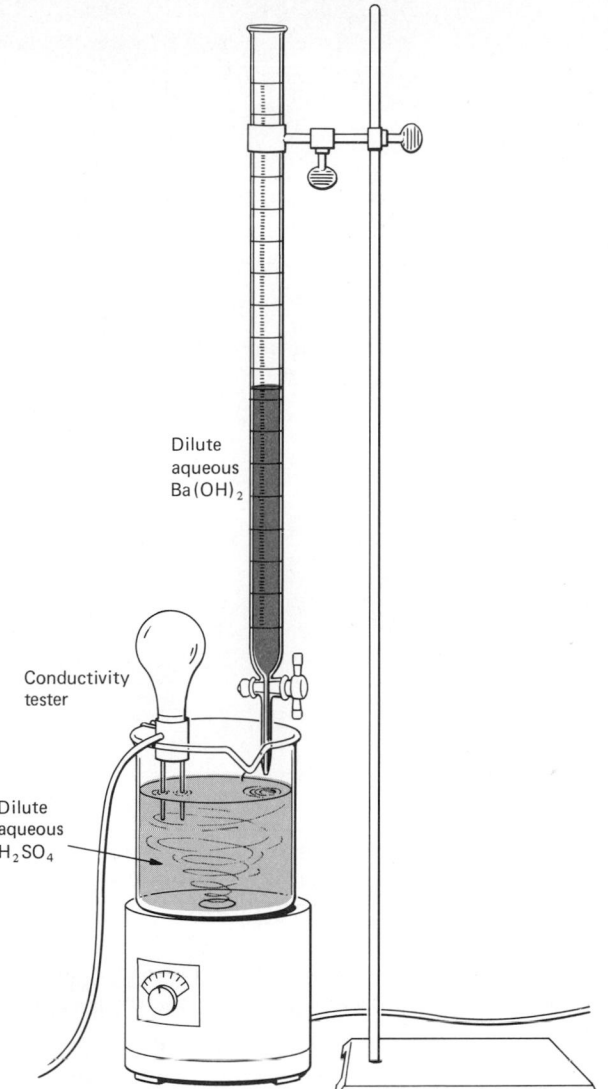

Dilute
aqueous
Ba(OH)$_2$

Conductivity
tester

Dilute
aqueous
H$_2$SO$_4$

Figure 11.5. Conductometric titration.

$$Ba^{2+}(aq) + HSO_4^-(aq) \longrightarrow BaSO_4(s) + H^+(aq)$$
$$2\,OH^-(aq) + H^+(aq) + H^+(aq) \longrightarrow 2\,H_2O(l)$$

You would expect the light to go out at the neutralization point and to come on again as excess base is added. The conductometric titration curve for the addition of 0.1 M Ba(OH)$_2$ to 0.1 M H$_2$SO$_4$ would have the form shown in Figure 11.6. The amount of H$_2$SO$_4$ in an unknown solution could be calculated from the molarity of the base and the volume required to reach the break in the curve.

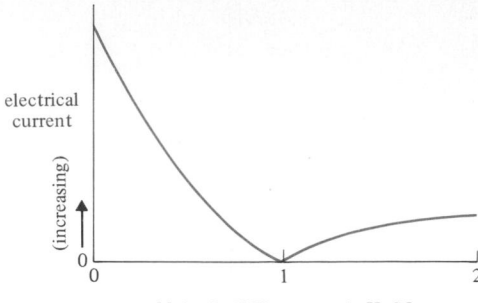

Moles Ba(OH)$_2$ per mole H$_2$SO$_4$

Figure 11.6. Conductometric titration, 0.1 M
Ba(OH)$_2$ added to 0.1 M H$_2$SO$_4$.

Exercise 1. Conductometric titration of an aqueous solution of H$_2$SO$_4$ shows
that the electric current reaches a minimum when 45.2 ml of
0.0520 M Ba(OH)$_2$ has been added. Calculate the number of grams of
H$_2$SO$_4$.

Answer: 0.231 g.

11.5 ACIDIC AND BASIC ANHYDRIDES

Before chemists reached general agreement that hydrogen is the key to the explanation of acid properties, it was believed that oxygen was the characteristic element
(Section 10.1). It will be interesting to examine the binary oxides for their relevance to
the problem of acid-base formation.

The oxides of the elements towards the upper right-hand corner of the periodic
table, the nonmetals, all tend to form acids when added to water, and that fact
probably caused some of the confusion previously mentioned. These oxides are called
acidic anhydrides and they produce some of our best known acids. A few examples
are given.

$$CO_2(g) \quad + H_2O \longrightarrow H_2CO_3(aq)$$

carbon dioxide   carbonic acid

Biological note: This reaction and its reverse are important in the human
body, serving to remove CO$_2$ from the tissues and release it in the lungs for
exhalation. The blood contains a special enzyme, carbonic anhydrase, that
catalyzes these reactions and makes them take place fast enough to permit
an adequate rate of metabolism.

$$SO_2(g) \quad + H_2O \longrightarrow H_2SO_3(aq)$$

sulfur dioxide   sulfurous acid

$$(P_2O_5)_2(s) \qquad + 6\,H_2O \longrightarrow \quad 4\,H_3PO_4(l)$$

phosphorus pentoxide   phosphoric acid

When NO_2 dissolves in water, both HNO_3 and NO are produced, a complication that will be dealt with in Chapter 12, Oxidation and Reduction. The cartoon in Figure 11.7 illustrates a number of gaseous pollutants, many of which are acidic, found in city air.

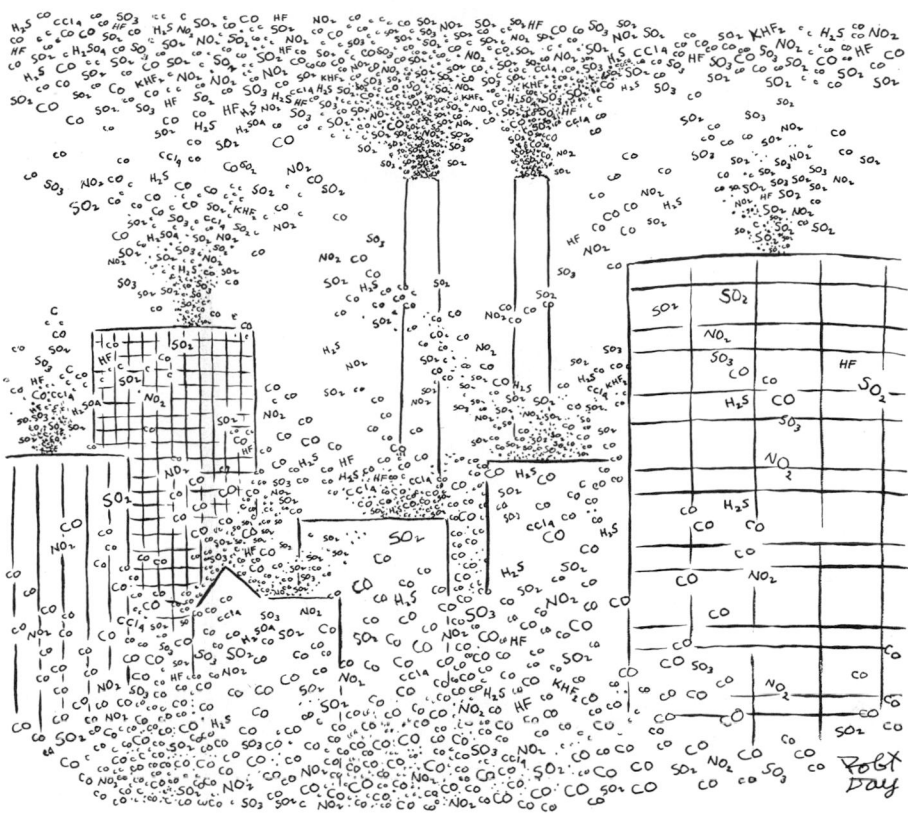

Figure 11.7. [Drawing by Robert Day; © 1967. Reprinted by permission from *The New Yorker Magazine.*]

Environmental note: The gaseous nonmetal oxides are harmful air pollutants partly because they react with fluids in the respiratory tissues to form acids. The emissions of NO (nitric oxide) from the internal combustion engine and SO_2 (sulfur dioxide) from coal burners have already been mentioned (see pages 89 and 163, respectively).

The oxides of the metals that form bases are called **basic anhydrides.** Those formed from the metals in the lower part of the second column of the periodic table, CaO, SrO, and BaO, were named alkaline earths by the ancients and are so stable to heat they were once thought to be elements. They are used today to line high temperature furnaces. Both MgO and RaO are also considered to be alkaline earths. In water they all react to form strong bases.

$$CaO(s) \quad + H_2O(l) \longrightarrow \quad Ca(OH)_2(s)$$

calcium oxide   calcium hydroxide
(lime)   (slaked lime)

The metals in the first column of the periodic table are called alkali metals. Their oxides also react with water to form strong bases.

$$Na_2O(s) \quad + H_2O(l) \longrightarrow \quad 2\,NaOH(s)$$

sodium oxide   sodium hydroxide

When acidic anhydrides and basic anhydrides react, they form salts but without forming water molecules.

$$CaO(s) + SO_2(g) \longrightarrow \quad CaSO_3(s)$$

calcium sulfite

$$Na_2O(s) + CO_2(g) \longrightarrow \quad Na_2CO_3(s)$$

sodium carbonate
(a salt which is also a base)

1. In the acid-base reactions shown, identify the Brønsted-Lowry acid-base pairs.
   (a) $HF + HCO_3^- \rightleftharpoons H_2CO_3 + F^-$
   (b) $HClO + H_2O \rightleftharpoons H_3O^+ + ClO^-$

   (c) $S^{2-} + 2\,H_3O^+ \rightleftharpoons H_2S + 2\,H_2O$

2. Same as problem 1.
   (a) $H_2SO_3 + H_2O \rightleftharpoons H_3O^+ + HSO_3^-$
   (b) $2\,CH_3COOH + CO_3^{2-} \rightleftharpoons$
   $$2\,CH_3COO^- + H_2CO_3$$
   (c) $NH_3 + H_3PO_4 \rightleftharpoons NH_4^+ + H_2PO_4^-$

3. Write balanced equations for possible acid-base reactions between the following pairs in aqueous solution. Use the values of K_a in Table 11.2 to predict whether the reactions actually occur.
   (a) $NO_2^- + HSO_4^- \longrightarrow$
   (b) $CO_3^{2-} + HPO_4^{2-} \longrightarrow$
   (c) $HS^- + H_2SO_4 \longrightarrow$

4. Same as problem 3.
   (a) $HCl + NH_3 \longrightarrow$
   (b) $NH_4^+ + PO_4^{3-} \longrightarrow$
   (c) $SO_4^{2-} + CH_3COOH \longrightarrow$

5. Calculate the pH on the titration curve in Figure 11.3 corresponding to the addition of 25.0 ml 0.20 M NaOH to 20.0 ml 0.20 M HCl.

6. Calculate the pH on the titration curve in Figure 11.3 corresponding to the addition of 15.0 ml 0.20 M NaOH to 20.0 ml 0.20 M HCl.

7. Methyl red is an indicator (Table 11.2) that has one color when protonated, HMR, and another color when deprotonated, MR$^-$. Use the fact that methyl red is red at pH < 4.4 and yellow at pH > 6.2 to choose which form of the methyl red molecule is responsible for the red color and which for the yellow. Write a balanced equation involving OH$^-$ or H_3O^+ and both methyl red species above.

8. Litmus is an indicator (Table 11.2) that has one color when protonated, HL, and another color when deprotonated, L$^-$. Use the fact that litmus is red at pH < 6 and blue at pH > 8 to choose which form of the litmus molecule is responsible for the red color and which for the blue. Write a balanced equation involving OH$^-$ or H_3O^+ and both litmus species above.

9. Persons with healthy stomachs do not find the presence of 0.02 M HCl at all painful, but persons with peptic ulcers do. One way to relieve the pain is to take sodium bicarbonate ($NaHCO_3$) to neutralize the acid. (a) Write a balanced equation for this reaction. (b) How many grams $NaHCO_3$ are needed to neutralize 50 ml 0.02 M HCl, the typical volume and concentration of stomach fluid?

10. Ordinary table vinegar sold in the United States contains about 5.0% acetic acid (CH_3COOH) by weight. (a) Assuming a density of 1.00 g ml^{-1} for the vinegar, calculate the volume (in ml) of 1.0 M NaOH required to neutralize 40 ml vinegar. (b) What is the molarity of CH_3COOH in vinegar?

11. The following metallic oxides are basic anhydrides. Write balanced equations for their reactions with water (a) K_2O; (b) BaO.

12. Write balanced equations in which the acidic anhydrides below react with water to form the indicated acids: (a) SO_3, H_2SO_4; (b) Cl_2O_7, $HClO_4$; (c) N_2O_3, HNO_2.

* 13. It is possible to calculate K_a for a weak monoprotic acid in the following manner. First, the pH is measured on a sample of the solution. Second, an accurately measured volume of the same solution is titrated against standard base to the end point. Explain.

* 14. Use K_a values in Table 11.1 to calculate the equilibrium constant for the following reaction and to predict whether the reaction occurs or not.

$$HF + CH_3COO^- \rightleftharpoons CH_3COOH + F^-$$

Hint: Apply the rules of algebra to the equilibrium expressions for the two acids in such a way as to give the required equilibrium expression.

12

OXIDATION AND
REDUCTION

Oxidation-reduction is another very broad classification of chemical reactions equally as important as acid-base reactions. Almost every chapter of this text contains an example of oxidation-reduction. Most chemical reactions in which elementary substances combine to form binary compounds, such as the combination of hydrogen and oxygen to form water (Section 4.6) or of sodium and chlorine to form salt (Section 6.4), are of this type. Often oxidation-reduction reactions are highly exothermic. The thermite reaction discussed on page 264 produces enough heat to melt iron, which is a product. The heat of combustion (oxidation) of gasoline produces gas pressures high enough to power the automobile and temperatures high enough to produce NO from N_2 and O_2 (page 89). The oxidation of carbohydrates such as cane sugar (Section 7.2) is united with certain endothermic biochemical reactions in the human body in such a way as to make the endothermic reactions take place. Other oxidation-reduction reactions can be harnessed to store various forms of energy such as that from sunlight, windmills, and waterfalls and release it later in the form of electricity.

Acid-base reactions and oxidation-reduction reactions comprise a large fraction of all known chemical processes. In acid-base reactions you were concerned with what happens to the proton; in oxidation-reduction the question is what happens to the electron.

12.1 TRANSFER OF ELECTRONS: OXIDATION-REDUCTION

A reaction between acids and bases can be thought of as a transfer of *protons;* a reaction between oxidizing and reducing agents can be thought of as a transfer of *electrons.*

Oxidation and reduction reactions were first described in terms of metallurgy. When metals were oxidized in air, they combined with oxygen to form metal oxides. The process of converting the metal to metal oxide was called *oxidation.* An example with Zn is

$$2\,Zn(s) + O_2(g) \longrightarrow 2\,ZnO(s). \tag{12.1}$$

If the Zn could be recovered by some chemical reaction, it was said that the ZnO had been *reduced.* For example by heating ZnO with a more active metal, Mg, the Zn is released.

$$ZnO + Mg \longrightarrow Zn + MgO$$

You will note that the Mg has become oxidized. Oxidation and reduction always go together. Whenever there is a reducing agent, such as Mg, there is an oxidizing agent, such as ZnO. This generalization might encourage you to look for a reducing agent in equation (12.1). Zinc is the reducing agent in that reaction; it reduces O_2 to ZnO. Since reductions of this type do not make much sense from a metallurgical point of view, another broader definition is needed. That definition is provided by the behavior of electrons in such reactions.

EXAMPLE 1. Since ZnO is an ionic compound, there must be a gain and loss of electrons when it is formed from its elements. Indicate this process by arrows connecting similar elements on both sides of the equation.

Solution: The number written over each atom represents its net electrical charge.

$$\overset{2(-2\,e^-)}{\overbrace{}}$$

$$\overset{0}{2\,Zn}\,(s) + \overset{0}{O_2}\,(g) \longrightarrow \overset{2+}{Zn}\;\overset{2-}{O}(s)$$

$$\underset{2(+2\,e^-)}{\underbrace{}}$$

(a) Each Zn loses $2\,e^-$. Zn is oxidized.
(b) Each O gains $2\,e^-$. O is reduced.

EXAMPLE 2. Show the gain and loss of electrons when ZnO is reduced by magnesium.

Solution: First write the net charge above every atom and then connect like atoms by arrows.

$$\overset{+2\,e^-}{\overbrace{\qquad\qquad}}$$

$$\underset{-2\,e^-}{\underbrace{\overset{2+\ 2-}{Zn\ O}\ +\ \overset{0}{Mg}\ \longrightarrow\ \overset{0}{Zn}\ +\ \overset{2+\ 2-}{Mg\ O}}}$$

(a) Zn^{2+} gains $2\,e^-$. Zn^{2+} is reduced.
(b) Mg loses $2\,e^-$. Mg is oxidized.
(c) O^{2-} is neither oxidized nor reduced.

The definition of oxidation and reduction that follows satisfies the old metallurgical viewpoint as well as our modern ideas.

DEFINITION: *During an oxidation-reduction reaction, the ion or atom that loses electrons is* **oxidized;** *the ion or atom that gains electrons is* **reduced.**

A mnemonic aid, which we hope will help you remember this important definition, is shown in Figure 12.1.

By this definition Example 3 illustrates an oxidation-reduction reaction even though no oxygen is involved.

EXAMPLE 3.   Show the gain and loss of electrons that occur when the ionic white solid $MgCl_2$ is formed from its elements.

Solution: Use the ionic charges predicted by the position of the elements in the periodic table (Table 6.3).

$$\overset{-2\,e^-}{\overbrace{\qquad\qquad}}$$

$$\underset{2(+1\,e^-)\,=\,+2\,e^-}{\underbrace{\overset{0}{Mg\,(s)}\ +\ \overset{0}{Cl_2\,(g)}\ \longrightarrow\ \overset{2+\ 1-}{Mg\,Cl_2\,(s)}}}$$

(a) Mg loses $2\,e^-$. Mg is oxidized.
(b) Cl gains $1\,e^-$. Cl is reduced.

In order to make the electrons balance in Example 3, each Cl gains one electron and two electrons are gained by Cl_2 because there are two atoms per molecule.

Figure 12.1.   A mnemonic lion.

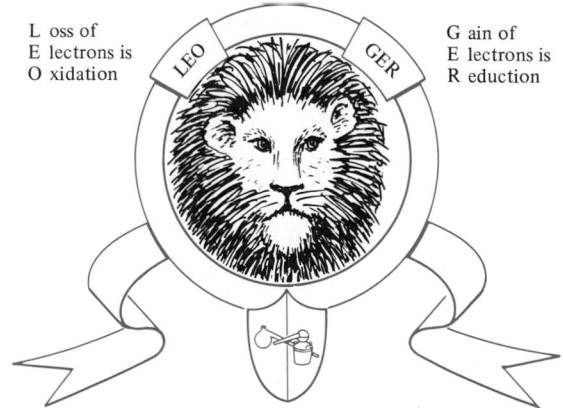

L oss of
E lectrons is
O xidation

G ain of
E lectrons is
R eduction

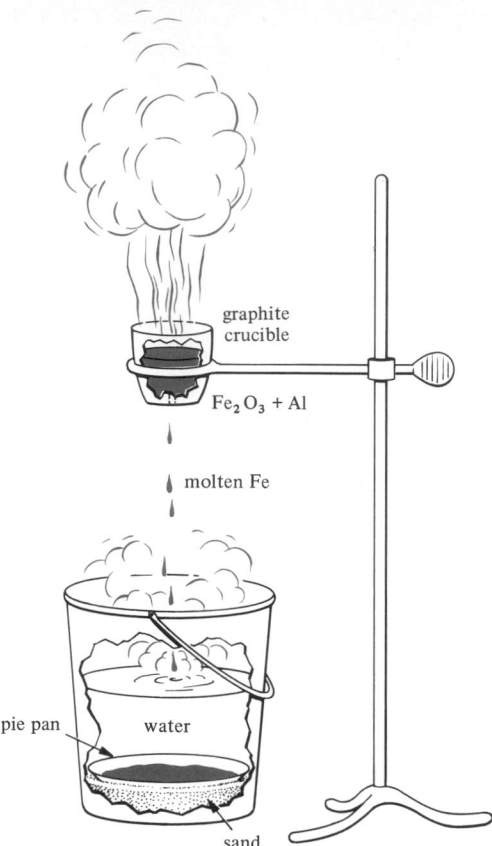

Figure 12.2. The thermite reaction, $2\,Al(powder) + Fe_2O_3(powder) \longrightarrow 2\,Fe(s) + Al_2O_3(s) + 208\,kcal$. The $Fe(s)$ melts and attains such a high temperature that it melts through the steel pie pan.

EXAMPLE 4. The thermite reaction (Figure 12.2) is such a highly exothermic oxidation-reduction reaction that it produces iron in the molten state. A mixture of the fine powder rouge (Fe_2O_3) and of powdered aluminum interact in the following manner.

$$Fe_2O_3(s) + 2\,Al(s) \longrightarrow 2\,Fe(l) + Al_2O_3(s)$$

Show that the number of electrons gained is equal to the number lost in this reaction.

Solution: The oxides are ionic and O and Al have their typical ionic charges, $2-$ and $3+$, respectively (Table 6.3). Use this fact and the principle of electroneutrality to calculate the ionic charge on Fe in its oxide.

$$2(+3\,e) = +6\,e$$

$$\underset{Fe_2\,O_3}{3+\ 2-} + \underset{2\,Al}{0} \longrightarrow \underset{2\,Fe(l)}{0} + \underset{Al_2\,O_3}{3+\ 2-}$$

$$2(-3\,e) = -6\,e$$

The total number of electrons gained (by Fe_2O_3) is 6 and the total number lost (by 2 Al) is 6. This principle of electron conservation will be very helpful later when you encounter more difficult equations of this type.

Exercise 1. In Figure 10.1 in Chapter 10, Acids and Bases, there is an equation for the production of $H_2(g)$ by the action of a metal on an acid. Show that in this reaction the number of electrons gained is equal to the number lost.

Answer:

$$2(+1 \ e^-) = +2 \ e^-$$

$$Zn + H_2SO_4 \longrightarrow H_2 + Zn \ SO_4$$
$$0 \quad\quad +1 \quad\quad\quad 0 \quad\quad +2$$

$$-2 \ e^-$$

12.2 RELATIVE TENDENCY TO LOSE ELECTRONS

If two dissimilar metals are brought into contact with one another, there is a greater tendency for one of them to lose electrons than the other. As a result an electric voltage or potential is produced. For example if Zn and Cu are brought together, as in Figure 12.3, there is a tendency for electrons to flow from the more active metal Zn to the relatively inert Cu. A slight separation of charge develops, but further movement of electrons is quickly halted as a result of the attractive and repulsive forces developed.

Figure 12.3. Different metals in contact. The more active metal has its electrons displaced toward the second metal.

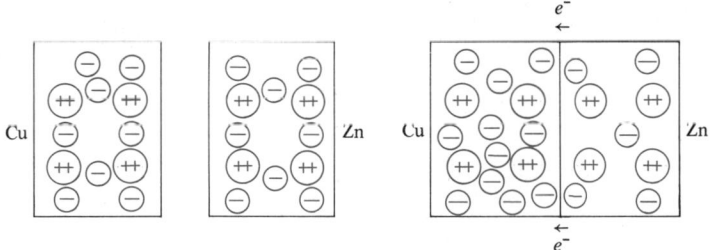

Note on Corrosion. A serious economic loss is incurred each year as a result of the oxidation of the fairly active metal Fe (iron). In air-saturated water, Fe serves as a source of electrons, and dissolves.

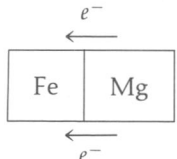

$$2(-2\,e^-)$$

$$\overset{0}{2\,Fe} + \overset{0}{O_2} + 2\,H_2O \longrightarrow \overset{2+}{2\,Fe}(aq) + \overset{2-}{4\,OH^-}(aq)$$

$$2(+2\,e^-)$$

In steel structures where corrosion is severe, such as in offshore drilling rigs, a more active metal such as magnesium is connected to the iron by a conductor. It furnishes the electrons needed by the iron and is oxidized sacrificially. The O_2 picks up its electrons on the iron surface, but the iron itself is not dissolved. The over-all reaction is

$$2(-2\,e^-)$$

$$\overset{0}{2\,Mg} + \overset{0}{O_2} + 2\,H_2O \longrightarrow \overset{2+}{2\,Mg^{2+}}(aq) + \overset{2-}{4\,OH^-}.$$

$$2(+2\,e^-)$$

The difference in tendency of Zn and Cu to lose electrons is observed if a shiny piece of Zn metal is dropped into a blue aqueous solution of $CuSO_4$. In this case extensive reaction takes place until one or the other of the reactants is consumed, as illustrated in Figure 12.4a. The Zn atoms lose electrons to the Cu^{2+} ions, by the following reaction.

$$Zn(s) + Cu^{2+}(aq) + SO_4^{2-}(aq) \longrightarrow Zn^{2+}(aq) + Cu(s) + SO_4^{2-}(aq)$$

Cu(s) precipitates out of solution and $ZnSO_4$ is formed in solution. Since the SO_4^{2-} ion does not enter into the reaction but is needed merely to maintain electroneutrality, it is usually left out of the ionic equation as before (Section 10.2).

$$Zn(s) + Cu^{2+}(aq) \longrightarrow Zn^{2+}(aq) + Cu(s)$$

This reaction can be thought of as two **half-reactions.** In one of these Zn loses electrons and is oxidized.

$$Zn(s) \longrightarrow Zn^{2+}(aq) + 2\,e^-$$

In the other, Cu^{2+} gains electrons and is reduced.

$$Cu^{2+}(aq) + 2\,e^- \longrightarrow Cu(s)$$

In the over-all reaction, electrons should cancel out. There is no measurable electric potential in this reaction because the electrons are neutralized as soon as they make contact with the Cu^{2+} ions.

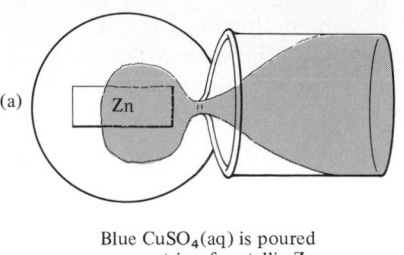

(a)

Blue $CuSO_4$(aq) is poured
over a strip of metallic Zn
in a Petri dish

A dark mass of metallic Cu
deposits on the Zn and the
solution becomes lighter

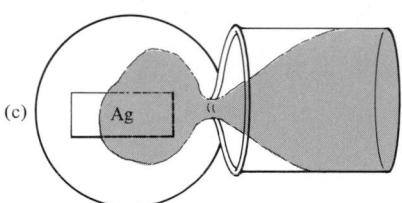

(b)

Colorless Ag_2SO_4(aq) is
poured over a strip of
metallic Cu

Crystals of Ag grow on the
Cu strip and the solution
becomes blue

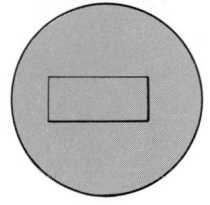

(c)

Blue $CuSO_4$(aq) is poured
over a strip of metallic Ag

No reaction is visible

Figure 12.4. Evidence that some metals have a greater tendency
than others to lose electrons and form aqueous
ions.

If you were to place metallic Cu in a solution of $ZnSO_4$, you would not expect a
reaction, although it is possible to write a balanced equation in which metallic Zn is
produced. You have just observed that metallic Zn has a greater tendency to lose
electrons than metallic Cu.

Other comparisons can be made by using other metals and their solutions. For
example Cu(s) and Zn(s) both are oxidized and dissolved in an aqueous solution of
Ag_2SO_4 (silver sulfate).

$$Cu(s) + 2\,Ag^+(aq) \longrightarrow Cu^{2+}(aq) + 2\,Ag(s) \quad \text{(see Figure 12.4b)}$$
$$Zn(s) + 2\,Ag^+(aq) \longrightarrow Zn^{2+}(aq) + 2\,Ag(s)$$

However, neither Cu^{2+}(aq) nor Zn^{2+}(aq) is reduced by Ag(s). This is illustrated for
copper in Figure 12.4c.

If a scale is devised in which the metals with the greatest tendency to lose electrons are at the top and those with the smallest tendency at the bottom, our three metals would be arranged in the following manner.

$$Zn \rightleftharpoons Zn^{2+} + 2\,e^-$$
$$Cu \rightleftharpoons Cu^{2+} + 2\,e^-$$
$$Ag \rightleftharpoons Ag^+ + e^-$$

It is even possible to place a nonmetal, such as chlorine, in this list but it should be compared in a form that is capable of losing electrons like the metals. The following reaction occurs between reddish metallic Cu and greenish-yellow $Cl_2(g)$.

$$Cu(s) + Cl_2(g) + aq \longrightarrow Cu^{2+}(aq) + 2\,Cl^-(aq)$$

Since the reverse reaction does not occur, $Cl^-(aq)$ should be located below $Cu(s)$ in our scale. The $Cl^-(aq)$ ion is also below $Ag(s)$, because it can be shown $Ag(s)$ is oxidized by $Cl_2(g)$, and that $Ag^+(aq)$ is not reduced by $Cl^-(aq)$. Actually $Ag^+(aq)$ and $Cl^-(aq)$ react with one another, but only to form an insoluble compound in which the ions retain their identity. This reaction, shown in equation (12.2), does not involve oxidation-reduction because there is no transfer of electrons.

$$\overset{+}{Ag}{}^+(aq) + \overset{-}{Cl}{}^-(aq) \longrightarrow Ag\,Cl\,(s) \tag{12.2}$$

An enlarged list including Cl^- can now be constructed.

$$Zn \rightleftharpoons Zn^{2+} + 2\,e^-$$
$$Cu \rightleftharpoons Cu^{2+} + 2\,e^-$$
$$Ag \rightleftharpoons Ag^+ + e^-$$
$$2\,Cl^- \rightleftharpoons Cl_2(g) + 2\,e^-$$

Exercise 1. A strip of metallic Zn dissolves in a solution of lead(II) nitrate, $Pb(NO_3)_2$. However, a strip of metallic Cu and a strip of metallic Ag are inert in the same solution. (a) Give the location of metallic Pb in the list above. (b) Would you expect an oxidation-reduction reaction between $Pb(s)$ and $Cl_2(g)$ in aqueous solution? If so, describe it.

Answer: (a) Between Zn and Cu. (b) Yes, to form $PbCl_2$.

You can proceed in the foregoing way to consider all possible pairs of oxidizing and reducing substances and construct a scale that permits you to predict whether reactions occur between any pair of substances in the list.

If the half-reactions are separated, it is usually possible to develop an electric potential and to measure it. This separation is the basis for the construction of an electrochemical cell or battery. If a half-reaction involves a metal that is a good conductor of electricity, the metal can serve as an electrode for that half of the cell. The metal ions, along with appropriate negative ions, will serve as the electrolyte which provides electrical conduction between it and the other half of the cell. A cell of this type is shown in Figure 12.5. It consists of Cu and Zn electrodes dipped in 1 M $CuSO_4$ and 1 M $ZnSO_4$, respectively. The voltmeter indicates that the electric potential of such a cell is 1.10 volt (V), with the Zn electrode more negative than the Cu electrode. If direct contact between the two electrolyte solutions is desired, such as in Figure 12.5, it is necessary to make the contact through some kind of porous barrier, such as unglazed ceramic or a gel. This barrier should permit the passage of ions under the influence of electrical forces, but it should prevent mechanical mixing due to agitation or differences in density. If mixing should happen, the $Cu^{2+}(aq)$ ions would attack the Zn(s) directly, as in Figure 12.4a, without producing a flow of electrons through the external circuit.

Figure 12.5. Completion of the circuit in an electrochemical cell. The voltmeter reading is for 1 M ionic concentrations.

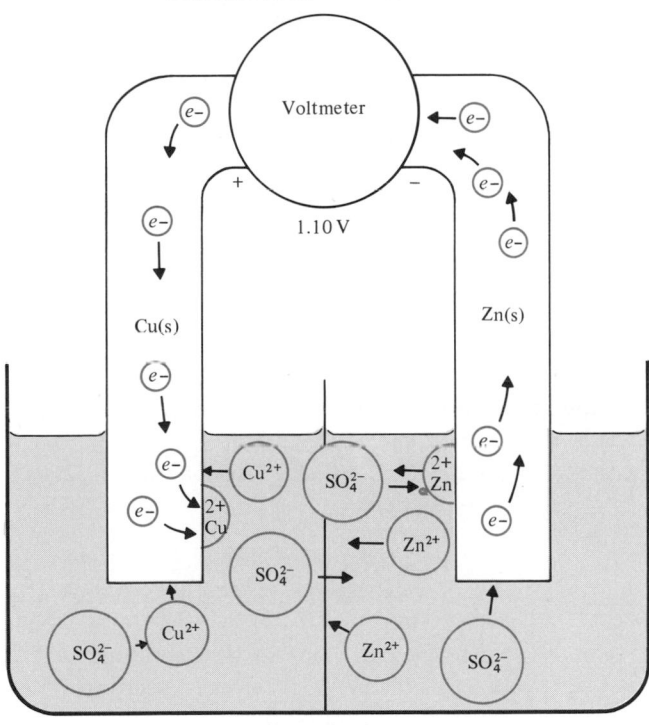

Porous barrier

When the circuit is closed, as shown in Figure 12.5, there is a flow of electricity. The voltmeter could be replaced by a flashlight bulb, which would permit even more current to be drawn. A number of processes begin simultaneously throughout the whole cell. You should understand how the circuit is provided in every region. Begin at one point, say at Zn(s), and move around the circuit with the electrons, back to the same point, to form a *complete circuit*. Unless the circuit is complete there will be no flow.

1. Electrons leave the $Zn(s)|Zn^{2+}(aq)$ electrode and pass through the external circuit to the $Cu(s)|Cu^{2+}(aq)$ electrode. The vertical lines represent the boundaries between the solids and the solutions of their sulfates.
2. Electrons are consumed at the $Cu(s)|Cu^{2+}(aq)$ electrode by the half-reaction $Cu^{2+}(aq) + 2\,e^- \longrightarrow Cu(s)$.
3. Electroneutrality is maintained around the $Cu(s)|Cu^{2+}(aq)$ electrode by moving $SO_4^{2-}(aq)$ out of the region and $Zn^{2+}(aq)$ ions into the region through the porous barrier.
4. Electroneutrality is maintained around the $Zn(s)|Zn^{2+}(aq)$ electrode by moving $Zn^{2+}(aq)$ out of the region and $SO_4^{2-}(aq)$ into the region.
5. Electrons are produced at the $Zn(s)|Zn^{2+}(aq)$ electrode by the half-reaction $Zn(s) \longrightarrow Zn^{2+}(aq) + 2\,e^-$. This brings you back to the original point, which completes the circuit.

A simpler diagram of this cell, which is more convenient to write, is

$$Cu(s) \mid Cu^{2+}(1\ M) \parallel Zn^{2+}(1\ M) \mid Zn(s).$$

Each single vertical line represents a boundary between liquid and solid, or in general any pair of different phases. The double line represents the completion of the electrical circuit internally, such as a porous barrier, without permitting a great amount of mixing of the two solutions.

Prediction of Oxidation-Reduction Reactions

Every pair of half-reactions in Table 12.1 is a possible oxidation-reduction reaction. The spontaneous direction for reaction is the one in which the half-reaction that is higher in the series goes in the direction that releases electrons. The other half-reaction must go in the direction that absorbs electrons.

EXAMPLE 1. Predict the over-all reaction that is a combination of the following half-reactions.

$$Co \rightleftharpoons Co^{2+} + 2\,e^-$$
$$Ag \rightleftharpoons Ag^+ + e^-$$

Solution: Since the first half-reaction is higher in Table 12.1 than the second, it must be the one that releases electrons. The half-reaction involving Ag^+ must be multiplied by two in order to balance the electrons.

$Co \longrightarrow Co^{2+} + 2\,e^-$		oxidation half-reaction
$2\,Ag^+ + 2\,e^- \longrightarrow 2\,Ag$		reduction half-reaction
$Co\ \ + 2\,Ag^+ \longrightarrow Co^{2+} + 2\,Ag$		over-all reaction

TABLE 12.1 Oxidation-reduction series. (All ions are aqueous and all elements are solids unless labeled otherwise.)

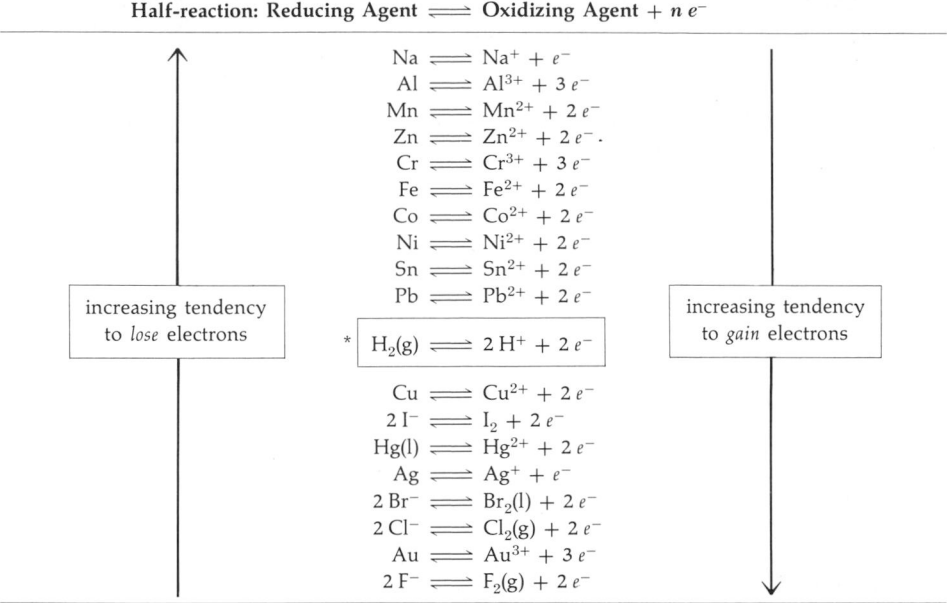

Half-reaction: Reducing Agent ⇌ Oxidizing Agent + $n\,e^-$

$$Na \rightleftharpoons Na^+ + e^-$$
$$Al \rightleftharpoons Al^{3+} + 3\,e^-$$
$$Mn \rightleftharpoons Mn^{2+} + 2\,e^-$$
$$Zn \rightleftharpoons Zn^{2+} + 2\,e^-.$$
$$Cr \rightleftharpoons Cr^{3+} + 3\,e^-$$
$$Fe \rightleftharpoons Fe^{2+} + 2\,e^-$$
$$Co \rightleftharpoons Co^{2+} + 2\,e^-$$
$$Ni \rightleftharpoons Ni^{2+} + 2\,e^-$$
$$Sn \rightleftharpoons Sn^{2+} + 2\,e^-$$
$$Pb \rightleftharpoons Pb^{2+} + 2\,e^-$$

increasing tendency to *lose* electrons

$$* \quad H_2(g) \rightleftharpoons 2\,H^+ + 2\,e^-$$

increasing tendency to *gain* electrons

$$Cu \rightleftharpoons Cu^{2+} + 2\,e^-$$
$$2\,I^- \rightleftharpoons I_2 + 2\,e^-$$
$$Hg(l) \rightleftharpoons Hg^{2+} + 2\,e^-$$
$$Ag \rightleftharpoons Ag^+ + e^-$$
$$2\,Br^- \rightleftharpoons Br_2(l) + 2\,e^-$$
$$2\,Cl^- \rightleftharpoons Cl_2(g) + 2\,e^-$$
$$Au \rightleftharpoons Au^{3+} + 3\,e^-$$
$$2\,F^- \rightleftharpoons F_2(g) + 2\,e^-$$

All metals higher than H_2, $2\,H^+$ in the series are considered "active."

EXAMPLE 2. What reaction do you expect between $Br_2(l)$ and a solution of NaCl?

Solution: First choose the possible half-reactions from Table 12.1.

$$2\,Br^- \rightleftharpoons Br_2(l) + 2\,e^-$$
$$2\,Cl^- \rightleftharpoons Cl_2(g) + 2\,e^-$$

The only conceivable reaction between the given reactants, since no Br^- ions are present, is one in which the first half-reaction goes from right to left. This requires that the second half-reaction goes from left to right.

$$Br_2(l) + 2e^- \longrightarrow 2\,Br^-$$
$$\underline{2\,Cl^- \longrightarrow Cl_2(g) + 2\,e^-}$$
$$Br_2(l) + 2\,Cl^- \longrightarrow 2\,Br^- + Cl_2(g)$$

However, this reaction cannot occur because ($2\,Br^- \rightleftharpoons Br_2 + 2\,e^-$) is above ($2\,Cl^- \rightleftharpoons Cl_2 + 2\,e^-$) in Table 12.1 and has a greater tendency to release electrons. Given an opportunity, this reaction would actually go in reverse! The answer is, "no reaction."

Exercise 2. In Figure 12.6 you will find several combinations of metals and salt solutions. Indicate the oxidation-reduction reaction that occurs in each, if any, and write a balanced ionic equation.

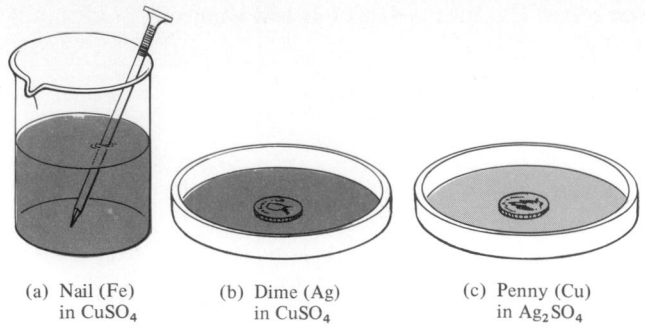

(a) Nail (Fe)   (b) Dime (Ag)   (c) Penny (Cu)
   in $CuSO_4$ in $CuSO_4$ in Ag_2SO_4

Figure 12.6. Predict the reactions in these oxidation-reduction pairs. See text for answer.

Answer:

Figure 12.6a: $Fe + Cu^{2+} \longrightarrow Fe^{2+} + Cu$

Figure 12.6b: $Ag + Cu^{2+} \longrightarrow$ N.R.

Figure 12.6c: $Cu + 2Ag^+ \longrightarrow Cu^{2+} + 2Ag$

Exercise 3. Predict which of the following conceivable reaction pairs give no reaction, "N.R."

(a) $Ni + Cu^{2+}$ _____   (d) $Br_2(l) + I^-$ _____
(b) $Ag + Au^{3+}$ _____   (e) $H_2 + F_2$ _____
(c) $Cu + Fe^{2+}$ _____   (f) $Hg^{2+} + Ag$ _____

Answer: (c) N.R., (f) N.R.

12.3 BALANCING OXIDATION-REDUCTION EQUATIONS

You have seen in the previous examples the advantage of knowing how many electrons are transferred in balancing an oxidation-reduction type equation. In fact, after you have decided which direction the reaction goes, the very next step is to balance the electrons.

Frequently a table of oxidation-reduction half-reactions may not contain the substances in which you are interested. In such cases there is another helpful method of balancing oxidation-reduction equations. This method is based on the oxidation numbers of the atoms that undergo change.

The Oxidation Number In those reactions involving only elementary substances (Zn, Mg, O_2, Cl_2, and so on) and binary ionic compounds ($Zn^{2+}O^{2-}$, $Mg^{2+}O^{2-}$, $Zn^{2+}Cl_2^-$, and so on) such as you have in Section 12.1, you may determine the number of electrons gained or lost by multiplying the change in electric charge by the number of atoms or ions in the reaction. In the following examples, the number written below each substance is the electric charge on *each* atom of that substance.

Positive

$$1(+3\ e^-) = +3\ e^-$$

$$Al^{3+} \longrightarrow Al$$
$$3+ \qquad\qquad 0$$

$3+$

$+3\ e^-$

0

Negative

Positive

$$2(+1\ e^-) = +2\ e^-$$

$$F_2 \longrightarrow 2\ F^-$$
$$0 \qquad\qquad 1-$$

0

$1-$ $\quad \downarrow\ +1\ e^-$

Negative

The number in parenthesis is the number of electrons required to change the charge *per atom* and it is multiplied by the number of atoms. Since the charges become more negative in both examples, downward on the adjoining number scale, they correspond to gaining electrons.

This method is so simple that it has been extended to atoms in more complicated compounds and ions. A positive or negative number is assigned to every atom that undergoes change as if it were an ion having that charge. In many cases the atom is only *partially* ionic, but the method works because it does not change the over-all number of electrons.

Take the molecule H_2SO_4 for example. In binary ionic compounds O is usually charged $2-$. If it is given this value in an H_2O molecule, each H must have the charge $1+$ to maintain electroneutrality. Using these values for O and H in H_2SO_4, what must be the "charge" on S?

$$H_2\ S\ O_4$$
$$1+\quad\ 2-$$

$$2(1+) + \boxed{} + 4(2-) = 0$$

$$2\ +\ \boxed{}\ -8\quad = 0$$

In order to account for all the electrons, the missing number must by $6+$.

$$H_2\quad S\quad O_4$$
$$6+$$

This is certainly not the actual charge on S in H_2SO_4. A positive charge of this magnitude would be too great for neighboring electrons to resist. Therefore, it is not called the ionic charge but the **oxidation number.** We may not be telling the truth about where the electrons are in the H_2SO_4 molecule, but we have neither destroyed nor created any electrons. We can use the oxidation numbers in a purely formal way to count the number of electrons gained and lost by reactants during reaction.

EXAMPLE 1.   Consider the reaction in which SO_2 is oxidized in the presence of H_2O and O_2 to become H_2SO_4. Show that the electrons lost by one atom are gained by another.

Solution: Every atom that undergoes change is located on the oxidation number scale. The oxidation number corresponds to the atom which is underlined. From these oxidation numbers, the numbers of electrons gained and lost are determined.

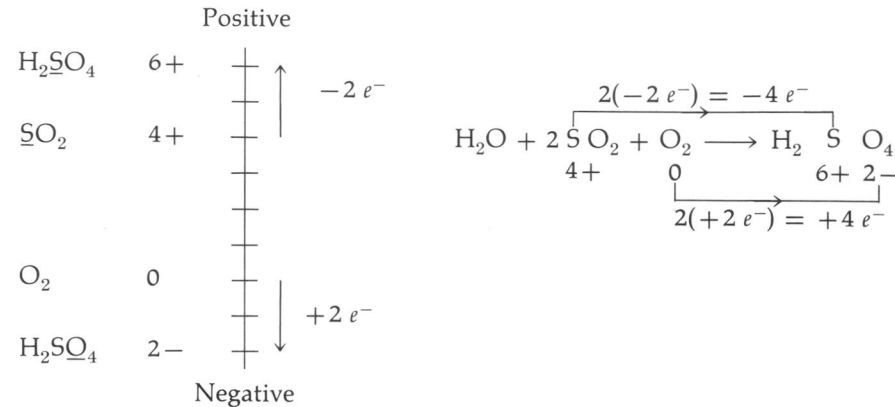

There are four electrons lost by the two S atoms in two molecules of SO_2 while changing from $4+$ to $6+$, and four electrons gained by the two O atoms in one molecule of O_2 while changing from 0 to $2-$. There would be no particular advantage in using the oxidation numbers to balance this equation because it is rather simply done by inspection, but it does give some useful information regarding the oxidizing capacity of an O_2 molecule and the reducing capacity of an SO_2 molecule. On the other hand some reactions are rather more difficult to balance. You might try your hand at the following exercise before learning how to use oxidation numbers.

Exercise 1.   Suppose you did not have access to a table of oxidation-reduction half-reactions that contained all the ions of interest. Try balancing the following equation by inspection, or trial and error.

$$MnO_4^- + SO_2 + H_2O \longrightarrow Mn^{2+} + SO_4^{2-} + H^+$$

Answer: Given in Example 2 (page 276).

By assigning oxidation numbers to each atom you could identify which atoms are oxidized and which are reduced and balance that part of the equation independently of the other parts. Having thereby fixed some of the coefficients, you should find it much easier to fix the others by inspection. But first you should learn a few simple rules for assigning oxidation numbers.

1. The oxidation number of each atom in an uncharged element is zero.

$$O_2 \quad Cl_2 \quad S_8$$
$$0 \qquad 0 \qquad 0$$

2. The oxidation number of a simple ion is equal to the charge on the ion.

$$Zn^{2+} \quad Fe^{3+} \quad Hg_2^{2+} \quad O_2^{2-}$$
$$2+ \qquad 3+ \qquad 1+ \qquad 1-$$

3. Oxygen is 2− except in peroxides where it is 1−.

$$H_2O \qquad H_2O_2 \text{ (hydrogen peroxide)}$$
$$2- \qquad\quad 1-$$
$$KClO_3 \qquad PbO_2 \qquad BaO_2 \text{ (barium peroxide)}$$
$$2- \qquad\quad 2- \qquad\quad 1-$$

4. Hydrogen is 1+ except in metal hydrides where it is 1−.

$$H_2O \quad HNO_3 \quad Ca(OH)_2 \quad CaH_2$$
$$1+ \qquad 1+ \qquad\quad 1+ \qquad\quad 1-$$

5. The alkali metals (principal group I) are always 1+ in compounds.

$$Li^+ \quad Na^+ \quad K^+ \quad Rb^+ \quad Cs^+$$

6. The alkaline earth metals (principal group II) are always 2+ in compounds.

$$Mg^{2+} \quad Ca^{2+} \quad Sr^{2+} \quad Ba^{2+} \quad Ra^{2+}$$

7. The principal group VII elements (F, Cl, Br, I) are always 1− in binary metallic compounds.

$$NaCl \quad FeBr_2 \quad GaI_3$$
$$1- \qquad\; 1- \qquad\; 1-$$

8. The sum of all the oxidation numbers in a molecule or ion must be equal to the charge on the molecule (zero) or ion.

$$H_2 \quad S \quad O_4$$
$$1+ \quad 6+ \quad 2- \qquad 2(+1) + (+6) + 4(-2) = 0$$

$$Mn \, O_4^-$$
$$7+ \quad 2- \qquad\qquad (+7) + 4(-2) = -1$$

$$Cr_2 \, O_7^{2-}$$
$$6+ \quad 2- \qquad\qquad 2(+6) + 7(-2) = -2$$

Exercise 2. Assign oxidation numbers to every atom shown. All peroxide oxygens are labeled.

$$P_4 \qquad NaH \qquad CrCl_3 \qquad Mg_3N_2 \qquad NH_3 \qquad Cl_2O \qquad Fe^{3+}$$

$$NaClO_4 \qquad HO_2^- \qquad PO_4^{3-}$$
$$\text{(peroxide)}$$

Answer:

$$\begin{array}{cccccccc}
& & 0 & 1+ \;\; 1- & 3+ \;\; 1- & 3- \;\; 2+ & 3- \;\; 1+ & 1+ \;\; 2- \\
& & P_4 & Na \;\; H & Cr \;\; Cl_3 & Mg_3 \;\; N_2 & N \;\; H_3 & Cl_2O \\
& & & 3+ & 1+ \;\; 7+ \;\; 2- & 1+ \;\; 1- & 5+ \;\; 2- & \\
& & & Fe^{3+} & Na \;\; Cl \;\; O_4 & H \;\; O_2^- & P \;\; O_4^{3-} & \\
\end{array}$$

EXAMPLE 2. Consider the reaction in Exercise 2 (page 274), in terms of oxidation numbers. Write a balanced equation.

Solution: The important oxidation numbers are assigned as follows.

$$MnO_4^- + SO_2 + H_2O \longrightarrow Mn^{2+} + SO_4^{2-} + H^+$$
$$\quad 7+ \qquad 4+ \qquad\qquad\qquad 2+ \qquad 6+$$

Investigation of the oxidation numbers of the other atoms shows that they do not change and need not be considered in the oxidation-reduction part of the reaction. The numbers of electrons transferred are

$$\overset{\overset{+5\,e^-}{\overbrace{\phantom{MnO_4^- + SO_2 + H_2O \longrightarrow Mn^{2+}}}}}{MnO_4^- + SO_2 + H_2O \longrightarrow Mn^{2+} + SO_4^{2-} + H^+.}$$

$$7+ \qquad 4+ \qquad\qquad 2+ \qquad 6+$$

$$\underset{-2\,e^-}{\underbrace{}}$$

Multiply $(+5\,e^-)$ by 2, *and* the Mn atoms on both sides by 2. Multiply $(-2\,e^-)$ by 5 *and* the S atoms by 5 on both sides. This completes the balancing of the oxidation-reduction part of the equation, with ten electrons gained and ten electrons lost.

$$\overset{2(+5\,e^-) = +10\,e^-}{\overbrace{\phantom{2\,MnO_4^- + 5\,SO_2 + H_2O \longrightarrow 2\,Mn^{2+}}}}$$

$$2\,MnO_4^- + 5\,SO_2 + H_2O \longrightarrow 2\,Mn^{2+} + 5\,SO_4^{2-} + H^+$$

$$7+ \qquad 4+ \qquad\qquad\qquad 2+ \qquad 6+$$

$$\underset{5(-2\,e^-) = -10\,e^-}{\underbrace{}}$$

Now that these numbers are fixed they should not be changed. You can of course multiply all of them by the same factor or start over. The remainder of the equation is balanced by balancing the O atoms with H_2O, or balancing the charge with H^+.

$$2\,MnO_4^- + 5\,SO_2 + 2\,H_2O \longrightarrow 2\,Mn^{2+} + 5\,SO_4^{2-} + 4\,H^+$$

Remember that in an ionic equation the charge must be balanced as well as the atoms.

charge: $2(-1)$ $\qquad\qquad = \qquad 2(+2) + 5(-2) + 4(+1)$
charge: $\qquad\qquad -2 \qquad = \qquad -2$

In Examples 3 and 4 you may use the oxidation number method, but you must also add H^+ ions if the solution is acidic and OH^- ions if it is basic.

Balancing Equations in Acidic and Basic Solutions

EXAMPLE 3. Balance the following reaction in *acidic* solution.

$$NO_2^- + Cr_2O_7^{2-} \longrightarrow NO_3^- + Cr^{3+}$$

Solution: The oxidation numbers are written, the number of electrons transferred is determined, and the oxidation-reduction part of the equation is balanced as shown.

$$\overset{3(-2\,e^-) = -6\,e^-}{\overbrace{\phantom{3\,NO_2^- + Cr_2O_7^{2-} \longrightarrow 3\,NO_3^-}}}$$

$$3\,NO_2^- + Cr_2O_7^{2-} \longrightarrow 3\,NO_3^- + 2\,Cr^{3+}$$

$$3+ \qquad 6+ \qquad\qquad 5+ \qquad 3+$$

$$\underset{2(+3\,e^-) = +6\,e^-}{\underbrace{}}$$

$$\text{charge: } 3(-1) + (-2) \quad \neq \quad 3(-1) + 2(+3)$$
$$\text{charge: } \qquad\qquad -5 \quad \neq \quad +3$$

Since the solution is *acidic*, add 8 H^+ on the left to make the total charge on that side $+3$. Then add 4 H_2O on the right to balance the H atoms on the left. Finally check the O atoms for balance.

$$3NO_2^- + Cr_2O_7^{2-} + 8H^+ \longrightarrow 3NO_3^- + 2Cr^{3+} + 4H_2O$$
$$\text{O atoms: } \quad 3(2) \quad + \quad 7 \qquad\qquad = \quad 3(3) \qquad\qquad + 4$$

EXAMPLE 4. Balance the following reaction in *basic* solution.

$$MnO_4^- + H_2O_2 \longrightarrow MnO_2 + O_2$$
$$\text{(hydrogen peroxide)}$$

Solution: Write the oxidation numbers and determine the number of electrons transferred by Mn.

$$\overset{\overset{\displaystyle +3\,e^-}{\longrightarrow}}{\underset{\underset{\displaystyle 7+ \qquad\quad 1- \qquad 4+2- \quad 0}{}}{Mn\,O_4^- + H_2O_2 \longrightarrow MnO_2 + O_2}}$$

Since Mn gains electrons the O atoms in H_2O_2 must lose electrons and become O_2. The oxidation-reduction part of the equation is balanced as follows.

$$\overset{\overset{\displaystyle 2(+3\,e^-)=\,+6\,e^-}{\longrightarrow}}{2MnO_4^- + 3H_2O_2 \longrightarrow 2MnO_2 + 3O_2}$$
$$7+ \qquad\qquad 1- \qquad\quad 4+ \qquad\quad 0$$
$$\underset{\underset{\displaystyle 3[2(-1\,e^-)]=\,-6\,e^-}{\longrightarrow}}{}$$

The charge on the right is 0 but the charge on the left is $2(-1)$ or -2.

Since the solution is *basic*, add 2 OH^- on the right to balance the -2 charge on the left.

$$2MnO_4^- + 3H_2O_2 \longrightarrow 2MnO_2 + 3O_2 + 2OH^-$$

Then add 2 H_2O on the right to balance the H atoms on the left, and finally check the O atom balance.

$$2MnO_4^- + 3H_2O_2 \longrightarrow 2MnO_2 + 3O_2 + 2OH^- + 2H_2O$$
$$\text{O atoms: } \quad 2(4) \quad + 3(2) \quad = \quad 2(2) \quad + 3(2) + 2 \qquad + 2$$

Exercise 3. Balance the following equation by the oxidation number method.

279

12.4
Practical Cells and Batteries

$$MnO_4^- + H_2O_2 \longrightarrow Mn^{2+} + O_2 \quad \text{in } \textit{acidic} \text{ solution}$$

(hydrogen peroxide)

Answer:

$$2\,MnO_4^- + 5\,H_2O_2 + 6\,H^+ \longrightarrow 2\,Mn^{2+} + 5\,O_2 + 8\,H_2O$$

oxidation numbers: $+7$, -1, $+2$, 0

$$2(+5\,e^-) = +10\,e^-$$
$$5[2(-1\,e^-)] = -10\,e^-$$

Exercise 4. Balance the following equation by the oxidation number method.

$$Cr(OH)_3 + IO_3^- \longrightarrow CrO_4^{2-} + I^- \quad \text{in } \textit{basic} \text{ solution}$$

Answer:

$$2\,Cr(OH)_3 + IO_3^- + 4\,OH^- \longrightarrow 2\,CrO_4^{2-} + I^- + 5\,H_2O$$

oxidation numbers: $+3$, $+5$, $+6$, -1

$$+6\,e^-$$
$$2(-3\,e^-) = -6\,e^-$$

12.4 PRACTICAL CELLS AND BATTERIES

The convenience of oxidation-reduction reactions as a portable source of electrical power is well known. Theoretically any oxidation reaction can produce electrons to be used in an external circuit. In the automobile battery it is lead (Pb) that is oxidized; in the dry cell used in a flashlight, zinc (Zn) is oxidized; in the electric toothbrush, the windless watch, and the pacemaker, cadmium (Cd) is oxidized; and in the fuel cell of a spaceship, $H_2(g)$ is oxidized. The automobile battery will be considered in detail in the next section.

An oxidation-reduction reaction which ordinarily would not be predicted from the series of half-reactions in Table 12.1 can be made to occur by applying a voltage from another cell. Direct current generators and rectifiers that convert alternating current to direct current can also be used to apply the voltage. An example of this type reaction is the dissociation of copper(II) iodide, CuI_2, into its elements. If a cell is constructed using the half-reactions $Cu \rightleftharpoons Cu^{2+} + 2\,e^-$ and $2\,I^- \rightleftharpoons I_2 + 2\,e^-$ the spontaneous reaction is

Electrolysis

$$Cu(s) + I_2(s) \longrightarrow Cu^{2+}(aq) + 2\,I^-(aq).$$

By forcing electrons in the opposite direction, a process called **electrolysis,** the reaction can be forced to go in reverse. This process is illustrated in Figure 12.7.

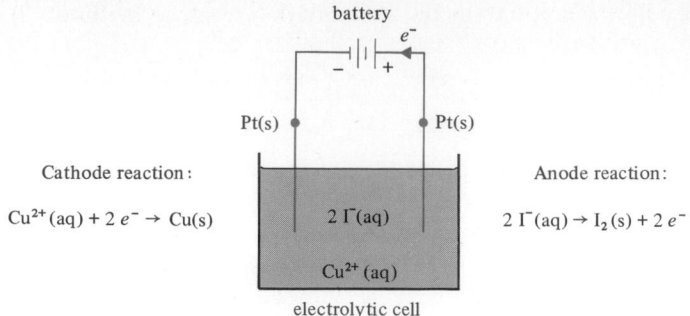

Figure 12.7. Electrolysis of $CuI_2(aq)$. (The platinum electrodes are inert under these conditions.)

DEFINITIONS: *Electrolysis is a chemical reaction caused by passing a direct current of electricity through a solution. During electrolysis or during discharge of a cell or battery, the electrode at which oxidation occurs is called the **anode**, and the electrode at which reduction takes place is called the **cathode**.*

In Figure 12.7, $Cu^{2+}(aq)$ is reduced to $Cu(s)$ at the cathode (left), and $I^-(aq)$ is oxidized to $I_2(s)$ at the anode (right).

DEFINITIONS: *Cations are **positively charged ions**. They are attracted to the cathode during electrolysis. **Anions** are **negatively charged ions**. They are attracted to the anode during electrolysis.*

The Cu^{2+} ions are cations and the I^- ions are anions in Figure 12.7.

The conduction of electricity by molten NaCl illustrated in Section 6.8 must be accompanied by electrolysis because that is the only way electrons can be transferred to the melt or removed from it. The reactions are

$$2\,Na^+ + 2\,e^- \longrightarrow 2\,Na(l) \qquad \text{cathode reaction}$$
$$2\,Cl^- \longrightarrow Cl_2(g) + 2\,e^- \qquad \text{anode reaction.}$$

However, alternating current causes a rapid reversal of anode and cathode from one electrode to the other, which prevents complete decomposition of sodium chloride. If direct current generators are employed, this electrolysis can be used for simultaneous production of metallic sodium and chlorine gas.

Environmental note: The factories that use this electrolytic process are called chloralkali plants although the electrolysis is usually performed in aqueous solution. Liquid mercury in which the sodium forms a solution or amalgam serves as the cathode. Wastes from these plants sometimes reach important bodies of water and contaminate the fish (See Section 8.3, Example 10). Japanese fishermen blockaded offending plants on the shores of Tokyo Bay in 1973.

In Chapter 7 you saw a number of spontaneous reactions that were exothermic. A very simple one was discussed in Section 7.3 and is repeated here.

$$2 H_2(g) + O_2(g) \longrightarrow 2 H_2O(l) + 136.6 \text{ kcal}$$

Since this is a typical oxidation-reduction reaction it can be reversed by an equivalent amount of electrical energy. However, pure water does not conduct electricity easily so that it cannot be electrolyzed alone. There are ions which will conduct electricity and yet are not discharged at the electrodes in the presence of water. This, of course, means that they do not enter into the reaction. All the metals in principal groups I and II of the periodic table are of this type as is Al^{3+}. Among the nonmetal ions only S^{2-} and those in principal group VII can be electrolyzed in aqueous solutions. A solution of Na_2SO_4 meets the requirements since neither ion electrolyzes. A few drops of a universal acid-base indicator, which is a mixture of indicators like those in Table 11.2, will give additional information regarding the course of the water electrolysis. This particular combination of indicators is red in acid, blue in base, and green in neutral solutions. In Figure 12.8 you can see the results of a brief period of electrolysis in a Petri dish using platinum (Pt) electrodes. The half-reactions at both electrodes are shown in equations (12.3) and (12.4). The cathode reaction is multiplied by two in order to use the same number of electrons that the anode reaction produces.

Cathode reaction:

$$4 H_2O + 4 e^- \longrightarrow 2 H_2 + 4 OH^- \tag{12.3}$$

Figure 12.8. Electrolysis of water.

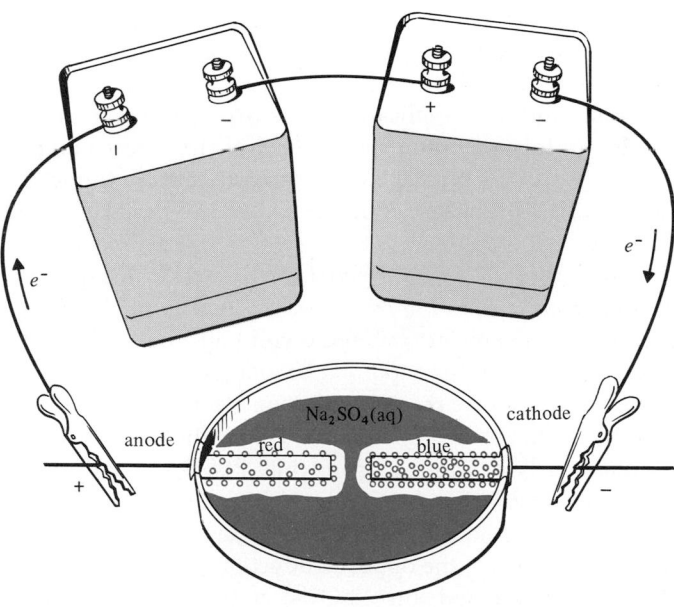

The solution in this region becomes basic. Some Na^+ ions migrate into the region and some SO_4^{2-} ions migrate out of the region to compensate for the new OH^- ions produced. Note that $H_2(g)$ bubbles form on and around the electrode.

Anode reaction:

$$2\,H_2O \longrightarrow O_2 + 4\,H^+ + 4\,e^- \tag{12.4}$$

The solution becomes acidic in this region. Some SO_4^{2-} ions migrate into the region and some Na^+ ions migrate out of the region to compensate for the new H^+ ions produced. Note that gas bubbles of O_2 form, but that their volume is somewhat less than that of the H_2. If electrolysis is discontinued and the solution is stirred, the mixture becomes neutral, as indicated by the return of the green color. The over-all reaction is determined by adding the cathode and anode reactions as shown in equation (12.5).

Over-all reaction:

$$6\,H_2O \longrightarrow 2\,H_2(g) + O_2(g) + 4\,H^+(aq) + 4\,OH^-(aq) \tag{12.5}$$

Canceling the H_2O molecules that are formed by neutralization on the right with the appropriate number on the left you obtain reaction (12.6).

Net over-all reaction:

$$2\,H_2O \longrightarrow 2\,H_2(g) + O_2(g) \tag{12.6}$$

The final result is that water is converted into its elementary components and the solution becomes more concentrated. Very pure samples of "electrolytic hydrogen" and "electrolytic oxygen" can be prepared in this way.

The Lead Storage Battery

The most common application of electrolysis is recharging a battery which has "run down." In this way the oxidized lead of the lead storage battery can be restored to metallic lead. The typical battery consists of from three to six cells arranged in series to increase the voltage. A single cell of the battery can be represented as follows.

$$Pb(s)\,|\,H_2SO_4(aq)\,|\,PbO_2(s)\,|\,Pb(s)$$

Each cell produces just under 2.0 volt with the electrode on the left being negative. There is no need for a porous barrier between electrode compartments of a lead storage cell because there is only one electrolyte, $H_2SO_4(aq)$. A cut-away view of a battery is shown in Figure 12.9. While the battery is discharging, metallic Pb is spontaneously oxidized to give white $PbSO_4(s)$, and red $PbO_2(s)$ is reduced to give $PbSO_4(s)$. The $PbSO_4(s)$ sticks to the plates where it is available for subsequent electrolysis. While the battery is charging, the reaction is made to go in reverse by forcing electrons in the opposite direction. The $PbSO_4(s)$ is converted back into $Pb(s)$ at one electrode and into $PbO_2(s)$ at the other. An over-all reaction is shown in equation (12.7).

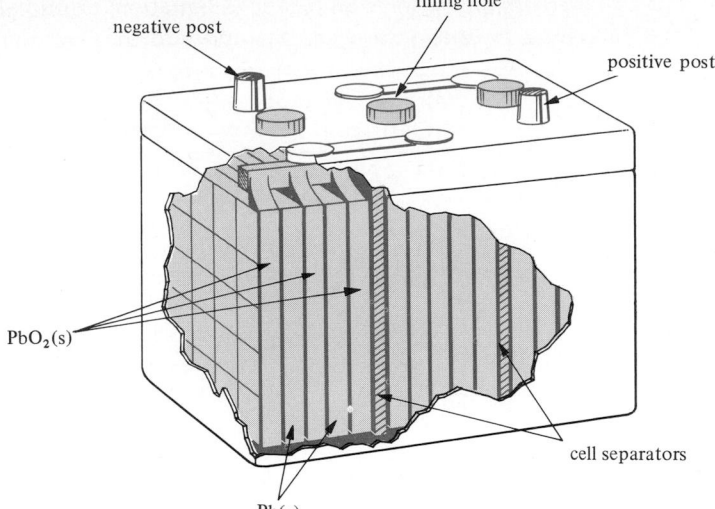

negative post

filling hole

positive post

PbO$_2$(s)

cell separators

Pb(s)

Figure 12.9. A three-cell lead storage battery.

$$Pb(s) + PbO_2(s) + 2\,H_2SO_4(aq) \underset{\text{charge}}{\overset{\text{discharge}}{\rightleftharpoons}} 2\,PbSO_4(s) + 2\,H_2O \qquad (12.7)$$

As the battery discharges, the H$_2$SO$_4$(aq) becomes more dilute, due to the formation of H$_2$O, and the solution becomes less dense. The level of the H$_2$SO$_4$(aq) may fall due to evaporation of H$_2$O, but H$_2$SO$_4$, which has a low vapor pressure, is not lost. A freshly charged battery has a solution density of approximately 1.290 g ml^{-1}.

Sealed Storage Batteries

The advantage of a sealed rechargeable battery is obvious. It has the portability and convenience of a dry cell and the long life of a lead storage battery. The latter advantage is particularly important in the case of the pacemaker for controlling the heart beat. In this case the cell is implanted by surgical operation. Sealed storage batteries are important in space travel as well where weightless conditions are encountered.

One such system is the nickel-cadmium (Ni-Cd) cell. In this cell the electrodes are constructed of metallic Cd (negative) and solid Ni(OH)$_3$ (positive), and the electrolyte is alkali. When discharging, the Cd(s) becomes Cd(OH)$_2$(s) and the Ni(OH)$_3$(s) becomes Ni(OH)$_2$(s).

EXAMPLE 1. Write balanced equations for (a) discharge and (b) charge of an alkaline Ni-Cd sealed storage cell.

Solution: (a) Write the half-reactions at each electrode during discharge.

$$Cd(s) + 2\,OH^-(aq) \longrightarrow Cd(OH)_2(s) + 2\,e^- \qquad \text{negative electrode}$$
$$Ni(OH)_3(s) + e^- \longrightarrow Ni(OH)_2(s) + OH^-(aq) \qquad \text{positive electrode}$$

To obtain the over-all balanced equation, multiply the second equation by 2 to cancel the electrons and add the two equations together.

$$Cd(s) + 2\,OH^-(aq) \longrightarrow Cd(OH)_2(s) + 2\,e^-$$
$$2\,Ni(OH)_3(s) + 2\,e^- \longrightarrow 2\,Ni(OH)_2(s) + 2\,OH^-(aq)$$
$$\overline{Cd(s) + 2\,Ni(OH)_3(s) \xrightarrow{\text{discharge}} Cd(OH)_2(s) + 2\,Ni(OH)_2(s)}$$

(b) Write the reverse of this equation for the over-all reaction that occurs during charge.

$$Cd(OH)_2(s) + 2\,Ni(OH)_2(s) \xrightarrow{\text{charge}} Cd(s) + 2\,Ni(OH)_3(s)$$

1. Balance the following oxidation-reduction reactions. Draw a *circle* around the element that is oxidized and a *line* under the element that is reduced.
   (a) $CO + O_2 \longrightarrow CO_2$
   (b) $Fe + H_2O \longrightarrow Fe_2O_3 + H_2$
   (c) $Al + H_2SO_4 \longrightarrow Al_2(SO_4)_3 + H_2$
   (d) $Fe^{3+} + Sn^{2+} \longrightarrow Fe^{2+} + Sn^{4+}$
   (e) $NO_2 + H_2O \longrightarrow NO + HNO_3$

2. Same as problem 1.

   (a) $C + O_2 \longrightarrow CO_2$
   (b) $Fe_2O_3 + CO \longrightarrow Fe + CO_2$
   (c) $HCl + Fe \longrightarrow FeCl_2 + H_2$
   (d) $Fe + Cu^{2+} \longrightarrow Fe^{3+} + Cu$
   (e) $H_2O_2 \longrightarrow H_2O + O_2$
   (hydrogen peroxide)

3. Balance the following oxidation-reduction reactions. How many electrons are gained and how many are lost by the element underlined for *each* atom or ion involved?
   (a) $\underline{Ag} + \underline{O}_2 \longrightarrow Ag_2O$

   (b) $\underline{Mg}\,\underline{Cl}_2 \xrightarrow{\text{electrolysis}} Mg + Cl_2$
   (c) $\underline{Sn} + \underline{O}_2 \longrightarrow SnO_2$
   (d) $\underline{Al} + \underline{H}_2 \longrightarrow AlH_3$
   (e) $\underline{Ba} + \underline{O}_2 \longrightarrow BaO_2$
   barium peroxide

4. Same as problem 3.

   (a) $\underline{Ca} + \underline{Cl}_2 \longrightarrow CaCl_2$

   (b) $\underline{Na}\,\underline{Cl} \xrightarrow{\text{electrolysis}} Na + Cl_2$
   (c) $\underline{Al} + \underline{O}_2 \longrightarrow Al_2O_3$
   (d) $\underline{Na} + \underline{H}_2 \longrightarrow NaH$
   (e) $\underline{Na} + \underline{O}_2 \longrightarrow Na_2O_2$
   sodium peroxide

5. The bismuth metal ion, $Bi^{3+}(aq)$, has a strong tendency to precipitate out of an aqueous solution as reduced Bi when metallic zinc, Zn, is added to it. The products of reaction are $Bi(s)$ and $Zn^{2+}(aq)$. Which of the following half-reactions is higher in the oxidation-reduction series in Table 12.1?

   $Bi \rightleftharpoons Bi^{3+} + 3\,e^-$, or
   $Zn \rightleftharpoons Zn^{2+} + 2\,e^-$.
   Explain.

6. The metal antimony, $Sb(s)$, has a strong tendency to dissolve in an aqueous solution containing $Ag^+(aq)$. The products of reaction are $Sb^{3+}(aq)$ and $Ag(s)$. Which of the following half-reactions is higher in the oxidation-reduction series in Table 12.1?

   $Sb \rightleftharpoons Sb^{3+} + 3\,e^-$, or
   $Ag \rightleftharpoons Ag^+ + e^-$.
   Explain.

7. The following reactions and nonreactions are applicable to aqueous ionic solution in which all elementary substances are solids unless otherwise labeled.
   (a) $Mg + Zn^{2+} \longrightarrow Mg^{2+} + Zn$
   (b) $Cu^{2+} + Zn \longrightarrow Cu + Zn^{2+}$
   (c) $Cu + H^+ \longrightarrow$ N.R.
   (d) $Zn + 2\,H^+ \longrightarrow Zn^{2+} + H_2(g)$

8. The following results are obtained from experiments with magnesium, Mg; platinum, Pt; and cadmium ion, $Cd^{2+}(aq)$.

(e) $Cu + 2\,Ag^+ \longrightarrow Cu^{2+} + 2\,Ag$

Arrange the following half-reactions with the strongest electron producers at the top and the weakest at the bottom, without referring to Table 12.1.

$$H_2(g) \rightleftharpoons 2\,H^+ + 2\,e^-$$
$$Ag \rightleftharpoons Ag^+ + e^-$$
$$Zn \rightleftharpoons Zn^{2+} + 2\,e^-$$
$$Cu \rightleftharpoons Cu^{2+} + 2\,e^-$$
$$Mg \rightleftharpoons Mg^{2+} + 2\,e^-$$

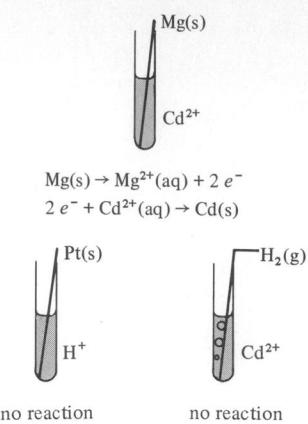

$$Mg(s) \rightarrow Mg^{2+}(aq) + 2\,e^-$$
$$2\,e^- + Cd^{2+}(aq) \rightarrow Cd(s)$$

no reaction   no reaction

How should the following substances be arranged in a line if those on the left in your list have the greatest tendency to release electrons? Mg, Pt, Cd, H_2

9. Use the oxidation-reduction series found in problem 7 to complete and balance the following equations. If no reaction occurs, write N.R.

(a) $Ag + H^+ \longrightarrow$
(b) $Mg + Cu^{2+} \longrightarrow$

10. Use the arrangement of elements based on their tendency to release electrons found in problem 8 to complete and balance the equations shown below. If no reaction occurs, write N.R.

(a) $Mg + H^+ \longrightarrow$
(b) $Pt + Cd^{2+} \longrightarrow$

11. Make a sketch of a cell like that in Figure 12.5 but in which the $Cu(s)|Cu^{2+}(aq)$ electrode is replaced by a $Ag(s)|Ag^+(aq)$ electrode. Show how the circuit is completed and indicate the direction of movement of all charged particles.

12. Make a sketch of the cell, $Pt(s)|H_2(g)|H^+(aq)\|Zn^{2+}(aq)|Zn(s)$, which shows the direction of movement of all charged particles (see Figure 12.5). Use $Cl^-(aq)$ ions to balance the positive ions in each half-cell.

13. Make a sketch of the sealed Ni-Cd storage cell described in Section 12.4. Use the abbreviated notation of a single vertical line for a phase boundary and a double vertical line for the separation of two solutions, if two are necessary.

14. The Edison storage cell consists of a $Ni(s)$ electrode that supports a mixture of $Ni(OH)_3(s)$ and $Ni(OH)_2(s)$ and is immersed in $OH^-(aq)$ solution and an iron electrode $Fe(s)$ that supports $Fe(OH)_2(s)$ and is also immersed in the $OH^-(aq)$ solution. Sketch this cell in the abbreviated form using a single vertical line for a phase boundary and a double vertical line for the separation of two solutions, if two are necessary.

15. Use the oxidation-reduction series in Table 12.1 to predict whether oxidation-reduction occurs in mixtures of the fol-

16. Same as problem 15.

lowing pairs of reactants and write a balanced equation. If no reaction occurs, write N.R.

(a) $H_2 + Cu^{2+} \longrightarrow$
(b) $Cl_2 + I^- \longrightarrow$
(c) $H^+ + Hg \longrightarrow$
(d) $Zn + Sn^{2+} \longrightarrow$

(a) $H_2 + Ag^+ \longrightarrow$
(b) $H^+ + Au \longrightarrow$
(c) $Hg^{2+} + Co \longrightarrow$
(d) $I_2 + Br^- \longrightarrow$

17. Write the ionic reactions that you would predict for the following mixtures in aqueous solution. If no reaction occurs, write N.R.

(a) $KBr(aq) + I_2 \longrightarrow$
(b) $CuI_2(aq) + Al \longrightarrow$
(c) $CuI_2(aq) + Br_2 \longrightarrow$

18. Same as problem 17.

(a) $NaCl(aq) + I_2 \longrightarrow$
(b) $CuI_2(aq) + Zn \longrightarrow$
(c) $CuI_2(aq) + Cl_2 \longrightarrow$

19. Assign an oxidation number to each atom in the ions and molecules shown below by writing it beneath the atom. Peroxide oxygens will be so labeled.

(a) Hg_2Cl_2 (e) NO_3^-
(b) K_2MnO_4 (f) LiH

(c) $Ca(ClO_4)_2$ (g) F_2
(d) H_3PO_3 (h) HSO_3^-

20. Same as problem 19.

(a) N_2 (e) $S_2O_3^{2-}$
(b) Na_2O_2 (f) $FeCl_3$
       sodium peroxide

(c) BaH_2 (g) $Mg(BrO_3)_2$
(d) SO_3^{2-} (h) $HClO$

21. Balance the following reactions, which are complete except for the coefficients. None of the oxides is a peroxide.

(a) $MnO_2 + KOH + O_2 \longrightarrow$
$$K_2MnO_4 + H_2O$$
(b) $U^{4+} + H_2O \longrightarrow$
$$UO_2^{2+} + U^{3+} + H^+$$
(c) $Br^- + BrO_3^- + H^+ \longrightarrow Br_2 + H_2O$

22. Same as problem 21.

(a) $HSO_3^- + IO_3^- \longrightarrow$
$$I_2 + SO_4^{2-} + H^+ + H_2O$$
(b) $Pb(OH)_3^- + ClO^- \longrightarrow$
$$Cl^- + PbO_2 + OH^- + H_2O$$
(c) $I_2 + HNO_3 \longrightarrow$
$$HIO_3 + NO_2 + H_2O$$

23. Complete and balance the following reactions which take place in acidic or basic solutions, as indicated.

(a) $MnO_2 + Cl^- \longrightarrow Cl_2 + Mn^{2+}$, in acidic solution
(b) $CrO_4^{2-} + SO_3^{2-} \longrightarrow CrO_2^- + SO_4^{2-}$, in basic solution

24. Same as problem 23.

(a) $Cr_2O_7^{2-} + H_2S \longrightarrow Cr^{3+} + SO_2$, in acidic solution
(b) $I^- + MnO_4^- \longrightarrow I_2 + MnO_2$, in basic solution

25. Write a balanced ionic equation for the electrolysis of aqueous $CuSO_4$, a strong electrolyte. Sketch the electrolytic cell and identify anode, cathode, anions, and cations, and indicate the direction of ion movement. (*Note:* The SO_4^{2-} ion is not dischargeable in aqueous solution.)

26. Write a balanced ionic equation for the electrolysis of aqueous NaCl. Sketch the electrolytic cell and identify anode, cathode, anions, and cations, and indicate the direction of ion movement. (*Note:* the Na^+ ion is not dischargeable in aqueous solution.)

27. Sealed Ag-Zn storage cells have been used in some spacecraft because of their high power output. The electrolyte solution is basic. Assume that during discharge the anode material changes from Zn to $Zn(OH)_4^{2-}(aq)$ and the cathode material changes from Ag_2O to Ag. (a) Write balanced equations for both discharge and charge of this cell. (b) Would you expect the concentration of $OH^-(aq)$ to change while the cell is discharging? Explain.

28. In the discharge of the Edison storage cell (see problem 14), oxidation occurs at the iron electrode. (a) Write balanced equations for both the discharge and the charge of this cell. Ni(s) does not enter into the reaction. (b) Would you expect the concentration of $OH^-(aq)$ to change while the cell is discharging? Explain.

*29. When two metals are brought into contact to protect one of them from corrosion, the procedure is called cathodic protection. If any electrochemical reaction occurs, the metal being protected acts like a cathode where reduction takes place. The other, more active, metal is called a sacrificial anode because it is consumed by oxidation. Assume that such a system is immersed in water containing traces of acid. There will be a tendency for the metals to form cations and for $H_2(g)$ to be released. Identify anode and cathode in the following systems during this process.

(a) A tin can (that is, a tinned iron can) has a scratch in it exposing the iron. It contains tomatoes, which are acidic.

(b) A galvanized (that is, zinc coated) iron fence post is in your backyard. The ground water around it is acidic due to the presence of carbon dioxide and other acid-forming substances.

(c) Discuss the merits of galvanizing the can and tinning the fence post.

*30. One disadvantage of electrochemical cells as sources of energy is their weight. Compare the amount of electricity (number of electrons) available per gram of starting material in the following cells.

(a) The lead storage cell (equation 12.7 on p. 283).

(b) A cell depending on the over-all reaction $2 H_2 + O_2 \longrightarrow 2 H_2O$ to produce electrons, such as that used in spacecraft. You may express your answer in moles of electrons, called faradays, or in coulombs per gram (1 mole electrons = 1 faraday = 96,500 coulomb, to three significant figures).

13

ATOMIC SPECTRA, ATOMIC ORBITALS, AND THE PERIODIC TABLE

We human beings take a lot of pleasure in the narrow band of electromagnetic radiation visible to the unaided eye, which we call light. We recognize more colors than we have names by which to call them. The rainbow, which is supposed to contain all the colors, seems to blend continuously from one shade to another. But the radiation that we can see is only a tiny fraction of the radiation that fills nature. If our eyes could see in the low energy infrared region, most substances that now appear colorless to us, like water and paraffin, would have colors for which we would probably want to invent names. A similar situation exists in the high energy ultraviolet region, which causes suntan and is invisible. X rays are a still higher energy form of this radiation. Radio waves are a very low energy form, beyond the infrared.

The discoveries about light made by the German scientists Max Planck and Albert Einstein at the beginning of this century launched completely novel fields of study in physics and chemistry that are still adding new insights into the nature of matter. From these studies we discovered not only the atomic numbers of the elements but also why they are arranged the way they are in the periodic table: 2 in the first period, 8 in the second, 8 in the third, 18 in the fourth, 18 in the fifth, 32 in the sixth, and the rest probably in the seventh.

13.1 THE COLOR OF LIGHT: WAVELENGTH, FREQUENCY, AND ENERGY

When white light, such as that from the hot tungsten (W) filament in an electric light bulb (Figure 13.1), is passed through a triangular glass prism, it is separated into its spectrum. This spectrum contains all the colors of the rainbow—violet, indigo, blue, green, yellow, orange, red, and shades in between. A beam of light striking the surface of the glass prism at an angle is bent as it enters the glass and again as it leaves. Violet-colored light is bent more than red-colored light causing violet to appear on the farthest end of the spectrum in Figure 13.1.

Light of different colors can be characterized in three different but equivalent ways, by wavelength, by frequency, and by energy.

Light as a Wave The characteristic measures of a wave are its wavelength and its frequency. All colors of light travel at the same velocity in air, $c = 3.00 \times 10^{10}$ cm sec^{-1}.

EXAMPLE 1. The velocity of light is considered the maximum velocity attainable in nature. Express this velocity in units of miles per second.

Solution: Depending on the one-factors with which you are familiar, the calculation could be made as follows:

$$x \frac{\text{mi}}{\text{sec}} = \left(3.00 \times 10^{10} \frac{\text{cm}}{\text{sec}}\right)\left(\frac{1}{2.54} \frac{\text{in.}}{\text{cm}}\right)\left(\frac{1}{12} \frac{\text{ft}}{\text{in.}}\right)\left(\frac{1}{5280} \frac{\text{mi}}{\text{ft}}\right)$$

$$= 186{,}000 \text{ mi sec}^{-1} \text{ (3 significant figures)}$$

The number of waves in monochromatic light, or single-color light, that pass a

Figure 13.1. The spectrum of white light.

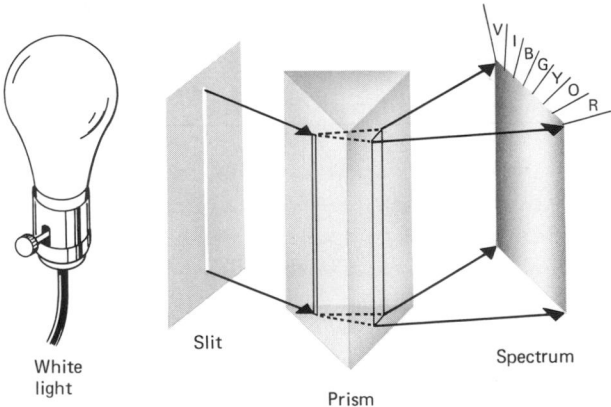

White light Slit Prism Spectrum

given point per second is called its frequency, ν sec^{-1}. Frequency is also expressed in units of cycles per second. For example, electricity in the United States is supplied as 60-cycle alternating current, which means that the flow of current in one direction reaches a maximum 60 times every second.

The **wavelength**, λ cm, is the distance in the wave from maximum to maximum or minimum to minimum as shown in Figure 13.2. The product of the length of the wave (λ) and the number of waves that go by per second (ν) must be equal to the velocity of the wave as shown below.

$$(\lambda \text{ cm})(\nu \text{ sec}^{-1}) = c \text{ cm sec}^{-1} \qquad (13.1)$$

Since c is fixed, you can calculate λ if you are given ν, and vice-versa. Red light has a wavelength of around 6500 Å and blue light around 4500 Å.

EXAMPLE 2. Calculate the approximate frequencies of (a) red 6500 Å and (b) blue 4500 Å light.

Solution: You will need the equality, 1 Å = 10^{-8} cm, from Table 2.3 in Section 2.1, to make the necessary calculations,
(a) Red light: λ = 6500 Å. From equation (13.1) you have

$$\lambda\nu = c.$$

Dividing both sides by λ gives

$$\nu = c/\lambda.$$

Substitute the known values for λ and c.

$$\nu = \left(\frac{3.00 \times 10^{10} \text{ cm sec}^{-1}}{6500 \quad \text{Å}} \right) \cdots$$

The calculation is complete except for a one-factor to provide the correct units. Change Ångstroms to centimeters using the equality given.

Figure 13.2. Light as a wave.

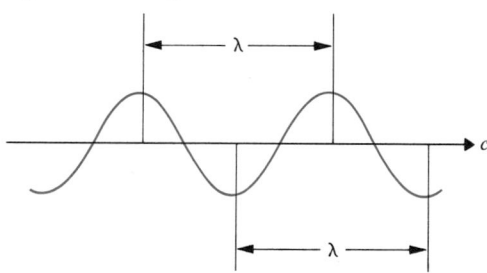

$$\nu = \left(\frac{3.00 \times 10^{10}\ \text{cm sec}^{-1}}{6500}\ \frac{}{\text{Å}}\right)\left(\frac{1}{10^{-8}}\ \frac{\text{Å}}{\text{cm}}\right)$$

$$= 4.62 \times 10^{14}\ \text{sec}^{-1}$$

(b) Blue light: $\lambda = 4500$ Å

$$\nu = \left(\frac{3.00 \times 10^{10}}{4500}\right)\left(\frac{1}{10^{-8}}\right)\ \text{sec}^{-1}$$

$$= 6.67 \times 10^{14}\ \text{sec}^{-1}$$

The blue and violet end of the visible spectrum has shorter wave-lengths and higher frequencies than the red end.

EXAMPLE 3. A certain FM station broadcasts at a frequency of 89.7 Mc ($89.7 \times 10^6\ \text{sec}^{-1}$). What is the wavelength of this transmission?

Solution: From equation (13.1) you have

$$\lambda\nu = c.$$

Substitute the known values of ν and c.

$$\lambda(89.7 \times 10^6\ \text{sec}^{-1}) = 3.00 \times 10^{10}\ \text{cm sec}^{-1}$$

Divide both sides by $89.7 \times 10^6\ \text{sec}^{-1}$.

$$\lambda = \frac{3.00 \times 10^{10}}{89.7 \times 10^6}\ \frac{\text{cm sec}^{-1}}{\text{sec}^{-1}}$$

$$= 3.34 \times 10^2\ \text{cm}$$

$$= 3.34\ \text{m}$$

This wave is long in comparison with those of visible light, but it is much shorter than an AM radio wave.

Exercise 1. Express the wavelength of 2800 Å ultraviolet light in nanometers (nm). The nanometer is frequently used as a unit of wavelength. You will need to use the equality, 1 nm = 10^{-9} m, from the list of prefixes in Table 2.2, p. 18.

Answer: ɯu 08Z

Light as
a Quantum
of Energy

The concept for which Planck found a good theory and for which Einstein found a good example was that light consists of **quanta,** or particles, of energy.

$$\Delta E = h\nu \tag{13.2}$$

Although a quantum is a given amount of energy, it represents the difference between two energy levels ΔE, where Δ means "the change in." (See page 295.) According to theory, light energy is not continuous but discrete, like atoms of matter. If the energy ΔE, is expressed in kilocalories per mole, h has the following constant value: $h = 9.54 \times 10^{-14}$ kcal mole^{-1} sec. The "mole^{-1}" refers to a mole of quanta. The energy per mole is often a more convenient unit than energy per quantum for the same reason that a mole of atoms is a more convenient unit than a single atom. The fixed value, h, is called **Planck's constant.**

EXAMPLE 4. Calculate the energy, ΔE, of a quantum of red light (6500 Å).

Solution: From equation (13.2) you have the following.

$$\Delta E = h\nu$$

Substitute the known value of h.

$$\Delta E = (9.54 \times 10^{-14} \text{ kcal mole}^{-1} \text{ sec})\nu$$

The value ν for 6500 Å light is calculated in Example 2, part (a), page 292. Substitute, $\nu = 4.62 \times 10^{14}$ sec^{-1} into equation (13.2).

$$\Delta E = (9.54 \times 10^{-14} \text{ kcal mole}^{-1} \text{ sec})(4.62 \times 10^{14} \text{ sec}^{-1})$$
$$= 44.1 \text{ kcal mole}^{-1}$$

A gas discharge tube can be made to produce light by passing an electric current through a gas at reduced pressures. A gas, such as H_2, is evacuated to a low pressure in a tube containing two electrodes, much like a neon sign or a fluorescent bulb. If you apply a voltage to these electrodes, they will produce an electric discharge. The excitation of the gas results in the emission of visible light. With hydrogen, for example, the tube glows red. If a beam of this light is passed through a prism, the previous rainbow of colors becomes a set of colored lines whose wavelengths, energies, or frequencies are very sharp and definite, as shown in Figure 13.3.

Figure 13.3. The spectrum of atomic H. How are these wavelengths related to the energy levels of the H atom?

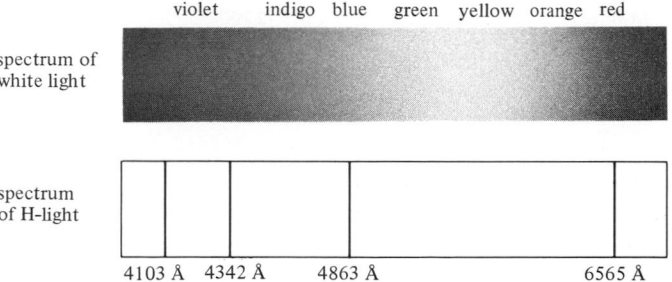

13.2 ATOMIC SPECTRA: THE H ATOM

Wavelength data are among the most precise measurements of which scientists are capable. One method of obtaining these data is to shine **monochromatic** light, that is, light consisting of a single wavelength, on a diffraction grating. This grating consists of a large number of carefully ruled parallel scratches on a metallic or other solid surface. The spacing between the scratched lines is similar in dimensions to that of the wavelength being measured. A characteristic diffraction pattern is produced which permits calculation of the wavelength of light in terms of the distances between the scratched lines. Mechanical devices have been developed for making scratches with very precise distances of separation.

> **Note on the metric system:** The wavelength of light emitted from an atom is a good standard of reference for other lengths because it never changes and it can be precisely measured. The centimeter in the metric system of measurement is defined as 16507.6373 times the wavelength of a certain line in the visible discharge spectrum of krypton (Kr).

The atomic spectrum of an element is the complete set of wavelengths emitted by its atoms in a discharge tube or some other instrument capable of exciting its electrons. A part of the line spectrum of the H atom is shown in comparison with the continuous spectrum of white light in Figure 13.3. The line spectrum is produced in a discharge tube containing H_2 molecules. The most intense of the lines is the red one at 6565 Å. It is this line that gives the discharge tube its red glow. The numerical values of these wavelengths and several others are listed in Table 13.1.

TABLE 13.1   A part of the spectrum of atomic hydrogen

Type of Radiation	λ(Å)	ΔE(kcal mole^{-1})	Δn
red	6565	43.6	3 ⟶ 2
blue	4863	58.8	4 ⟶ 2
violet	4342	65.9	5 ⟶ 2
violet	4103	69.7	6 ⟶ 2
. . .	. . .	. . .	. . .
ultraviolet	1216	235.2	2 ⟶ 1
ultraviolet	1026	278.8	3 ⟶ 1
ultraviolet	973	294.0	4 ⟶ 1
. . .	. . .	. . .	. . .
		313.6	∞ ⟶ 1

Exercise 1.   Show by calculation that the blue line in the atomic spectrum of H (4863 Å) corresponds to an energy value of 58.8 kcal mole^{-1} (see Table 13.1).

The explanation for these lines is simple if you think of an excited hydrogen atom as occupying a certain step on a stairway of energy levels, as shown in Figure 13.4. Every line in the **visible series** (red, blue, violet, . . .) is explained by jumps ending on step number 2. The lowest energy line, the red line, is a result of the atom jumping from the third step to the second step, the blue line as a jump from the fourth to the second step, and so on. Now you can see why the energy associated with a line is called ΔE, the *change* in energy, and not just E. Each line represents the difference between two energy levels. The energy of the light quantum emitted is just enough to balance the decrease in energy of the atom. Scientists have assigned a whole number to each energy level because the ΔE values are given by a single equation in which these level numbers are involved. The general equation (13.3) and two examples are shown.

$$\binom{\text{final}}{\text{energy}} - \binom{\text{initial}}{\text{energy}} = -313.6\left(\frac{1}{\left[\begin{smallmatrix}\text{final}\\\text{level}\\\text{number}\end{smallmatrix}\right]^2} - \frac{1}{\left[\begin{smallmatrix}\text{initial}\\\text{level}\\\text{number}\end{smallmatrix}\right]^2}\right) \quad (13.3)$$

$$3 \longrightarrow 2: \quad E_2 - E_3 = -313.6\left(\frac{1}{2^2} - \frac{1}{3^2}\right) = -313.6\left(\frac{1}{4} - \frac{1}{9}\right) = -313.6\left(\frac{5}{36}\right)$$
$$= -43.6 \text{ kcal mole}^{-1}$$

$$4 \longrightarrow 2: \quad E_2 - E_4 = -313.6\left(\frac{1}{2^2} - \frac{1}{4^2}\right) = -313.6\left(\frac{1}{4} - \frac{1}{16}\right) = -313.6\left(\frac{3}{16}\right)$$
$$= -58.8 \text{ kcal mole}^{-1}$$

These energy differences represent the energy changes in the atom, which are negative. They are negative because the energies of the light quanta emitted are positive, and this provides for conservation of energy. The energies of the light quanta are the ones listed in Table 13.1.

Exercise 2. Calculate the next value of ΔE in this series. Compare your answer with the value given in Table 13.1.

Answer: 5 ⟶ 2: −65.9 kcal mole⁻¹

Figure 13.4. A portion of the energy levels in atomic H.

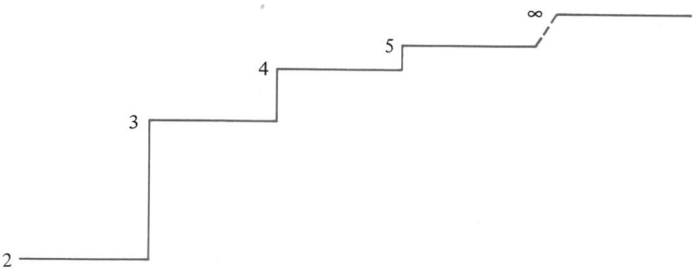

The visible series of lines emitted by hydrogen was discovered first, but its discovery led to the observation of other series in ranges of wavelengths invisible to the human eye. The ultraviolet series is explained by introducing one more step at the bottom of the previous set, as shown in Figure 13.5. The last step is a big one! The equation used for the visible series (equation 13.3) gives the observed energy lines in the ultraviolet series. Several examples are given.

$$2 \longrightarrow 1: \quad E_1 - E_2 = -313.6\left(\frac{1}{1^2} - \frac{1}{2^2}\right) = -313.6\left(\frac{1}{1} - \frac{1}{4}\right) = -313.6\left(\frac{3}{4}\right)$$

$$= -235.2 \text{ kcal mole}^{-1}$$

$$3 \longrightarrow 1: \quad E_1 - E_3 = -313.6\left(\frac{1}{1^2} - \frac{1}{3^2}\right) = -313.6\left(\frac{1}{1} - \frac{1}{9}\right) = -313.6\left(\frac{8}{9}\right)$$

$$= -278.8 \text{ kcal mole}^{-1}$$

$$\infty \longrightarrow 1: \quad E_1 - E_\infty = -313.6\left(\frac{1}{1} - \frac{1}{\infty}\right) = -313.6(1 - 0) = -313.6(1)$$

$$= -313.6 \text{ kcal mole}^{-1}$$

Ultraviolet light has high enough energy to break chemical bonds. As you can see from the last calculation, ultraviolet light is capable of lifting the electron in hydrogen from $n = 1$ to $n = \infty$, or, in other words, of completely removing the electron. Atoms in molecules are held together by chemical bonds having energies on the order of

Figure 13.5. Energy levels of atomic H in kilocalories per mole.

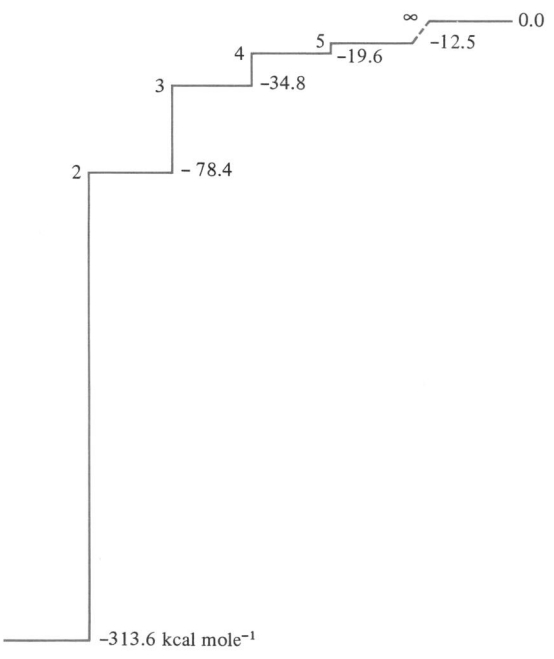

100 kcal mole^{-1}. A quantum containing several hundred kilocalories per mole colliding with a molecule can easily destroy one of its chemical bonds and cause the molecule to dissociate. For this reason prolonged exposure of the skin to the sun or looking directly at the sun, which produces considerable radiation in the ultraviolet range, can be extremely harmful. Fortunately, ultraviolet light is not very penetrating and most of it is absorbed by the O_3 (ozone) in the atmosphere before it reaches earth. Microbiologists work under ultraviolet lamps with some pathogenic microorganisms in order to lessen the chance of airborne infection.

Every step in Figure 13.5 can be assigned an energy value if one step is arbitrarily given the value zero. It is customary to assign zero to the highest energy step. Of course this requires that the lower steps be given negative values. Physically the zero energy level represents the complete separation of the electron and the proton, under which conditions the force of attraction is zero (equation 5.2, p. 99). At lower energy levels the electron is closer to the proton. All the energy levels are given by equation (13.4).

Zero on the Energy Diagram

$$E = \frac{-313.6}{n^2} \qquad \text{where } n = 1, 2, 3, \ldots \qquad (13.4)$$

Each of the steps in Figure 13.5 is labeled with the energy calculated using equation (13.4) and the corresponding step number. Note that all the energies are negative except the highest, which has the value zero.

1. An excited H atom occupies one of the energy levels defined by the whole number n. The normal H atom is in energy level 1.
2. By jumping from one energy level to a lower one, an excited H atom emits a quantum of light, whose energy is given by subtracting the lower energy level from the higher.

Summary: Atomic Spectrum of the H Atom

EXAMPLE 1. Show that the ultraviolet line in the H atom spectrum corresponding to the transition, $4 \longrightarrow 1$, has (a) energy 294.0 kcal mole^{-1} and (b) wavelength 973 Å (Table 13.1).

Solution: The energy (ΔE) is most easily calculated using the values in Figure 13.5. Then use the algebraic equations in Section 13.1 to convert ΔE to λ.
(a) From Figure 13.5, the energy of level 4 is seen to be -19.6 kcal mole^{-1} and that of level 1 to be -313.6 kcal mole^{-1}. Write the definition of ΔE.

$$\Delta E = E_{\text{final}} - E_{\text{initial}}$$

Substitute the initial and final values from the figure.

$$\Delta E = -313.6 - (-19.6) \text{ kcal mole}^{-1}$$
$$= -313.6 + 19.6$$
$$= -294.0 \text{ kcal mole}^{-1}$$

This is the energy change of the *atom;* the energy of the *light quantum* emitted has the positive value

$$\Delta E = +294.0 \text{ kcal mole}^{-1}.$$

(b) Write equations (13.2) and (13.1) and eliminate ν, whose value is unknown, in the following algebraic steps.

$$\Delta E = h\nu \qquad (13.2)$$

$$\lambda\nu = c \qquad (13.1)$$

Divide both sides of equation (13.1) by λ.

$$\nu = c/\lambda$$

Substitute this expression for ν into equation (13.2).

$$\Delta E = h\left(\frac{c}{\lambda}\right)$$

Multiply both sides by $\lambda/\Delta E$.

$$\lambda = \frac{hc}{\Delta E}$$

Substitute into this equation the known constants, $h = 9.54 \times 10^{-14}$ kcal mole^{-1} sec, and $c = 3.00 \times 10^{10}$ cm sec^{-1}, and the positive value of ΔE, the light quantum calculated in part (a).

$$\lambda = \frac{(9.54 \times 10^{-14} \text{ kcal mole}^{-1} \text{ sec})(3.00 \times 10^{10} \text{ cm sec}^{-1})}{294.0 \text{ kcal mole}^{-1}}$$

$$= \frac{(9.54 \times 10^{-14})(3.00 \times 10^{10})}{294.0} \text{ cm}$$

Use the one-factor method to convert centimeters into Ångstroms.

$$\lambda = \left[\frac{(9.54 \times 10^{-14})(3.00 \times 10^{10})}{294.0} \text{ cm}\right]\left[\frac{1}{10^{-8}} \frac{\text{Å}}{\text{cm}}\right]$$

$$\lambda = 973 \text{ Å}$$

Exercise 3. In the atomic spectrum of H, there is a series of lines in the infrared region corresponding to excited atoms falling to the $n = 3$ level from higher levels. Use Figure 13.5 to calculate the energies of the quanta corresponding to the first two lines in this series.

299

13.3
Quantum Numbers

Answer: 4 ⟶ 3:   15.2 kcal mole^{-1}, 5 ⟶ 3:   22.3 kcal mole^{-1}

13.3 QUANTUM NUMBERS

The integers $n = 1, 2, 3, \ldots, \infty$ in equation (13.4) are called the **principal quantum numbers.** They are also called the energy shell numbers. The word "shell" in everyday language usually implies something about space or location and not about energy. However, the location of an electron in an atom is *never* precisely known, nor is its orbit or path of motion. Only the energy level is exact. Of course, qualitatively the higher the value of n the farther the electron is from the nucleus.

From a closer analysis of atomic spectra, other quantum numbers in addition to the principle quantum numbers have been discovered. The mathematical treatment of the energies and locations of electrons in atoms is called **quantum mechanics.** Scientists have used all the quantum numbers that have been discovered through the study of atomic spectra and the known shapes of molecules to develop a set of models that explain much of the chemical behavior of all the elements. These models will be described in the following paragraphs.

Orbitals

Each energy shell in the H atom consists of a certain number of orbitals. A maximum of two electrons can occupy each of these orbitals, but of course the orbital does not have to be occupied. The first shell consists of one orbital, the second shell consists of four orbitals, the third shell consists of nine orbitals, and so on. If n is the shell number, the number of available orbitals in that shell is n^2.

EXAMPLE 1. Calculate the number of orbitals available for electron occupancy in the first, second, third, and fourth shells of an atom. What is the maximum number of electrons permitted in each shell?

Solution:

	n	n^2	Maximum Number of Electrons $2n^2$
first shell	1	$(1)^2 = 1$	$2 \times 1 = 2$
second shell	2	$(2)^2 = 4$	$2 \times 4 = 8$
third shell	3	$(3)^2 = 9$	$2 \times 9 = 18$
fourth shell	4	$(4)^2 = 16$	$2 \times 16 = 32$

Each orbital has a different shape, that is, the electron or electrons which occupy that orbital fill space in a characteristic geometric form. The first shell consists of an s orbital, which is spherically symmetric. This orbital has already been described (Figure 5.3, p. 103) for the normal H atom. As shown in Figure 5.3 the most probable *location as a point* of an **s electron** is a position that coincides with the proton or nucleus, but the most probable *distance* from the nucleus is represented by the surface of a sphere having a radius of 0.5 Å with the nucleus at the center (Figure 5.3b). The locations of the electron at any two points on the surface of the sphere in Figure 13.6a are equally probable. Thus the diagram represents a kind of contour map for electron density or probability.

The second shell consists of an *s* orbital, which you may designate as 2s to distinguish it from the 1s orbital in the first shell, and three **p orbitals.** The 2s orbital is spherically symmetric, like the 1s orbital, but the *p* orbitals are directed along the three mutually perpendicular axes, *x*, *y*, and *z*. To differentiate among the *p* orbitals you may designate them as $2p_x$, $2p_y$ and $2p_z$. The positions of maximum probability for an electron in the $2p_x$ orbital are on the *x* axis a short distance to the right of the nucleus and an equal distance to the left. The electron spends equal time at each point and no statement can be made about how it gets from one point to the other. The p_y orbital has the same symmetry along the *y* axis, and the p_z orbital along the *z* axis. The two-lobed surfaces shown in Figure 13.6 represent points of equal probability within the given *p* orbitals.

The 2s, $2p_x$, $2p_y$, and $2p_z$ orbitals comprise the four orbitals in the second shell. In the H atom, all four are on the same energy level, represented by step 2 in Figure 13.5.

In the third shell there are one 3s orbital, three 3p orbitals, and five **3d orbitals,** making nine all together. The *d* orbitals are typically four lobed.

The fourth shell has the usual number of *s*, *p*, and *d* orbitals plus seven **4f orbitals,** for a total of 16. The *f* orbitals vary considerably in their geometry.

EXAMPLE 2.   Show that the maximum number of electrons in the fourth shell of any atom is 32.

Solution: The fourth shell contains four types of orbitals, namely 4s, 4p, 4d, and 4f. The total number of orbitals is given by the equation

$$n^2 = 4^2 = 16 \text{ orbitals.}$$

Figure 13.6.   The 1s and 2p orbitals of the H atom. These surfaces represent points of equal probability for the location of an electron in the given orbital.

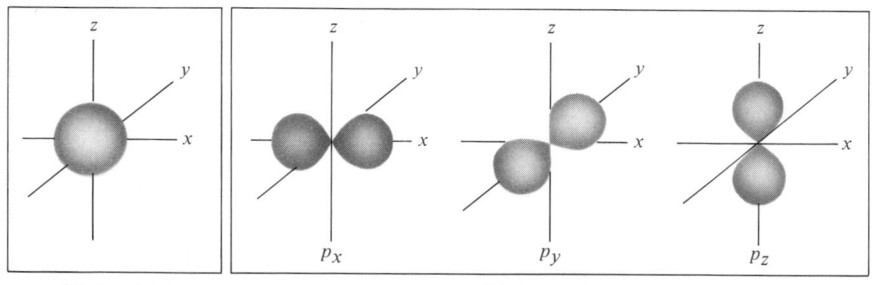

(a) 1s–orbital   (b) 2p-orbitals

$$4s \ 4p4p4p \ 4d4d4d4d4d \ 4f4f4f4f4f4f4f.$$

Each orbital has a maximum occupancy of two electrons, which gives the following maximum occupancies by orbital type.

$$4s^2 \quad 4p^6 \quad 4d^{10} \quad 4f^{14}$$

By adding the superscripts you may calculate the maximum number of electrons in the fourth shell.

$$\text{number of electrons} = 2 + 6 + 10 + 14 = 32$$

There are no elements in the periodic table for which the s, p, d, and f orbitals are not sufficient to describe the unexcited atom.

Filling Orbitals in Multi-electron Atoms

The geometric shapes of the orbitals of the same type (s, p, d, or f) in all atoms are similar to those for the H atom. Of course the greater attractive force of the more positive nuclei and the repulsive forces between electrons in the same atom cause considerable changes in the energy levels of these orbitals. Orbitals of the same type, for example, $2p_x$, $2p_y$, $2p_z$, are still on the same energy level, but different types of orbitals, for example, $2s$, $2p$, are on different energy levels. There are **subshells** of energy within each shell; the s orbital comprises one subshell, the p orbitals are in another subshell, the d orbitals in still another subshell, and so on. The energy levels of the orbitals in the same energy shell (n) have the following order: $s < p < d < f$, where the first is on the lowest energy level. Furthermore, $1s < 2s < 3s \ldots$, and $2p < 3p < 4p \ldots$, but there is no requirement for $3d < 4s$ or $4f < 6s$, and so on.

If you will imagine constructing the whole periodic table by adding one proton and one electron at a time, with whatever neutrons are necessary for the stability of the nucleus, the order of filling the orbitals, starting with the lowest energy first, is shown in Table 13.2. You will notice that there are a number of irregularities in the filling order. For example, the $4s$ orbital is introduced before the third shell is completed. Likewise before the fourth shell is completed, $5s$, $5p$, and $6s$ orbitals are introduced.

This order of filling is shown more clearly on the relative energy step diagram in Figure 13.7. The general rule is that electrons must fill the lower energy levels before they begin to occupy orbitals at higher energy levels.

The electronic structures of the first few elements are given in Table 13.3, where the superscript represents the number of electrons in each orbital. The subscript on the symbol of each element represents its atomic number and consequently the total of its electrons.

TABLE 13.2   The order of filling the atomic orbitals in the periodic table

$1s$	$2s$	$2p$	$3s$	$3p$	$4s$	$3d$	$4p$	$5s$	$4d$	$5p$	$6s$	$4f$	$5d$	$6p$	$7s$	$5f$

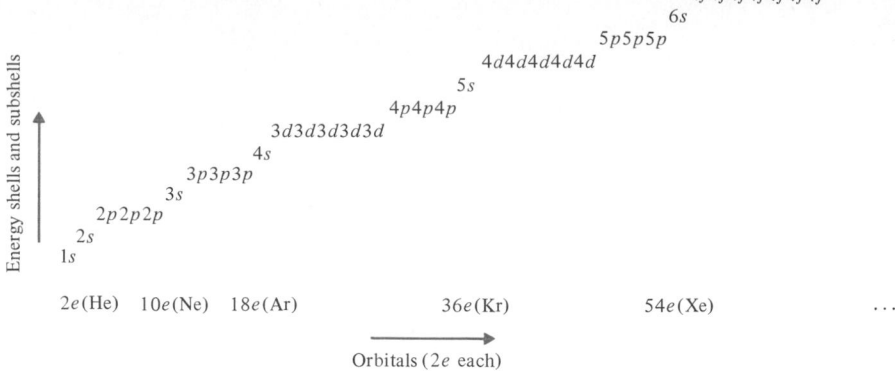

Figure 13.7. The relative energy levels that show the order of filling the atomic orbitals in the periodic table.

Exercise 1. Give the electronic structures of Xe and Sr. You may combine all the orbitals of a given subshell.

Answer:

$$_{38}Sr(1s^2 2s^2 2p^6 3s^2 3p^6 4s^2 3d^{10} 4p^6 5s^2)$$

$$_{54}Xe(1s^2 2s^2 2p^6 3s^2 3p^6 4s^2 3d^{10} 4p^6 5s^2 4d^{10} 5p^6)$$

You will notice that after each element located in the last group of the periodic table (He, Ne, Ar) a new shell is introduced, occupied by a single electron. Naturally these elements in the first group are all very similar chemically, since their chemical reactions involve only the outermost electrons (Section 6.1). The order of filling the orbitals becomes obvious when you superimpose it on the periodic table as shown in Table 13.4.

For example, there are ten elements in the fourth period after $_{20}$Ca, which

TABLE 13.3 Electronic structures of the first twenty elements in the periodic table

$_1$H	$1s^1$	*$_{11}$Na	$[Ne]3s^1$
$_2$He	$1s^2$	$_{12}$Mg	$[Ne]3s^2$
$_3$Li	$1s^2 2s^1$	$_{13}$Al	$[Ne]3s^2 3p_x^1$
$_4$Be	$1s^2 2s^2$	$_{14}$Si	$[Ne]3s^2 3p_x^1 3p_y^1$
$_5$B	$1s^2 2s^2 2p_x^1$	$_{15}$P	$[Ne]3s^2 3p_x^1 3p_y^1 3p_z^1$
$_6$C	$1s^2 2s^2 2p_x^1 2p_y^1$	$_{16}$S	$[Ne]3s^2 3p_x^2 3p_y^1 3p_z^1$
$_7$N	$1s^2 2s^2 2p_x^1 2p_y^1 2p_z^1$	$_{17}$Cl	$[Ne]3s^2 3p_x^2 3p_y^2 3p_z^1$
$_8$O	$1s^2 2s^2 2p_x^2 2p_y^1 2p_z^1$	$_{18}$Ar	$[Ne]3s^2 3p_x^2 3p_y^2 3p_z^2$
$_9$F	$1s^2 2s^2 2p_x^2 2p_y^2 2p_z^1$	$_{19}$K	$[Ne]3s^2 3p_x^2 3p_y^2 3p_z^2 4s^1$
$_{10}$Ne	$1s^2 2s^2 2p_x^2 2p_y^2 2p_z^2$	$_{20}$Ca	$[Ne]3s^2 3p_x^2 3p_y^2 3p_z^2 4s^2$

*[Ne] *represents the electronic structure of Ne.*

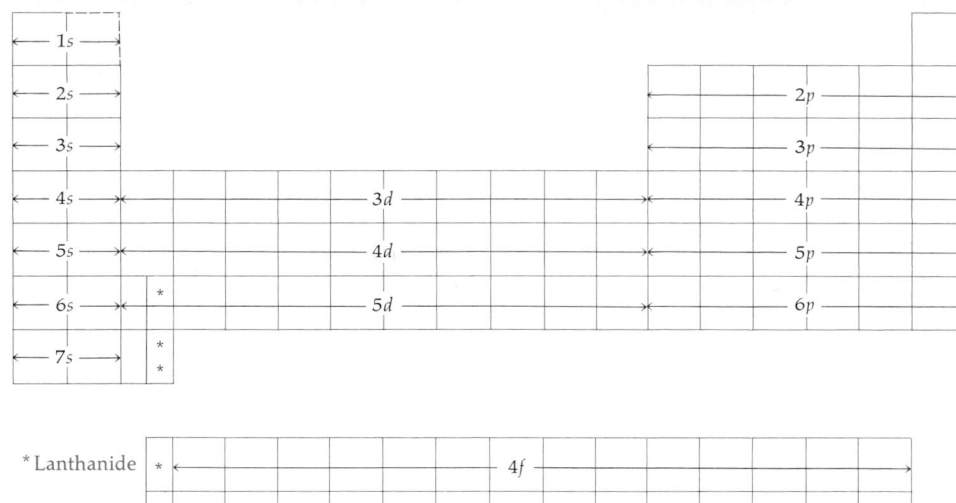

correspond to filling the 3d orbitals: $3d^1$, $3d^2$, . . . , $3d^{10}$. This relates to the chemical fact that $_{20}$Ca is much like $_{12}$Mg above it in principal group II, but it is not until you reach $_{31}$Ga that you find another element like $_{13}$Al in principal group III.

There are minor exceptions to the rules of filling orbitals, especially while the d orbitals and f orbitals are being filled, but the general pattern is very strong and compels scientists to consider the application of quantum mechanics to chemistry as extremely important.

This brings us back to the introductory paragraphs of this chapter in which we declared that the numbers of elements in each of the first seven rows of the periodic table, namely 2, 8, 8, 18, 18, and 32, could be explained by a study of atomic spectra. By adding up the number of subshell electrons in each period of Table 13.4, you obtain the appropriate numbers of elements, as shown in Table 13.5.

TABLE 13.5 Filling of orbitals to form new periods in the periodic table

Period Number	Electronic Structure	Total Number of Elements
1	$1s^2$	2
2	$2s^2 2p^6$	8
3	$3s^2 3p^6$	8
4	$4s^2 3d^{10} 4p^6$	18
5	$5s^2 4d^{10} 5p^6$	18
6	$6s^2 4f^{14} 5d^{10} 6p^6$	32
7	$7s^2 5f^{14}$ . . .	incomplete

When electrons are removed from neutral atoms to form positive ions, the order of emptying the orbitals is often the reverse of the order of filling, but there are a great many exceptions. The effect of adding protons to the nucleus in the building-up process is to shift the subshell energy levels toward the levels held by the other subshells in the same shell. Within the subshells you still find $s < p < d < f$, but you may also find that $4s < 3d$ has changed to $3d < 4s$, and other similar changes.

EXAMPLE 3.   Give the electronic structures of the following pairs of atoms and their positive ions. (a) Cl, Cl^+ (b) K, K^+ (c) Fe, Fe^{2+}, Fe^{3+}.

Solution: In parts (a) and (b), the order of emptying the orbitals is the reverse of the order of filling. In part (c) an exception is found.

(a) $_{17}Cl$: $1s^2 2s^2 2p^6 3s^2 \underline{3p^5}$ (b) $_{19}K$: $1s^2 2s^2 2p^6 3s^2 3p^6 \underline{4s^1}$
   $_{17}Cl^+$: $1s^2 2s^2 2p^6 3s^2 \underline{3p^4}$ $_{19}K^+$: $1s^2 2s^2 2p^6 3s^2 3p^6$

(c) $_{26}Fe$: $1s^2 2s^2 2p^6 3s^2 3p^6 \underline{4s^2} 3d^6$
   $_{26}Fe^{2+}$: $1s^2 2s^2 2p^6 3s^2 3p^6$ $\underline{3d^6}$
   $_{26}Fe^{3+}$: $1s^2 2s^2 2p^6 3s^2 3p^6$ $\underline{3d^5}$

Note: For emptying the orbitals, $4s > 3d$, which is the opposite for filling.

The rule for determining the electronic structures of positive ions can be stated in the following manner.

EMPTYING RULE: *The rule for emptying orbitals in an atom or ion, holding the number of protons constant, is that orbitals having higher shell numbers are emptied first. Within a given shell the subshells are emptied in the following order: f, d, p, s.*

An interesting result of the emptying rule is that several long series of elements in Table 13.4 have very similar chemical and physical properties because they empty in the same way. The metals in periods 4, 5, 6, and 7 lying between principal groups II and III are all somewhat similar. The three series of ten elements each, corresponding to filling the $3d$, $4d$, and $5d$ subshells are called **transition metals.** The two series corresponding to filling the $4f$ and $5f$ subshells are called the **rare-earth metals** or **lanthanides,** and the **actinides,** respectively. Many of the actinides must be produced by nuclear reactions and disintegrate almost immediately. The **rare earths** as found in nature are difficult to separate because they have almost identical properties.

Exercise 2.   Give the electronic structures of the rare earth metal europium, Eu, and its ion, Eu^{3+}.

Answer: $_{63}Eu^{3+}$: $1s^2 2s^2 2p^6 3s^2 3p^6 4s^2 3d^{10} 4p^6 5s^2 4d^{10} 5p^6$ $\underline{4f^6}$
            $_{63}Eu$: $1s^2 2s^2 2p^6 3s^2 3p^6 4s^2 3d^{10} 4p^6 5s^2 4d^{10} 5p^6 6s^2 4f^7$

1. Match the following:
   (a) a quantum of variable frequency __   (1) λ/ν
   (b) frequency __   (2) $h\nu$
   (c) velocity of light __   (3) $\lambda\nu$
   (d) ultraviolet __   (4) 100 kcal mole^{-1}
      (5) sec^{-1}
      (6) 6500 Å

2. Match the following:
   (a) frequency __   (1) 4863 Å
   (b) (wave length) × (frequency) __   (2) c/λ
   (c) energy __   (3) $h\nu$
   (d) infrared __   (4) 3×10^{10} cm sec^{-1}
      (5) Δn
      (6) 10,000 Å

3. The waves on a certain beach come in at a velocity of 3.0 ft sec^{-1}. A person observing them estimates that there is one wave every 10 sec. (a) What is the wave frequency? (b) What is the wave length?

4. The waves on a certain beach come in at a velocity of 2.0 ft sec^{-1} and they are 10 ft apart (crest to crest). (a) What is their frequency? (b) How could you measure their frequency if you had only a stopwatch?

5. Infrared spectroscopists still express wavelength in units of μ (micron, 10^{-6} m) although, strictly speaking, it is not a metric unit. (a) What is the metric unit for 10^{-6} m? (b) Express the infrared wavelength, $8\,\mu$, in Ångstrom units (Å).

6. Many spectral instruments are graduated in wavelength units of "$m\mu$" (millimicron, 1 micron = 10^{-6} m) although strictly speaking, it is not a metric unit. (a) What is the metric equivalent of "$m\mu$"? (b) Express the visible wavelength, $650\,m\mu$, in units of Ångstroms and nanometers.

7. A certain x ray has wavelength 1 Å. What is its frequency?

8. A certain radio wave has wavelength of 250 m. (a) What is its frequency expressed in sec^{-1}? (b) What is its frequency in kilocycle?

9. The radiation from a certain toaster has wavelength 10 μm. (a) Identify it as infrared, visible, or ultraviolet. (b) Calculate the corresponding energy in kilocalories per mole. (c) A quantum of this light collides with a diatomic molecule held together by a bond having energy 40 kcal mole^{-1}. Is it likely that the bond will break?

10. (a) What is the energy of light having wavelength 100 Å, expressed in kilocalories per mole? (b) How does this compare with the amount of energy required to ionize H atoms? (c) What effect would a light quantum of this wavelength have if it collided with a hydrogen atom?

11. A series of lines in the H spectrum correspond to excited atoms falling to the fifth level from higher levels. (a) Are these lines in the infrared or ultraviolet region? (b) Identify the lowest energy line in this series and, using equation (13.3), obtain a numerical value for the energy.

12. A series of lines in the H spectrum correspond to excited atoms falling to the fourth level from higher levels. (a) Are these lines in the infrared or ultraviolet region? (b) Identify the lowest energy line in this series and, using the steps in Figure 13.5, obtain a numerical value for the energy.

13. There are some elements in the periodic table that contain electrons in the sixth shell, but none in which the sixth shell is full. Calculate the maximum number of subshells, the maximum number of orbitals, and the maximum number of electrons possible in the sixth shell.

14. In the first shell of an atom only the $1s$ orbital is available for electron occupancy, whereas in the second shell four orbitals are available: $2s$, $2p_x$, $2p_y$ and $2p_z$. (a) How many orbitals are available in each of the third, fourth, and fifth shells? (b) How many orbitals are there in the g subshell, which follows the f subshell?

15. Predict the electronic structures of each of the following atoms and ions by shell number (1, 2, 3, . . .), subshell (s, p, d, or f), and number of electrons. (a) Rb; (b) Rn; (c) H⁻; (d) Ca^{2+}; (e) S^{2-}; (f) Cr^{3+}.

16. Predict the electronic structures of each of the following atoms and ions by shell number (1, 2, 3, . . .), subshell (s, p, d, or f), and number of electrons. (a) Ra; (b) I; (c) Na⁺; (d) Cl⁻; (e) Br⁺; (f) Ni^{2+}.

*17. The centimeter is defined as 16507.6373 wavelengths of light emitted by the inert gas krypton at a certain frequency. Calculate the wavelength of this light expressing your answer in as many significant figures as permitted. Predict its color.

More on the One-Factor Method

The direct proportion is one of the simplest mathematical relationships found in science. For example, every problem in the conversion of units involves a direct proportion. Consider the problem of converting yards into feet. The length of an object in feet is directly proportional to its length in yards.

$$1 \text{ yd} = 1 \times 3 \text{ ft} = 3 \text{ ft}$$
$$2 \text{ yd} = 2 \times 3 \text{ ft} = 6 \text{ ft}$$
$$3 \text{ yd} = 3 \times 3 \text{ ft} = 9 \text{ ft}$$

If you double the number of yards, you double the number of feet; if you triple the number of yards, you triple the number of feet; and so on.

Mathematically the relationship is stated in the following way, where y = number of feet and x = number of yards.

(a) $y \propto x$ ($\propto$ means "is proportional to")

(b) $y = 3x$ (The number "3" is called a **proportionality constant**.)

A characteristic of proportional relationships is that, if you plot one quantity (y)

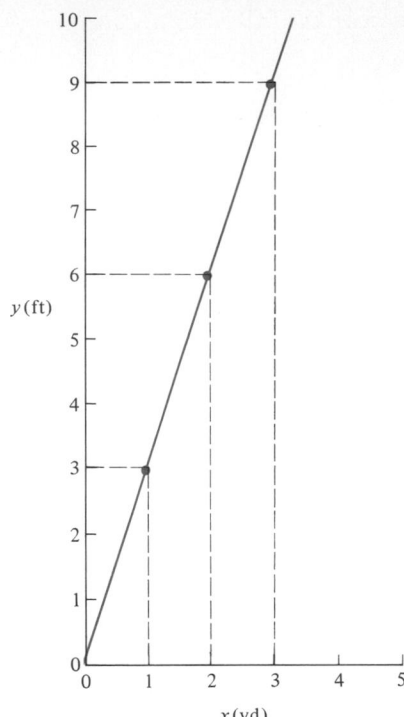

against the other quantity (x) on a graph, you obtain a straight line, as shown in the figure above. You will note that the straight line passes through the origin, which is the point where the number of yards and the number of feet are both equal to zero. Some straight-line relationships are not proportional because they do not pass through the origin. An example of nonproportionality is the relationship between Fahrenheit and Celsius temperatures (Figure 2.5, p. 26), in which 0°F is not equal to 0°C.

2   INVERSE PROPORTION

In some mathematical relationships one quantity decreases when the other increases. If one quantity is cut in half when the other is doubled, or cut in one third when the other is tripled, and so on, the relationship is called an **inverse proportion.** An example follows.

If an automobile travels 100 miles at one speed and then another 100 miles at another speed, the length of time required for each is inversely proportional to the speed, as described in the following sentence.

If 100 miles at <u>50 MPH</u> takes <u>2 hours</u>, then
100 miles at <u>25 MPH</u> takes <u>4 hours</u>, and
100 miles at <u>10 MPH</u> takes <u>10 hours</u>, and so on.

Mathematically the relationship is stated in one of the following ways.

(a)
$$\text{time} \propto \frac{1}{\text{speed}}$$

(b)
$$\text{time} = \frac{\text{distance}}{\text{speed}}$$

In this case "distance" is a proportionality constant.

3   THE MEANING OF "EQUIVALENT TO" IN CONSTRUCTING ONE-FACTORS

One-factors can be set up with numerators and denominators equal to one another in simple problems, such as conversion of units. For example $\frac{1000 \text{ g}}{1 \text{ kg}}$, $\frac{10^{-2} \text{ m}}{1 \text{ cm}}$, and $\frac{10^{-3} \text{ }\ell}{1 \text{ ml}}$, and their reciprocals, can be used as one-factors because their numerators and denominators are *equal*.

In other problems it may be desirable to use quantities that are only *equivalent* to one another to construct the needed one-factors.

EXAMPLE 1.   We know 1 mole of Na weighs 23 g; therefore, 23 g Na is equivalent to 1.0 mole Na, and $\frac{23 \text{ g}}{1 \text{ mole}}$ is a one-factor. You could say that 1.0 mole Na *equals* 6.0×10^{23} atoms of Na, but it is *not* correct to say that 1.0 mole Na *equals* 23 g Na.

EXAMPLE 2.   The density of gold is 19.3 g ml^{-1}. You could *not* say that 19.3 g gold *equals* 1.00 ml gold, but they *are equivalent*. The reason is that solid gold always has the same density and a formula exists ($d = \text{weight/volume}$) with which weight and volume could be calculated from one another. To have 1.00 ml gold means that you must have 19.3 g gold, and vice versa; therefore, the quantities are equivalent. Both $\frac{19.3 \text{ g}}{1 \text{ ml}}$ and $\frac{1 \text{ ml}}{19.3 \text{ g}}$ are satisfactory one-factors.

EXAMPLE 3.   In a chemical reaction the number of moles of reactant and product are all equivalent to one another. The following reaction is an example.

$$2 \text{ H}_2 + \text{O}_2 \longrightarrow 2 \text{ H}_2\text{O}$$

The balanced equation means that 2 mole H$_2$, 1 mole O$_2$, and 2 mole H$_2$O are all equivalent to one another, so that $\frac{2 \text{ mole H}_2}{1 \text{ mole O}_2}$, $\frac{1 \text{ mole O}_2}{2 \text{ mole H}_2\text{O}}$, and so on, are one-factors. Of course, the one-factors are applicable only to the number of moles that are actually consumed or produced during the chemical reaction (see Section 4.6, Limiting Reactants).

EXAMPLE 4.   We know that 1 mole phosphoric acid, H_3PO_4, contains 3 mole H. If all the H is consumed in a chemical reaction, you could say that 1 mole H_3PO_4 is equivalent to 3 mole H. The ratio $\dfrac{3 \ \text{mole H}}{1 \ \text{mole } H_3PO_4}$ and its reciprocal are both satisfactory one-factors for this kind of reaction.

2. (c).
4. light *small* objects sink.
6. (a) mass (inertia) of the bicyclist   (b) weight   (c) mass (inertia)
8. (a) potassium, group I, period 4   (b) iodine, group VII, period 5
    (c) neon, group 0, period 2   (d) silicon, group IV, period 3
    (e) oxygen, group VI, period 2.
10. (a) Ca, group II, period 4   (b) Kr, group 0, period 4
    (c) Al, group III, period 3   (d) N, group V, period 2
    (e) Br, group VII, period 4.
12. (a) chemical   (b) physical.
14. (a) heterogeneous   (b) homogeneous   (c) heterogeneous.
16. (a) boiling   (b) sublimation   (c) condensation.
18. (a) sodium bromide   (b) magnesium chloride   (c) aluminum sulfide
    (d) calcium hydride.
20. (a) carbon tetrafluoride   (b) nitrogen triiodide   (c) sulfur dioxide.

2. (a) 10^{-2} (b) 10^{-4} (c) 10^{-7} (d) 10^{7}.
4. (a) 63360 in.
    (b) It is exact and has as many significant figures as you like after
    the decimal.

6. (a) 1000 cm^3 (b) 10 pl   (c) 2 × 10^6 cm   (d) 5 × 10^8 mg
 (e) 4 × 10^{-2} m.
8. (a) 50 mi hr^{-1} (b) 168 lb, "VIKT" means "WEIGHT" in Swedish.
10. (a) 37.8°C (compare 38°C)   (b) 546°K.
12. 5.6° on the Celsius scale.
14. w, 0.84 cm 2 s.f.; x, 3.65 cm 3 s.f.; y, 8.01 cm 3 s.f.; z, 13.81 cm 4 s.f.
16. (a) 7 × 10^{-5} g   (b) 0.001%   (c) 30%.
18. (a) 4   (b) 5   (c) 3   (d) 3   (e) 4.
20. (a) 4 × 10^{-3} (b) 3.6 × 10^{-2} (c) 6.023 × 10^{23} (d) 2.001 × 10^3
 (e) 2.9 × 10^{-2}.
22. (a) 2.736 × 10^2 (b) 9.42 × 10^{-5} (c) 2.75 × 10^7 (d) 2.3 × 10^{10}
 (e) 2.6 × 10^6.
24. 2.7 tons.
★25. (a) 4.6 m   (b) 13.7 m   (c) 3.0   (d) 0.1, 3%, 3 s.f.
 (e) No. The sea may not have been perfectly round or the measurements
 may have been reported to the nearest cubit only.
★26. (a) sink   (b) float   (c) float.

Chapter 3   2. 1.5 × 10^4 lb in.$^{-2}$
4. (a) yes   (b) Too low because the air exerts pressure on top of the column.
6. 1.5 ft^3.
8. 0.15 ml.
10. Slower gas particles hit the walls *less frequently* and with *less force* each time.
12. (a) 1.0 cm^3 (b) 0.005 cm^3.
14. 133 cm^3.
16. 1.87 × 10^5 m^3.
★17. II lower than position I, III lower than positions I *and* II. Pressure in
 space above column is greater in III than in II by Boyle's law.
★18. (a) barometric pressure   (b) It has air in it.
 (c) As the water is removed, more air leaks into the bottle.
★19. 68 ft.

Chapter 4   2. (a) 300 cm^3 NO$_2$(g)   (b) 200 cm^3 CO$_2$(g).
4. (a) 4, 5, 4, 6   (b) 4, 4, 1, 2   (c) 1, 2, 3, 1   (d) 3, 4, 1, 4
 (e) 1, 2, 1, 2.
6. (a) 864 shoes   (b) 3.6 × 10^{24} H atoms.
8. (a) 3 × 10^{11} bees   (b) 5 × 10^{-13} mole of bees.
10. 1.7 × 10^{21} H$_2$O molecules.
12. (a) 1.20 × 10^{24} atoms Cl   (b) 1.20 × 10^{24} atoms Cl
 (c) 6.02 × 10^{23} atoms Cl.
14. (a) 3.0 g H   (b) 4.0 g H   (c) 0.50 g H.
16. (a) 0.50 mole Al$_2$O$_3$ (b) 0.15 mole O$_2$.
18. (a) 0.37 mole O$_2$ (b) 0.73 mole O$_2$.
20. 8 lb O$_2$.
22. (a) 62.5% Ca   (b) 72.3% Fe.

24. SnO_2.
26. (a) CH   (b) C_6H_6.
★27. N_2H_4.
★28. 55.5 mole ℓ^{-1}.

Chapter 5

2. (a) 4F (R)   (b) 4F (R)   (c) 3F (A)   (d) 6F (A).
4. (a) 100F   (b) 0.01F.
6. (a) 0.5F (R)   (b) F (R)   (c) 0.5F (A)   (d) 2F/9 (A)   (e) 16F (A).
8. (a) 2.66×10^{-23} g   (b) 2.67×10^{-23}, approximately equal.
10. 0.8 in.
12. (a) endothermic, separation of unlike charges
    (b) exothermic, addition of a single electron to a neutral atom
    (c) endothermic, see part (a)   (d) exothermic, see part (b)
    (e) endothermic, combination of like charges.
14. group VII: period 2, F, 2 shells; period 3, Cl, 3 shells;
    period 4, Br, 4 shells; period 5, I, 5 shells; period 6, ·At, 6 shells.
16. calculated: 6.9 g mole^{-1}; table: 6.94 g mole^{-1}.
18. Cs^+: 55, 55, 54, 78; Xe: 54, 132, 54, 54;
    He^{2+}: 2, 3; I^-: 53, 53, 54, 74; Li^+: 6, 3.
★19. (b) Force must decrease from left to right but it must
    increase without limit as distance approaches zero.
★20. (a) 9.1×10^{-28} g   (b) 6.0×10^{23}.
★21. 10^{15} times.

Chapter 6

2. (a) 2   (b) 2   (c) 1   (d) 3   (e) 1.
4. (a) 7   (b) 1   (c) 2   (d) 6   (e) 3.

6. (a) H:P:H with H   (b) :Br:Br: (c) :Cl:O:⁻
   (He)(Ar)   (Kr)   (Ar)(Ne)

   (d) :C::N:⁻   (e) :O:H H:O:
   (Ne)(Ne)   (He)(Ne)

8. (a) :N=O: , H—O: (b) :O:²⁻ , C with O and O
   (c) :O—S—O with :O: and H   (d) :O=S: with :O:

10. (a) covalent   (b) covalent   (c) ionic   (d) ionic.

12. (a) $:\overset{..}{\underset{..}{Ra}}:^{2+}$ $:\overset{..}{\underset{..}{S}}:^{2-}$, radium sulfide
     (Rn)   (Ar)

(b) $Li:^+$ $:\overset{..}{\underset{..}{S}}:^{2-}$ $Li:^+$, lithium sulfide
    (He)   (Ar)

(c) $:\overset{..}{\underset{..}{Cl}}:^-$ $:\overset{..}{\underset{..}{Mg}}:^{2+}$ $:\overset{..}{\underset{..}{Cl}}:^-$, magnesium chloride
    (Ar)   (Ne)

14.

	MgO magnesium oxide	MgH$_2$ magnesium hydride
Na$_3$P sodium phosphide	Na$_2$O sodium oxide	NaH sodium hydride
AlP aluminum phosphide	Al$_2$O$_3$ aluminum oxide	AlH$_3$ aluminum hydride

16. (a) tin(II) chloride, tin(IV) chloride   (b) iron(II) oxide, iron(III) oxide.

18. (a) MnO, MnO_2 (b) CoI_2, CoI_3.

20. (a) $BaCl_2$ (b) Cs_2O (c) AlF_3.

22. (a) $2\,Na + Br_2 \longrightarrow 2\,NaBr$ (b) $6\,Ca + P_4 \longrightarrow 2\,Ca_3P_2$
(c) $2\,Al + 3\,I_2 \longrightarrow 2\,AlI_3$.

24. (a)   109° 28′   (b)   109° 28′
          (tetrahedral)   (bent)

(c) 180° (linear)   (d)   120° (planar)

(e) H   H   109° 28° (pyramidal)

26. (a) nonpolar   (b) yes, B   (c) yes, S   (d) yes, P.

28. (a) yes, H—F, bond is polar, center of positive charge is located to left of center of negative charge

(b) yes,

All bonds are polar. The center of the positive charge is above the center of the negative charge.

(c) no, $\overset{\longrightarrow}{H-H}\overset{\longleftarrow}{\equiv C-H}$. The bonds are slightly polar but the center of the positive charge and the center of the negative charge are both at the center of the molecule.

(d) no, a nonpolar bond.

★ 29. CaC_2 $:\overset{..}{\underset{..}{Ca}}:^{2+}$ $:C\equiv C:^{2-}$ carbide ion

H_2O $:\overset{..}{O}-H$
            |
            H

$Ca(OH)_2$ $:\overset{..}{\underset{..}{Ca}}:^{2+}$ $2:\overset{..}{\underset{..}{O}}-H^-$ hydroxide ion

C_2H_2 $H-C\equiv C-H$

★30.

H $:\overset{..}{O}-H$
 | |
H$-$C$-$C$-$H
 | |
 H H

ethyl alcohol

$\overset{..}{\underset{..}{O}}$
H$-$C   C$-$H
 / | | \
H H   H H

dimethyl ether

★31.

H
 \
  C$=\overset{..}{O}:$ (I)
 /
H

     H
     |
$:C=\overset{..}{O}:$ (II)
     |
     H

Structure (I), because in (II) carbon and oxygen have three bonds each.

★32. Each Cs^+ ion is surrounded by Cl^- ions at each of the eight corners of a cube. Since 8 cubes meet at each corner, each Cl^- ion belongs only $\frac{1}{8}$ to the given Cs^+ ion. $\frac{1}{8} \times 8 = 1\ Cl^-$ ion per Cs^+ ion.

★33. $NaCl(g)$ is polar because it is ionic; $NaCl(s)$ is not polar because, although it is ionic the symmetry of the fixed arrangement of ions in the crystalline solid (see Figure 6.1) makes the center of the positive charge coincide with the center of the negative charge. However, the crystal is polar at its surface and attracts other polar substances such as water.

2. 18°F change.   Chapter 7

4. (a) 5×10^4 cal   (b) 1×10^4 cal.

6. 3.33 kcal g^{-1}.

8. 7.13 kcal (g ethanol)$^{-1}$; 3.95 kcal (g sucrose)$^{-1}$.

10. 47.4 kcal mole^{-1} exothermic; yes, because it is exothermic.

12. (a) Steam, or $H_2O(g)$, causes a worse burn because it gives off heat of vaporization when it condenses.

    (b) Same temperature, no heat flows from one to the other when they are brought in contact.

14. (a) 225 mm

    (b) No, the vapor would force the Hg completely out of the tube.

16. (a) 5.1×10^2 mm (table value 522 mm)   (b) 4 mm, right side higher.

18. 34°C (table value 34.6°C).
20. (a) 93°C (table value 93.5°C)   (b) 89°C (table value 88.7°C).
22. 397 mm Hg.
★**23.** 625 g (final temperature 0°C).
★**24.** (c) 2.82 Å.
★**25.** 3.4 mi.

Chapter 8   **2.** *similarities:* homogeneous, more than one element, cannot be separated by filtration; *differences:* solutions have variable composition, solutions can be separated by fractional distillation or crystallization, solutions have variable melting and boiling points.

4. *avoid* water and other *polar* solvents; *recommend* carbon tetrachloride, chloroform and other *nonpolar* or *weakly polar* solvents.

6. *attraction* of the polar water molecules for one another makes dissolving *endothermic.*

8. (a) increase   (b) decrease.

10. 50 ml volumetric buret, graduated every 0.1 ml up to 50 ml—not at 1 ml intervals like 100 ml graduated cylinder.

12. 36 g.

14. (a) 4.00 g NaOH(s) weighed, dissolved in water, and diluted to 500 ml in volumetric flask
(b) 2.00×10^{-3} mole.

16. 70 ml.

18. (a) 15.8 M   (b) 19.0 ml conc. HNO_3 diluted to 100 ml in a volumetric flask.

20. 8 ppm.

★**21.** (95 g bact. sol'n) $\left(\dfrac{70}{100} \dfrac{\text{g pure alc.}}{\text{g bact. sol'n}} \right) \left(\dfrac{100}{95} \dfrac{\text{g conc. sol'n}}{\text{g pure alc.}} \right) = 70$ g conc. sol'n.

★**22.** 75.5% N_2, 23.2% O_2, 1.2% Ar.

★**23.** 0.5 ppm.

★**24.** 4 kg.

★**25.** 0.08 ppm; the toxic level of Pb in blood is 0.07 ppm.

Chapter 9   **2.** (a) closed   (b) open   (c) open.

4. not *reversible* because the drop of water is not at equilibrium; must be carried out slowly as in a cylinder with a piston.

6. closed, but not at equilibrium; the temperature is not the same throughout, e.g., the air around the cooling coils of the air conditioner is colder than the air in the other parts of the house.

8. Under reduced pressure (about 1 mm Hg) a process occurs that tends to increase pressure, $H_2O(s) \longrightarrow H_2O(g)$. This sublimation process, which is endothermic, keeps the ice frozen.

10. closed; reversible because heat is exchanged slowly.

12.

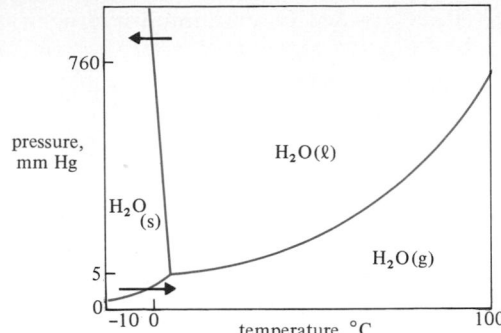

14. no; K^+, NO_3^- contain no ions in common with Na^+, Cl^-.

16. (a) 7×10^{-7} M   (b) 5×10^{-10} M.

18. (a) decreased pressure favors the direction giving higher pressure: larger volume; therefore more $N_2(g)$ is formed
(b) too rapid an increase in altitude.

20. (a), two gas molecules on left and two gas molecules on right, i.e., no volume change at constant pressure.

22. (a) $[CH_4]/[H_2]^2$ (b) yes, large K
(c) less, increase in temperature favors the direction giving lower temperature: the endothermic reaction.

24. 2×10^{-9}.

26. (a) $[Mg^{2+}][F^-]^2$ (b) $[H_3O^+][OAc^-]/[HOAc]$ (c) $[I^-]^2/[Br^-]^2$
(d) $[CrO_4^{2-}]^2[I^-]/[IO_3^-][OH^-]^4$.

★27. (a) NaCl at 20°C, KCl at 60°C   (b) 50 g.

★28. $[H_2O(l)]$ is *decreased* slightly by salt dissolving in it. Melting occurs in order to *increase* $[H_2O(l)]$. Since melting is endothermic, the temperature falls until a new freezing point is reached.

2. $2\,OH^- + H_2PO_4^- \longrightarrow 2\,H_2O + PO_4^{3-}$.   Chapter 10

4. $3\,Mg + 2\,H_3PO_4 \longrightarrow 3\,H_2 + Mg_3(PO_4)_2$; magnesium phosphate.

6. yes; $CO_3^{2-} + 2\,H_3PO_4 \longrightarrow 2\,H_2PO_4^- + H_2O + CO_2$.

C: 1			=				1
O: 3	+	2(4)	=	2(4)	+	1	+ 2
H:		2(3)	=	2(2)	+	2	
P:		2(1)	=	2(1)			
charge: (−2)			=	2(−1)			

8. (a) $H^+ + NH_3 \longrightarrow NH_4^+$ (b) $HClO + OH^- \longrightarrow ClO^- + H_2O$.

10. (a) 1 mole H_2O, $H^+ + OH^- \longrightarrow H_2O$
(b) 2 mole H_2O, $2\,H^+ + 3\,OH^- \longrightarrow 2\,H_2O + OH^-$
(c) 3 mole H_2O, $3\,H^+ + 4\,OH^- \longrightarrow 3\,H_2O + OH^-$.

12. (a) $Al(OH)_3 + 3 H_2SO_4 \longrightarrow Al(HSO_4)_3 + 3 H_2O$, aluminum hydrogen sulfate

(b) $2 Al(OH)_3 + 3 H_2SO_4 \longrightarrow Al_2(SO_4)_3 + 6 H_2O$, aluminum sulfate.

14.

NH_4I ammonium iodide	NH_4HSO_3 ammonium hydrogen sulfite	NH_4ClO ammonium hypo-chlorite	$(NH_4)_3PO_4$ ammonium phosphate
AlI_3 aluminum iodide	$Al(HSO_3)_3$ aluminum hydrogen sulfite	$Al(ClO)_3$ aluminum hypo-chlorite	$AlPO_4$ aluminum phosphate
RaI_2 radium iodide	$Ra(HSO_3)_2$ radium hydrogen sulfite	$Ra(ClO)_2$ radium hypo-chlorite	$Ra_3(PO_4)_2$ radium phosphate

16. (a) 2   (b) 3   (c) 12   (d) 12.3.

18. 1.7 before, 1.1 after.

20. 1.6×10^{-8} M (pH = 7.8).

★21. $Ca_3(PO_4)_2 + 2 H_2SO_4 \longrightarrow Ca(H_2PO_4)_2 + 2 CaSO_4$, 12.2 wt.% P; $Ca_3(PO_4)_2 + 4 H_3PO_4 \longrightarrow 3 Ca(H_2PO_4)_2$, 26.5 wt.% P, better.

★22. 13.36 kcal (mole HNO_3)$^{-1}$, 13.35 kcal (mole HCl)$^{-1}$.

Chapter 11   2. (a) $a + b \rightleftharpoons a + b$ (b) $a + b \rightleftharpoons b + a$ (c) $b + a \rightleftharpoons a + b$.

4. (a) $HCl + NH_3 \longrightarrow Cl^- + NH_4^+$, yes, $K_{HCl} > 5.7 \times 10^{-10}$

(b) $NH_4^+ + PO_4^{3-} \longrightarrow NH_3 + HPO_4^{2-}$, yes, $5.7 \times 10^{-10} > 4.4 \times 10^{-13}$

(c) $SO_4^{2-} + CH_3COOH \longrightarrow HSO_4^- + CH_3COO^-$, no, $1.8 \times 10^{-5} < 1.3 \times 10^{-2}$.

6. 1.5.

8. $HL + OH^- \longrightarrow H_2O + L^-$(blue); $L^- + H_3O^+ \longrightarrow HL$(red) $+ H_2O$.

10. (a) 33.3 ml   (b) 0.83 M.

12. (a) $SO_3 + H_2O \longrightarrow H_2SO_4$ (b) $Cl_2O_7 + H_2O \longrightarrow 2HClO_4$

(c) $N_2O_3 + H_2O \longrightarrow 2HNO_2$.

★13. $K_a = (10^{-pH})^2/(\text{conc.} - 10^{-pH})$.

★14. $K = 37$, reaction occurs.

Chapter 12   2. (a) 1, 1, 1, Ⓒ, $\underline{O_2}$ (b) 1, 3, 2, 3, ⒸO, $\underline{Fe_2O_3}$

(c) 2, 1, 1, 1, Ⓕⓔ, $\underline{HCl}$ (d) 2, 3, 2, 3, Ⓕⓔ, $\underline{Cu^{2+}}$

(e) 2, 2, 1, $H_2\text{Ⓞ}_2$, $\underline{H_2O_2}$.

4. (a) 1, 1, 1, 1; 1 e^- gained by Cl, 2 e^- lost by Ca

(b) 2, 2, 1; 1 e^- gained by Na^+, 1 e^- lost by Cl^-

(c) 4, 3, 2; 2 e^- gained by O, 3 e^- lost by Al

(d) 2, 1, 2; 1 e^- gained by H, 1 e^- lost by Na

(e) 2, 1, 1; 1 e^- gained by O, 1 e^- lost by Na.

6. $Sb(s) \rightleftharpoons Sb^{3+}(aq) + 3\,e^-$, loses electrons more easily.

8. Mg, Cd, H_2, Pt.

10. (a) $Mg + 2\,H^+ \longrightarrow Mg^{2+} + H_2$ (b) N.R.

12.

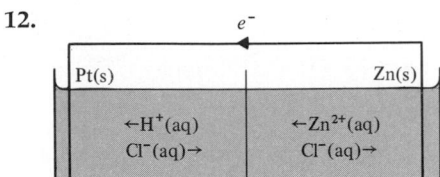

14. $Ni(s)\,|\,Ni(OH)_3(s)\,|\,Ni(OH)_2(s)\,|\,OH^-(aq)\,|\,Fe(OH)_2(s)\,|\,Fe(s)$.

16. (a) $H_2 + 2\,Ag^+ \longrightarrow 2\,H^+ + 2\,Ag$ (b) N.R.

(c) $Hg^{2+} + Co \longrightarrow Hg + Co^{2+}$ (d) N.R.

18. (a) N.R.   (b) $Cu^{2+}(aq) + Zn \longrightarrow Cu + Zn^{2+}(aq)$

(c) $2\,I^-(aq) + Cl_2 \longrightarrow I_2 + 2\,Cl^-(aq)$.

20. (a) N_2 (b) Na_2O_2 (c) $Ba\,H_2$ (d) $S\,O_3^{2-}$ (e) $S_2\,O_3^{2-}$
       0       $1+1-$ $2+1-$ $4+2-$ $2+2-$

(f) $Fe\,\,Cl_3$ (g) $Mg\,(Br\,O_3)_2$ (h) $H\,\,Cl\,O$
   $3+\,\,1-$ $2+5+2-$ $1+1+2-$

22. (a) 5, 2, 1, 5, 3, 1   (b) 1, 1, 1, 1, 1, 1   (c) 1, 10, 2, 10, 4.

24. (a) $Cr_2O_7^{2-} + H_2S + 8\,H^+ \longrightarrow 2\,Cr^{3+} + SO_2 + 5\,H_2O$

(b) $6\,I^- + 2\,MnO_4^- + 4\,H_2O \longrightarrow 3\,I_2 + 2\,MnO_2 + 8\,OH^-$.

26. Compare cathode reaction with equation (12.3) on page 281.

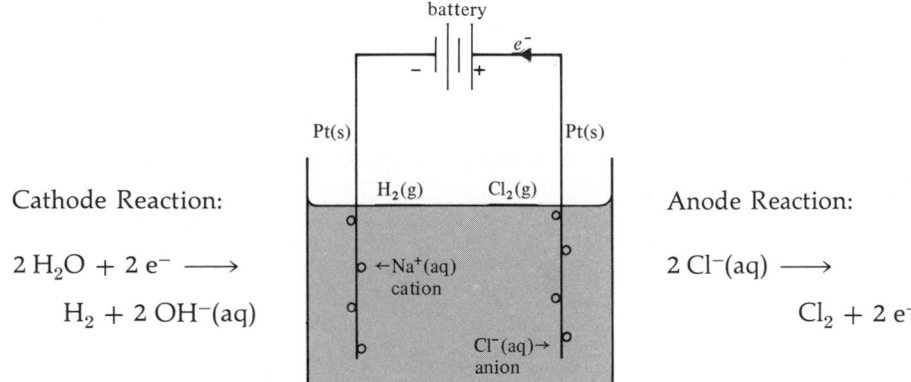

Cathode Reaction:

$2\,H_2O + 2\,e^- \longrightarrow$

   $H_2 + 2\,OH^-(aq)$

Anode Reaction:

$2\,Cl^-(aq) \longrightarrow$

     $Cl_2 + 2\,e^-$

Over-all Reaction:

$$2\,Cl^-(aq) + 2\,H_2O \longrightarrow Cl_2 + H_2 + 2\,OH^-(aq)$$

28. $Fe(s) + 2 Ni(OH)_3(s) \underset{\text{charge}}{\overset{\text{discharge}}{\rightleftharpoons}} Fe(OH)_2(s) + 2 Ni(OH)_2(s)$; no; it does not appear as a reactant or product.

⋆29. (a) Fe anode, Sn cathode   (b) Zn anode, Fe cathode

(c) dissolved $Zn^{2+}(aq)$ would contaminate the tomatoes; a scratch on the fence post would cause Fe to oxidize more rapidly than ever.

⋆30. (a) 3.11×10^{-3} faraday g^{-1}, 3.00×10^2 coulomb g^{-1}

(b) 1.11×10^{-1} faraday g^{-1}, 1.07×10^4 coulomb g^{-1}.

Chapter 13

2. (a) (2)   (b) (4)   (c) (3)   (d) (6).

4. (a) 0.2 sec^{-1}

(b) Divide the number of waves by the number of seconds.

6. (a) nm   (b) 6500 Å, 650 nm.

8. (a) $1.2 \times 10^6 \text{ sec}^{-1}$ (b) 1200 kilocycle.

10. (a) 2.86×10^3 kcal mole^{-1} (b) larger than IE

(c) it would ionize the H atom.

12. (a) infrared   (b) Δn: 5 $\longrightarrow$ 4, 7.1 kcal mole^{-1}.

14. (a) 9, 16, 25   (b) 9.

16. (a) $1s^2 2s^2 2p^6 3s^2 3p^6 4s^2 3d^{10} 4p^6 5s^2 4d^{10} 5p^6 6s^2 4f^{14} 5d^{10} 6p^6 7s^2$

(b) $1s^2 2s^2 2p^6 3s^2 3p^6 4s^2 3d^{10} 4p^6 5s^2 4d^{10} 5p^5$ (c) $1s^2 2s^2 2p^6$

(d) $1s^2 2s^2 2p^6 3s^2 3p^6$ (e) $1s^2 2s^2 2p^6 3s^2 3p^6 4s^2 3d^{10} 4p^4$ (f) $1s^2 2s^2 2p^6 3s^2 3p^6 3d^8$.

⋆17. 6057.80211 Å, orange to red.